**Dynamic Pulse Buckling**

**MECHANICS OF ELASTIC STABILITY**
*Editors:* H.H.E. Leipholz and G.Æ. Oravas

H H.E. Leipholz, Theory of elasticity. 1974.
ISBN 90-286-0193-7
L. Librescu, Elastostatics and kinetics of anisotropic and heterogeneous shell-type structures. 1975.
ISBN 90-286-0035-3
C.L. Dym, Stability theory and its application to structural mechanics. 1974.
ISBN 90-286-0094-9
K. Huseyin, Nonlinear theory of elastic stability. 1975.
ISBN 90-286-0344-1
H.H.E. Leipholz, Direct variational methods and eigenvalue problems in engineering. 1977.
ISBN 90-286-0106-6
K. Huseyin, Vibrations and stability of multiple parameter systems. 1978.
ISBN 90-286-0136-8
H.H.E. Leipholz, Stability of elastic systems. 1980.
ISBN 90-286-0050-7
V.V. Bolotin, Random vibrations of elastic systems. 1984.
ISBN 90-247-2981-5
D. Bushnell, Computerized buckling analysis of shells. 1985.
ISBN 90-247-3099-6
L.M. Kachanov, Introduction to continuum damage mechanics. 1986.
ISBN 90-247-3319-7
H.H.E. Leipholz and M. Abdel-Rohman, Control of structures. 1986.
ISBN 90-247-3321-9
H.E. Lindberg and A.L. Florence, Dynamic pulse buckling. 1987.
ISBN 90-247-3566-1

# Dynamic Pulse Buckling

*Theory and Experiment*

*By*

Herbert E. Lindberg

*APTEK, Inc., San Jose, CA, USA*

Alexander L. Florence

*SRI International, Menlo Park, CA, USA*

This work was sponsored by the Defense Nuclear Agency under RDT&E RMSS Code B342078464 N99QAXAH30103 H2590D and published as report number DNA 6503H, under contract number DNA 001-78-C-0287.

1987 **MARTINUS NIJHOFF PUBLISHERS**
a member of the KLUWER ACADEMIC PUBLISHERS GROUP
DORDRECHT / BOSTON / LANCASTER

**Distributors**

*for the United States and Canada*: Kluwer Academic Publishers, P.O. Box 358, Accord Station, Hingham, MA 02018-0358, USA
*for the UK and Ireland*: Kluwer Academic Publishers, MTP Press Limited, Falcon House, Queen Square, Lancaster LA1 1RN, UK
*for all other countries*: Kluwer Academic Publishers Group, Distribution Center, P.O. Box 322, 3300 AH Dordrecht, The Netherlands

**Library of Congress Cataloging in Publication Data**

Lindberg, Herbert E., 1930-
Dynamic pulse buckling.

(Mechanics of elastic stability ; 12)
Bibliography: p.
Includes index.
1. Buckling (Mechanics) 2. Elasticity. 3. Stability of structures. I. Florence, Alexander L., 1928- II. Title. III. Series.
TA656.2.L53 1987 624.1'6 87-14049

ISBN 90-247-3566-1 (this volume)
ISBN 90-247-2743-X (series)

# FOREWORD

This book originally appeared as a text prepared for the Defense Nuclear Agency to summarize research on dynamic pulse buckling, by the authors and their colleagues at SRI International, during the period from 1960 to 1980. The original printing of 300 copies by the DNA Press was followed shortly by a small second printing to meet the demand by readers who heard of the book from the primary recipients. This supply was also quickly exhausted, to researchers and practicing engineers outside the DNA community and to academics who wanted to include the material in courses on elastic and plastic stability of structures. Commercial publication by Martinus Nijhoff Publishers was therefore undertaken to meet the needs of this broader community.

The objective of the book was to gather into a cohesive whole material that had been published in reports and the open literature during the two decade period. In the process of knitting this material together, a substantial amount of new work was done. The book therefore contains many new results never published in the open literature.

These include the systematic development of the concepts of pulse buckling in Chapter 2, with a simple bar as an example, and a demonstration of the relationships between pulse buckling and the more familiar features of static buckling. In particular, the representation of imperfections that trigger static buckling, developed a century ago, is shown to be useful, in an appropriate random form, for pulse buckling. New experimental results are given for equivalent imperfections in thin bars made of sheet material typically used in thin-walled shells.

Similar development from classical static buckling theory is given in Chapter 3 for shells under radial impulse. For convenience of the reader, the Donnell shell equations needed for both static and dynamic analysis under both radial and axial loads are developed from first principles. The theory for radial impulse is developed from low impulse levels, for which instability is in autoparametric resonance, to intermediate levels, for which higher order terms must be included in the equations of motion in order to analyze autoparametric resonance, to high levels, at which buckling takes place during a single inward motion in the hoop mode. This again illustrates the relationship between more familiar types of buckling and single-pulse buckling, which is the main topic of the monograph. As a supplement to application of the theory for buckling during circumferential plastic flow, new experimental results and a unique constitutive relation are given for plastic stress-strain behavior of aerospace metals.

These ideas, developed to explain and extend experimental observations for ideally impulsive loading, are then extended to radial pressure pulses with finite pulse durations. Formulas are derived to estimate critical combinations of peak pressure and impulse intensity that cause buckling for loads over the entire range from ideally impulsive to step pressures of infinite duration. As for impulsive loads, extensive experimental results are given as a guide to the theoretical development and for confirmation of the accuracy of the theoretical results.

Chapter 4 gives advanced theory for plastic-flow buckling under radial impulse. The first examples are of buckling in shells of various lengths, in which the axial boundary conditions allow bi-axial plastic flow. This introduces a new buckle-resisting bending moment, called the directional moment, which occurs because of a kinematic constraint that forces yield points for material at various positions through the shell thickness to move along the yield surface in such a way as to resist bending. A new theory for viscoplastic flow buckling with directional moments is also developed. These theories are applied to prediction of threshold buckling in the conventional sense, and also to threshold buckling for shells thrown in at very high velocities such that the shell material becomes a solid mass, as used in explosive pipe closures and shaped charges for oil well pipe perforation.

In Chapter 5, buckling theory is developed for cylindrical shells under axial impact. As in Chapter 3, relationships are demonstrated among static buckling, buckling under long duration step loads, and pulse buckling which occurs during a single transit of the axial stress wave up and down the shell. These relationships show that the theory for step loading cannot be applied for impulsive loading in the manner suggested in the literature. As for radial impulse, thicker-walled shells buckle under very high velocity impact, which results in bi-axial plastic flow. Again, directional moments play an important roll in this buckling. Both the elastic and plastic-flow buckling theories are guided and confirmed by experimental results summarized in the text.

Experimental results are also given for gross buckling collapse of moderately thick shells, and a simple kinematic energy theory is derived to predict the average collapse force. These results demonstrate the order-of-magnitude change in time scales between the dynamics that initiate buckles and the quasi-static energy absorption from later stretching and folding. The last chapter gives theoretical and experimental results for plastic flow buckling of rectangular plates.

*H. E. Lindberg*
*A. L. Florence*
*March, 1987*

# CONTENTS

# PREFACE

This monograph brings together research results on dynamic buckling from reports and technical papers produced during the past two decades by the authors and their co-workers at SRI International (formerly Stanford Research Institute). Work on the monograph was supported by the Defense Nuclear Agency (DNA), under the technical direction of Mr. D.J. Kohler in the Shock Physics Directorate, Aerospace Systems Division (SPAS). DNA also sponsored much of the original research, either directly or through the Air Force Weapons Laboratory. Air Force sponsorship of original research was also provided by the Space and Missile Systems Office (now the Ballistic Missile Office).

The need to design structures to resist static buckling, particularly structures made from high strength alloys, has been recognized for more than a century. Buckling from dynamic loads has received serious attention only since World War II, and only within the last two decades has a basic understanding been developed for buckling under explosive and impact loads. This development followed closely the introduction of high-speed electronic and photographic instrumentation to observe such buckling, which can occur in a fraction of a millisecond. The monograph makes liberal use of experimental results in establishing the physical basis for dynamic pulse buckling theory.

We focus on buckling from intense loads, well above static buckling loads, but of short enough duration that the buckle amplitudes are small and cause no serious damage. A common example is a nail, which is driven by very high loads but does not buckle because the hammer and nail are in contact for only a short time. In large structures, the intense loads cause the modes of dynamic buckling to be of much higher order than in static buckling. Practical applications in which dynamic buckling plays an important role are evolving with the advancing technology of high speed, light weight military and civilian vehicles and with the renewed emphasis on safety in transportation and industry.

As in many technologies, early motivation for dynamic buckling research stemmed from military needs, such as in the design of aircraft landing struts and the design of ballistic missiles to resist launch and attack loads. Launch and reentry vehicles were found to buckle in uniquely high-order wrinkling patterns under X-ray blowoff (surface impulse) and blast loads. The new feature in ballistic missile skin buckling was that the wrinkle patterns had to be determined in addition to the critical loads.

Dynamic wrinkling also occurs in thin structures under impact, as in automobile or railway crashes. An idealized form of such wrinkling is the crushup of a cylindrical shell under axial impact. It is often desirable that the wrinkles be tightly packed, in order to absorb energy and protect passengers sitting farther back in the vehicle. This takes unique advantage of the natural tendency toward high-order wrinkling in dynamic buckling.

Industrial safety often requires that facilities survive accidental explosions without release of toxic gases. A current example is found in nuclear reactors, where safety receives great emphasis. Within an explosive environment, high pressure and impact conditions are felt by many structural elements, and dynamic buckling is an important mode of energy absorption. One example is the above-core structure support columns, whose dynamic buckling protects the reactor cover from impact by the lower internal structure.

Dynamic buckling is an important design consideration in implosion devices, in which metal cylindrical or conical shells are collapsed inwardly by explosives. These devices are used for rapid closing of pipes and for shaped-charge weapons and oil well perforators. To work effectively, shaped-charge liners should collapse without buckling.

New problems in dynamic buckling will undoubtedly appear as space structures are built, because these structures are very large and very thin. Any slight impact will lead to local dynamic buckling because the wave transit time through the structure is long compared with the time for buckles to form.

An important new phenomenon of pulse buckling is that the mode of buckling is determined by the load, so the perspective for analysis is reversed from that in static buckling. In static analysis, the buckling mode is known (the fundamental mode) and the maximum safe load is to be determined. In pulse buckling, the load amplitude is specified and dictates the buckling modes. The design quantity to be determined is the maximum safe duration of the load. This inverted viewpoint is needed in order to recognize and analyze dynamic pulse buckling. As pulse buckling becomes familiar to engineers, buckling that could otherwise cause serious problems will become a standard response feature to be considered in the design of thinner and lighter weight structures.

The monograph presents a systematic development of dynamic pulse buckling, from simple elastic buckling of bars to the most recent developments in biaxial plastic flow buckling of shells. Emphasis is on developing an understanding of the buckling processes and on making available practical theory that can be used for estimating buckling strengths of structural elements (bars, plates, rings, shells) under a variety of pulse loadings. Familiarity with static buckling in these elements is assumed, but most derivations are made from fundamental principles so that the monograph can be used as a graduate level textbook. Each chapter is written to be understood independently so that the practicing engineer can go directly to the theory and experiment most appropriate to a specific problem. For student use, Chapters 1 and 2 should be read first because they describe in more detail fundamental ideas that are used throughout the remaining text.

The authors are indebted to the staff of the Poulter Laboratory at SRI International for contributions to both theory and experiment. Those who worked directly on the research are D.L. Anderson, B.P. Bain, J.H. Busma, J.D. Colton, L.J. Dary, J.K. Gran, G.R. Greenfield, R.E. Herbert, T.C. Kennedy, C.M. Romander, P. Schwindt, G.R. Sliter, H. Vaughan, and W. Zietzke. Manuscripts for papers, reports, and this text were prepared with the assistance of M. Adams and J. Berry and their staff, and N. Hall, V.A. Jercha, D.M. Phillips, K. Reeds, and N.J. Smith.

The research was made possible by the guidance and support of G.R. Abrahamson, Director of Poulter Laboratory and Vice President of SRI International, J.N. Goodier of Stanford University, and M.C. Atkins of Defense Nuclear Agency. Dr. Abrahamson performed the first experiments on plastic flow buckling in 1959 and with J.N. Goodier developed a theoretical explanation of the observed buckles. Section 3.2 is their paper reproduced with little change. In 1966 Goodier made another pioneering contribution to the theory by discovering the directional moment. Dr. Atkins was the technical monitor on this early research while he was with the Air Force Weapons Laboratory and continued his encouragement and support while at Avco Corporation and DNA. Dr. Atkins suggested that this monograph be written and with D.J. Kohler saw us through its preparation.

# 1. INTRODUCTION

## 1.1 FORMS OF DYNAMIC BUCKLING

Dynamic stability of structures is a very broad subject that includes not only dynamic buckling from transient and vibratory loads, but also interaction of structures with other media, such as in aircraft flutter, and interaction with active control systems that have their own dynamic characteristics. A review of many forms of dynamic stability of structures is available in the proceedings of an international conference on the subject held at Northwestern University in 1965.[1]

This monograph is concerned with buckling from prescribed dynamic loads acting on structural elements (bars, plates, rings, shells). To more precisely define the scope and areas of practical application, it is useful to further distinguish among several types of dynamic buckling based on the physical phenomena of the buckling processes. The first distinction is between buckling from oscillatory loads and buckling from transient loads consisting of a single pulse characterized by its amplitude, shape, and duration. The first type we call vibration buckling and the second we call pulse buckling.

In vibration buckling, the amplitudes of vibration caused by an oscillating load become unacceptably large at critical combinations of load amplitude, load frequency, and structure damping. A simple example is a column supporting an oscillating axial load as shown in Figure 1.1a. Inevitable imperfections in the column give rise to bending moments that excite lateral motion. The column vibrates laterally at large amplitude when the loading frequency is twice the natural bending frequency of the column: each time the column bows out to one side or the other, the axial loading force reaches its maximum and produces bending moments. The term vibration buckling describes the similarity to vibration resonance. The difference is that in vibration resonance the load is in the same direction as the motion (lateral to the column in the example) and excites the motion directly as an applied force in the equation of motion. Simple resonance occurs when the lateral loading frequency coincides with the natural frequency of vibration. By contrast, in vibration buckling the bending moment induced by the axial force introduces the force as a parameter multiplying the displacement in the equation of motion. A mathematical description of vibration buckling is therefore: *dynamic stability of vibrations induced by oscillating parametric loading.* This type of resonance is therefore called *parametric resonance.* An extensive treatment of this subject is given in a book by W. W. Bolotin.[2]

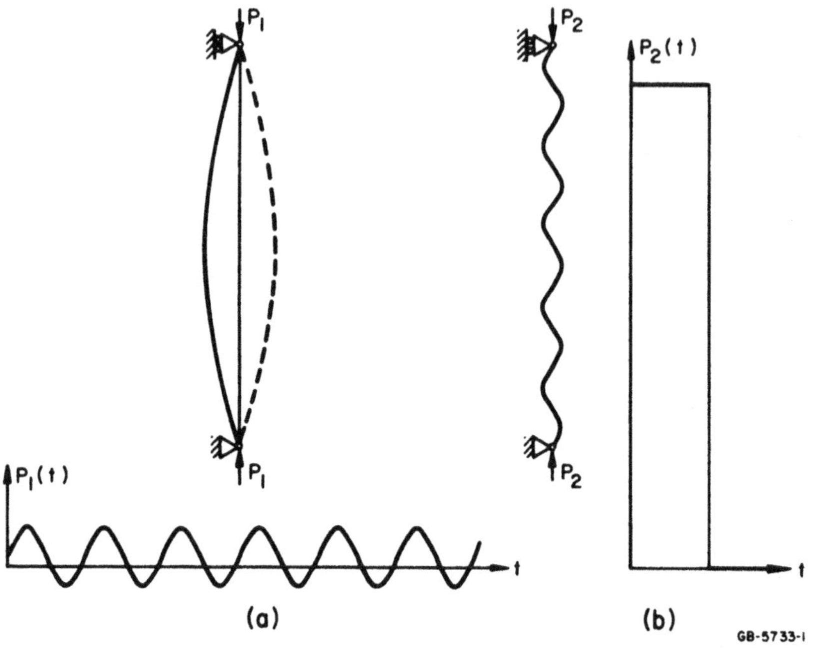


FIGURE 1.1 VIBRATION BUCKLING AND PULSE BUCKLING

In pulse buckling, the structure deforms to an unacceptably large amplitude as the result of transient response to the applied load. The deformation can be permanent, as a result of plastic response or snap through to a large-deformation postbuckled state, or the structure can return to its undeformed state. A simple example is a long column with a suddenly applied axial load many times greater than the static Euler load, as shown in Figure 1.1b. Motion grows exponentially in all modes with wavelengths longer than the Euler wavelength for the given load. The critical modes are those with greatest total growth during the time of load application. The critical condition for buckling is an unacceptably large deformation or stress. The column can survive a large axial load before reaching this condition if the load duration is short enough. Under an intense, short duration load, the column buckles into a very high-order mode as shown in Figure 1.1b.

This type of buckling might also be called response buckling because it is similar in form and analytical approach to simple dynamic response from prescribed loading histories. Again, the difference from simple dynamic response is that the load appears as a parameter multiplying the displacement in the equations of motion; therefore, pulse buckling falls under the mathematical definition: *dynamic response of structural systems induced by time-varying parametric loading.*

The final distinction to be made relates to the modes in which the pulse buckling takes place. In this monograph, we focus mainly on response in high-order modes as in the example in Figure 1.1b. In all the problems to be addressed, the load amplitude is higher than the static buckling load; in most of the problems, the amplitude is *much* higher than the static buckling load. We use the term pulse buckling because it tends to emphasize the high amplitude, short duration loading of primary concern here. We distinguish between high and low order buckling because the extensive published work on low-order buckling is beyond the scope of the monograph. The extent of this published work, however, emphasizes the importance of low-order dynamic buckling in conventional structural design, in which dynamic loads often do not differ widely from static buckling loads.

Three problems in which a great deal of low-order dynamic buckling work has been done are (1) buckling of columns, for example, in aircraft landing struts, (2) buckling of arches and spherical caps, and (3) buckling of cylindrical shells under axial load. The last two problems have in common that the structures are extremely sensitive to initial imperfections in shape. As a result, critical dynamic buckling loads of long duration (e.g., step loads) can be smaller in amplitude than the corresponding slowly applied load.[3] Problems in which this feature of dynamic buckling is important are not addressed in this monograph. The fundamentals of the low-order column buckling problem are included in the monograph, but the many elaborations available in the literature are not. Emphasis is on high-order buckling because it relates more closely to high-order buckling of plates, rings, and shells, the subject that has received the most attention of the authors. Dynamic buckling of cylindrical shells under axial impact is treated in considerable detail for both elastic and plastic axial waves, but always with emphasis on high-order response.

## 1.2 EXAMPLES OF DYNAMIC PULSE BUCKLING

A few examples illustrate the features of dynamic pulse buckling. The descriptions here are qualitative, to give the nature and scope of dynamic pulse buckling. Detailed interpretations and analyses are given in the following chapters. The simplest example is again a long column as in Figure 1.1b. In practice, the large amplitude load is typically applied by impact at one end. Figure 1.2 gives a sequence of high speed photographs of a thin strip buckling under axial impact at its lower end. The impact produced an elastic compression wave that traveled up the strip at velocity 0.2 inch/$\mu$s (5 mm/$\mu$s), starting at the lower end at $t = 0$. As is typical in such impact buckling, the wave front moved through a distance much greater than the buckle wavelength before any visible buckling took place. At the time of first perceptible buckling, near the impacted end at $t = 24$ $\mu$s, the wave front had traveled 5 inches (127 mm), about twice the length of the portion of the strip in the photographs. Thus, each buckle sees essentially a constant, suddenly applied axial thrust. The buckles therefore remain fixed in position and merely

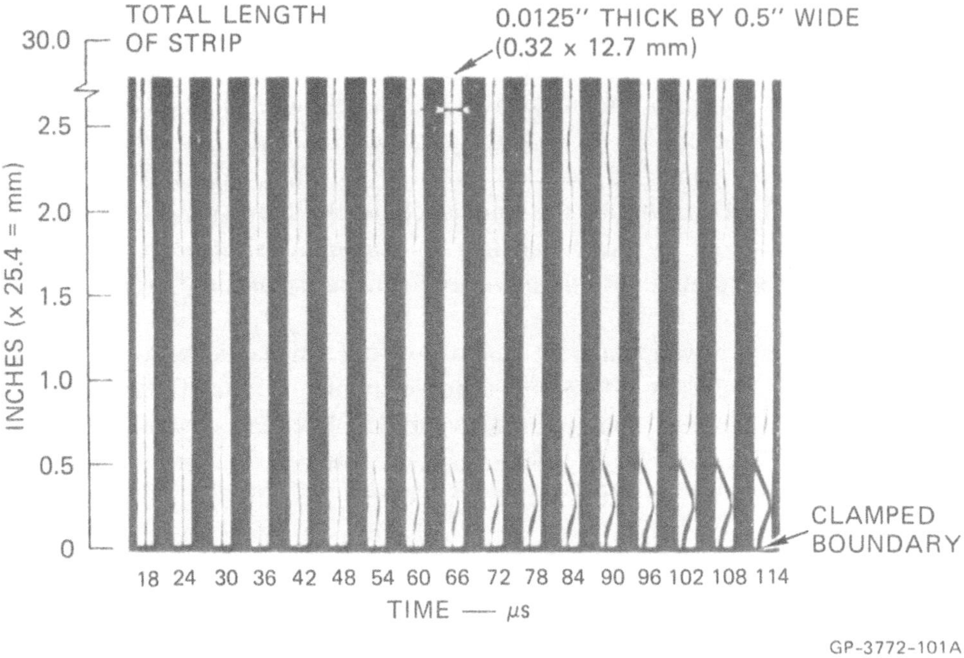


FIGURE 1.2 WAVES FORMING IN A 6061-T6 ALUMINUM STRIP

Time is measured from instant 40,000 psi (276 MPa) compressive wave reflects from clamp support.

grow in amplitude with time. The amplitude of buckles near the impacted end is larger than the amplitude of buckles farther up the strip because of unavoidable eccentricity of the impact and because the duration of axial loading is longest at the end. In this example and those that follow, the buckling motion was allowed to proceed well beyond any threshold of acceptable motion in order to see the buckling process clearly.

Similar high speed photographs of buckling in a thin cylindrical shell under axial impact at its lower end are given in Figure 1.3. The shell is very thin (radius-to-thickness ratio $a/h = 550$), so the buckling process is similar to that in the thin strip. The amplitude of the axial stress wave is 1.5 times the classical elastic axial buckling stress of the shell. The large-amplitude buckles at late time are in the familiar diamond pattern, but the buckle wavelengths are much shorter than those of the static large deflection post-buckled state. This difference is illustrated in Figure 1.4, which compares post-test photographs of a shell buckled dynamically as in Figure 1.3 and an identical shell buckled by a slowly applied load in a testing machine. (In all the shells, a clamped end condition was provided by a thick internal ring and a thin exterior ring whose upper edge can be seen in the photographs; the narrow buckle in the dynamically buckled shell in Figure 1.4 is just above this

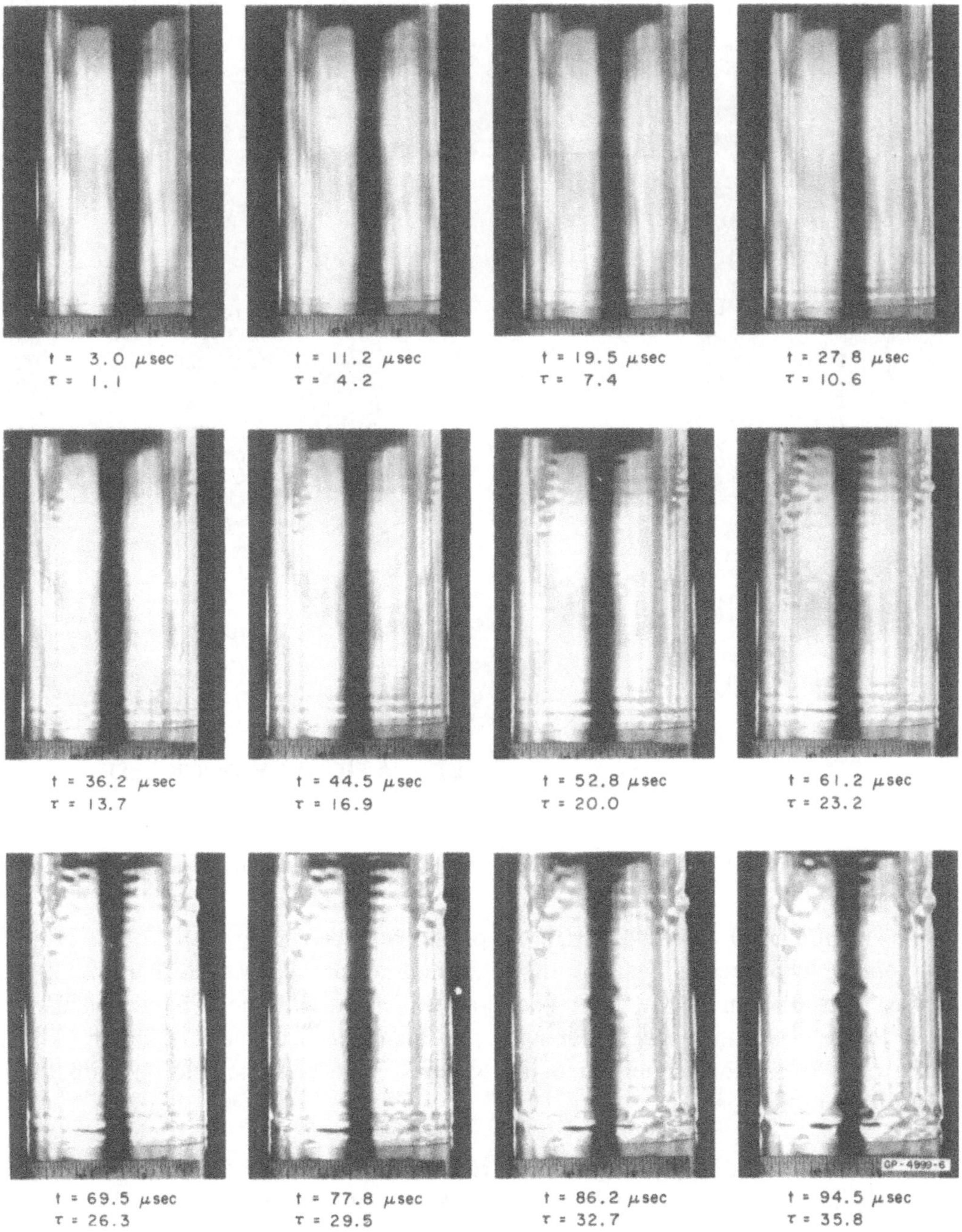


FIGURE 1.3 HIGH SPEED PHOTOGRAPHS OF BUCKLING IN A THIN CYLINDRICAL SHELL UNDER AXIAL IMPACT

Time is from initial impact at rigid end ring; a/h = **550**, $\sigma/\sigma_s$ = 1.5.

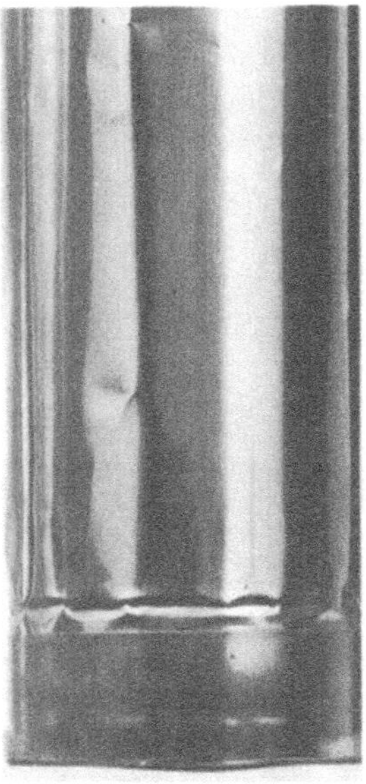

(a) DYNAMICALLY BUCKLED

(b) STATICALLY BUCKLED

FIGURE 1.4 COMPARISON OF STATIC AND DYNAMIC POST-BUCKLED PATTERNS

clamped boundary.) Observe that most of the buckling in Figure 1.3 was elastic; only one or two buckles near the impacted end are permanently formed by plastic hinges, as shown in Figure 1.4. The axial wavelength of the dynamic buckles is close to the classical buckle wavelength. A theoretical treatment of this buckling, based on a dynamic counterpart to the classical theory, agrees closely with these experimental observations.

Dynamic buckling of shells under radial pulse loads is illustrated in Figure 1.5. The very thin shell in Figure 1.5a ($a/h = 480$) was loaded by an impulsive radial pressure over one side of the shell. The resulting hoop stress as the shell moved inward produced a destabilizing thrust similar to that of the axial compression wave in the thin strip in Figure 1.2. Because the peak value of the hoop stress was close to the value of the compressive stress in the strip, the wavelength of buckles is also similar. This is shown in Figure 1.5b, in which a buckled strip from the experiments of Figure 1.2 is placed near a group of buckles in the shell in Figure 1.5a.

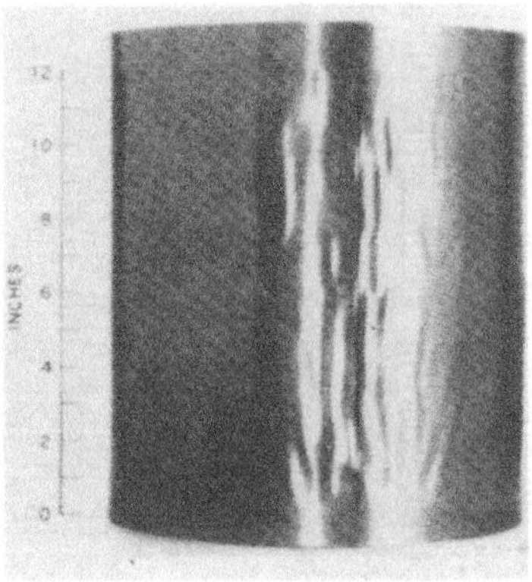


(a) THIN SHELL (a/h = 480)

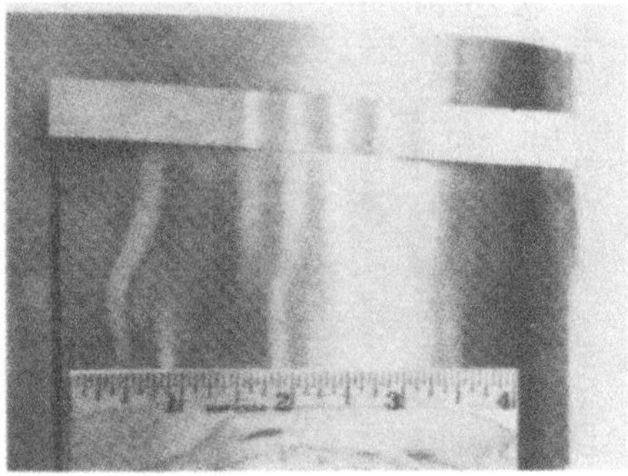


(b) THIN SHELL AND STRIP

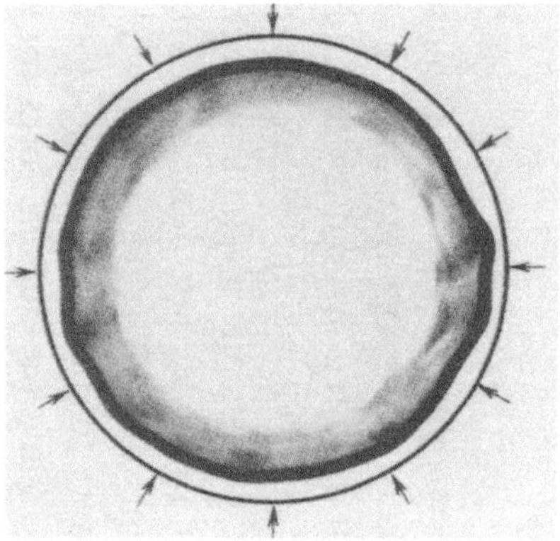

(c) THICK SHELL

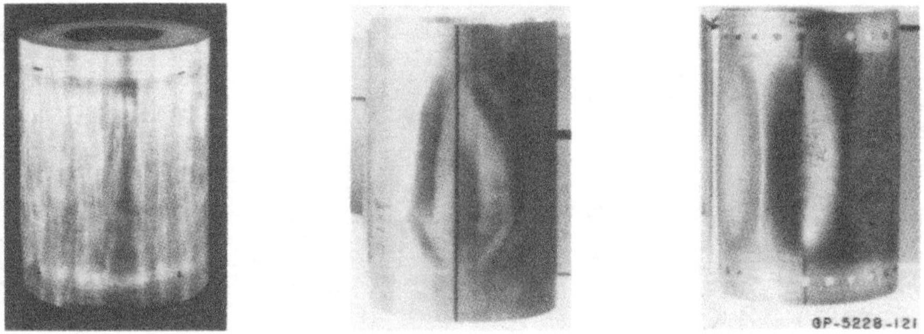


(d) THIN SHELLS (a/h = 100), INCREASING LOAD DURATION

GP-5228-121A

FIGURE 1.5 DYNAMIC PULSE BUCKLING OF CYLINDRICAL SHELLS UNDER RADIAL LOADS (Aluminum 6061-T6)

The mode number of buckling is very high; the wavelengths in Figure 1.5a correspond to mode numbers of $n = 50$ to 100 buckles around the circumference. Part of the reponse analysis for pulse buckling is to determine these "preferred" modes of buckling.

The wall of the shell in Figure 1.5c was projected inward by a radial impulse uniform around the circumference. It is much thicker ($a/h = 19$) than the wall of the shell in Figure 1.5a, so the initial radial velocity required to produce buckling was much higher. The hoop strain was several percent, well into the plastic range, which results in another form of pulse buckling called dynamic plastic flow buckling. In this simple form it is similar to the elastic examples just discussed, with the elastic modulus replaced by the plastic tangent modulus. The permanent wrinkles in the elastic examples in Figures 1.4 and 1.5 were formed by late-time plastic hinges following elastic buckling. In plastic flow buckling the buckles form while the material throughout the entire wall thickness is flowing plastically from hoop compression. Resistance to buckling therefore comes from the plastic tangent modulus rather than from the elastic modulus. The tangent modulus in metals is typically about 100 times smaller than the elastic modulus. Buckle wavelengths, which are proportional to the square root of the modulus, are therefore about 10 times shorter (in comparison to wall thickness) than in elastic buckling. This can be seen by comparing the edge view of the buckling thin strip in Figure 1.2 with the end view of the buckled thick shell in Figure 1.5c. A similar situation occurs in thick bars under high velocity axial impact.

Figure 1.5d illustrates the effect of changing the pulse duration of radial pressure loading. All three shells have $a/h = 100$ and were clamped at their ends. The shell on the left was loaded by an impulsive pressure as in Figures 1.5a and b and is buckled into a very high-order mode ($n \approx 45$). The shell on the right was loaded by a long duration blast wave and is buckled on its loaded side into wavelengths corresponding to $n \approx 7$ waves around the complete circumference. This is close to the $n = 6$ pattern for static radial pressure uniform around the shell. The shell in the center was loaded by a blast wave of intermediate duration and is buckled into an $n \approx 13$ wave pattern. The changing wave pattern is the result of the increase in peak pressure (and hence hoop stress) required to cause buckling as the pulse duration is reduced. This illustrates again that the mode number of buckling is an important response feature that must be determined in dynamic pulse buckling analysis.

Figure 1.6 shows the symmetric form of buckling in thick-walled tubes under high velocity axial impact. The tubes were 25 mm O.D. with a wall thickness of 2.5 mm (mean radius-to-thickness ratio $a/h = 4.5$). As in the thick shell under radial impulse, buckling took place during sustained plastic flow, this time with the unperturbed flow mainly in the axial direction. A new feature in this flow buckling is

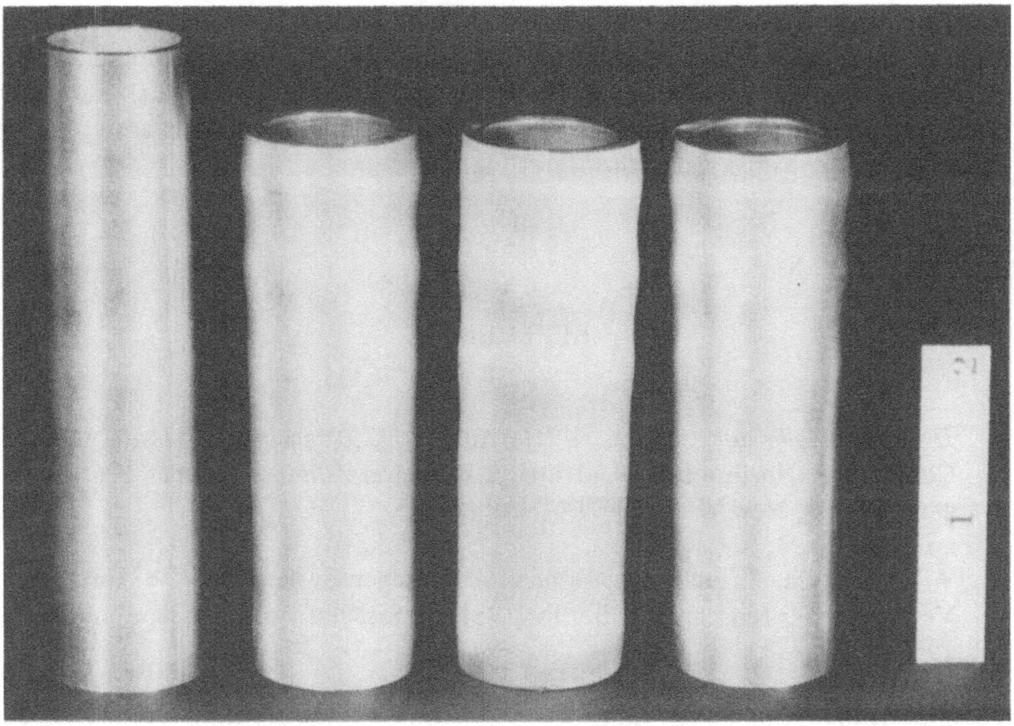

MP-7504-17

FIGURE 1.6 AXISYMMETRIC PLASTIC FLOW BUCKLING IN CYLINDRICAL SHELLS FROM IMPACT AT 342 ft/sec (104 m/s)

Tube at left shows shape before impact.

that there is plastic flow in both in-plane directions (axially and circumferentially) and that the flow kinematics require that stress states through the wall thickness lie at different positions on the yield surface. Flow kinematics dictate the direction of the strain rate vector, and the associated flow rule (the Mises yield function is assumed) therefore dictates the position of the stress state on the current yield surface. During flow perturbed by buckling motion, these stress states are such that a bending moment is induced that resists buckling even in the absence of strain hardening. This new bending moment is called the directional moment, because it is the result of the kinematic constraint on the direction of the strain rate vector. A carefully developed explanation of directional moments is given in the text.

Directional moments also occur in plastic flow buckling of intermediate-length cylindrical shells loaded by radial impulse. These shells are long enough to impose a kinematic constraint on axial flow (constant through the thickness) but short enough that there is axial flow and hence a directional moment. For long shells (Figure 1.5c) or rings, there is no directional moment because the strain rate vector is either fixed in direction (long shell, no axial flow) or there is no axial flow restraint (ring). In some materials, a third type of bending moment results from an increase in stress with an increase in strain rate--the viscoplastic moment. The effects and relative importance of strain hardening (tangent modulus) moments, directional moments, and viscoplastic moments in various situations are crucial in the analysis of dynamic plastic flow buckling.

## REFERENCES

1. *Dynamic Stability of Structures*, G. Herrmann, Ed., Proceedings of International Conference, Northwestern University, Evanston, Illinois, October 1965 (Pergamon Press, New York, 1967).

2. W.W. Bolotin, "Kinetische Stabilität Elastischer Systeme" (VEB Deutscher Verlag der Wissenschaften, Berlin, 1961), translated from Russian edition of 1956.

3. J.W. Hutchinson and B. Budiansky, "Dynamic Buckling Estimates," AIAA J., **4**, 3, pp. 525-530 (March 1966).

# 2. IMPACT BUCKLING OF BARS

## 2.1 INTRODUCTION

Most of the features unique to dynamic pulse buckling are illustrated in detail by buckling of bars under axial impact. An important feature that does not occur is directional moments, because all stresses are axial. Study of buckling in long bars shows that many high-order modes are excited and that they grow exponentially in amplitude with time as long as the load is maintained. These are called the *buckling modes.* The shape into which the bar deforms depends on the relative amplitude of growth in the buckling modes and on the bar imperfections that initiate growth. A plot of relative amplitude versus mode number is called the *amplification function.* The most amplified mode is called the *preferred mode* of dynamic buckling.

Imperfections are treated as either discrete or random. The bar is a convenient structural element in which to examine the relative importance of these two imperfection types. Eccentricity at the point of impact provides a discrete imperfection and associated unique buckled shape. Deviations from perfect straightness along the length of the bar are treated as random. Experiments on bars are inexpensive enough that many tests were performed to compare the theoretical distribution of wavelengths calculated using this assumption with the experimental distribution observed in a collection of bars. The random imperfection assumption is found to give a reasonably accurate description of the experimentally observed results. This demonstration is used to formulate a critical condition for buckling based on amplification of random imperfections. In later chapters this same buckling criterion is used for pulse buckling of cylindrical shells from radial and axial loads.

The bar is also a convenient structural element in which to examine the change in buckling patterns and critical load duration as the load amplitude is varied. Experiments were performed on aluminum bars with elastic impact stresses ranging from 30% to 80% of the yield stress, all with nominally a rectangular pulse shape. At each value of stress the pulse duration was increased until permanent buckles were just perceptible. The resulting curve of critical pulse duration versus axial stress (actually, a band of overlapping data points from buckled and unbuckled bars, owing to the random nature of the imperfections) is compared with theoretical curves for several assumed imperfection amplitudes. These critical curves are simple examples that anticipate similar critical curves of peak pressure versus impulse (given in the next chapter) for cylindrical shells under radial pressure pulses.

In one respect, impact buckling of bars is a more complex problem than buckling of shells under radial pressure. In dynamic buckling under axial load, the load is communicated to the bar by axial stress waves that move up and down the bar. For the high-order buckling considered in this monograph, buckling takes place during one transit of the compressive wave up the bar. The effect of this moving wave is not treated explicitly in the theoretical analysis given here, but it is shown that the amount of lateral buckling motion that takes place during the time to traverse a buckle wavelength is small compared with the total buckling motion. Thus, from a practical standpoint the complex problem of buckle motion as the wave passes can be neglected.

The effect of moving waves on a simply supported bar buckling in its fundamental mode was treated by Sevin.[1] In this case, buckling takes place as the axial wave moves up and down the bar many times. Sevin's approach is similar to that given here: he neglects motion as the wave traverses the bar and considers only that the bar is subjected to a load that increases uniformly along its length in a series of steps that add to the load with each arrival of the wave at the ends of the bar. The ends are taken to move toward each other at fixed velocity, as in a testing machine. His conclusion is the same as that here, namely, that because the motion during each step is small compared with the total buckling motion, this simplification is reasonable.

The final form of pulse buckling demonstrated by bar impact is dynamic plastic flow buckling under very high velocity, short duration impact. This is the subject of the final section of this chapter and is introduced more completely at that point.

## 2.2 ELASTIC BUCKLING OF LONG BARS

This section is concerned mainly with dynamic elastic buckling of long bars from axial loads well in excess of the static Euler load of the bar considered as a simply supported column. In fact, in bar impact experiments of the type given in Figure 1.2, the bars are so long that they buckle before any signal is received from the free end, so that there is no bar length and hence no physical Euler load. Nevertheless, it is useful to formulate the theoretical problem as though the bar were a column with supports at both ends because of the familiarity of this formulation and because it allows direct use of a statistical response analysis available from communication theory. Also, before we consider dynamic pulse buckling of this bar, it is useful to present the theory of static buckling. This helps in relating the static and dynamic problems and introduces, in terms of long established static buckling experimental results, the later treatment of discrete and random imperfections in dynamic buckling. Both the static and dynamic theories are presented before the dynamic experiments are described in detail so that the experiments can be better understood.

### 2.2.1 Equations of Motion

We consider elastic buckling of a simply supported uniform bar under axial compression, as in Figure 2.1a. The bar is of length L and supports an axial compressive force P. Its cross section is uniform with axial distance x, measured from one end. Deflection y is taken positive downward and is measured from an unstressed initial deflection $y_o(x)$. An element of length dx between two cross sectons taken normal to the original (undeflected) axis of the bar is shown in Figure 2.1b. The shearing force Q and bending moment M acting on the sides of the element are taken positive in the directions shown. The inertia force acting on the element is $\rho A(\partial^2 y/\partial t^2)\,dx$, where $\rho$ is the density of the bar material, A is the area of the cross section, and t is time.

The basic equations for the analysis of bar buckling are derived from dynamic equilibrium of the element in Figure 2.1b and the moment-curvature relation for the bar. Summing forces in the y direction gives

$$-Q - \rho A \frac{\partial^2 y}{\partial t^2} dx + (Q + dQ) = 0$$

or

$$\rho A \frac{\partial^2 y}{\partial t^2} = \frac{dQ}{dx} \qquad (2.2.1)$$

Taking moments about point B and neglecting rotary inertia of the element results in

$$M - \rho A \frac{\partial^2 y}{\partial t^2} dx \frac{dx}{2} + (Q + dQ)\,dx$$

$$- (M + dM) + P \frac{\partial}{\partial x}(y + y_o)\,dx = 0$$

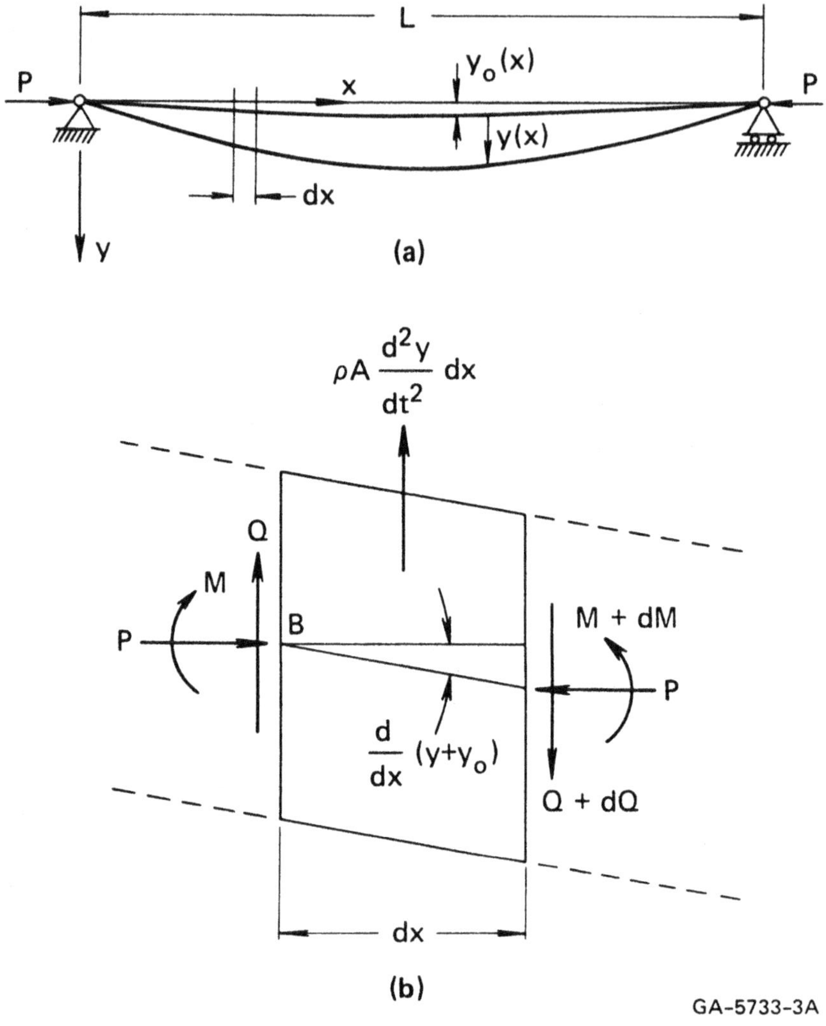


FIGURE 2.1 BAR NOMENCLATURE AND ELEMENT OF LENGTH

Terms of second order are neglected, reducing this equation to

$$Q = \frac{\partial M}{\partial x} - P \frac{\partial}{\partial x}(y + y_o) \tag{2.2.2}$$

If the effects of shear deformations and shortening of the bar axis are neglected, the curvature of the bar axis is related to the bending moment by

$$EI \frac{\partial^2 y}{\partial x^2} = -M \tag{2.2.3}$$

in which E is Young's modulus and I is the moment of inertia of the bar section, assumed symmetric about the xy plane (otherwise the bar would twist in addition to bending). The differential equation for the deflection of the bar axis is found by differentiating (2.2.2) and then eliminating Q by means of (2.2.1) and M by means of (2.2.3) twice differentiated. The result is

$$EI \frac{\partial^4 y}{\partial x^4} + P \frac{\partial^2}{\partial x^2}(y + y_o) + \rho A \frac{\partial^2 y}{\partial t^2} = 0 \tag{2.2.4}$$

### 2.2.2 Static Elastic Buckling of a Simply Supported Bar

For static buckling, the inertia term is neglected and (2.2.4) becomes

$$EI \frac{d^4 y}{dx^4} + P \frac{d^2 y}{dx^2} = -P \frac{d^2 y_o}{dx^2}$$

or, substituting $k^2 = P/EI$,

$$\frac{d^4 y}{dx^4} - k^2 \frac{d^2 y}{dx^2} = -k^2 \frac{d^2 y_o}{dx^2} \tag{2.2.5}$$

If we consider first a bar with no initial deflection, we need only the general solution to the homogeneous equation [with $y_o(x) \equiv 0$]. This solution is

$$y = A \sin kx + B \cos kx + Cx + D \tag{2.2.6}$$

For a simply supported bar, the deflection and bending moment are zero at the ends, and the boundary conditions are therefore

$$y = \frac{d^2 y}{dx^2} = 0 \text{ at } x = 0 \quad \text{and} \quad x = L \tag{2.2.7}$$

Applying these to (2.2.6) gives

$$B = C = D = 0, \quad \sin kL = 0$$

and therefore

$$kL = \pm n\pi$$

where n is an integer. Using the definition of k, this becomes an equation for P.

$$P_n = \frac{\pi^2 EI}{L^2} \cdot n^2 \tag{2.2.8}$$

Thus, with no initial deflection, only discrete values of P give a nontrivial solution, and the magnitude A of the deflection is undetermined.

Before discussing these solutions further, let us treat the bar having an initial shape $y_o(x)$. The solution for the perfectly straight bar suggests that $y_o(x)$ should be expressed by the Fourier sine series

$$y_o(x) = \sum_{n=1}^{\infty} a_n \sin \frac{n\pi x}{L} \tag{2.2.9}$$

The coefficients in this series are found from

$$a_n = \frac{2}{L} \int_0^L y_o(x) \sin \frac{n\pi x}{L}\, dx \tag{2.2.10}$$

Substitution of (2.2.9) into (2.2.5) gives the following differential equation for the imperfect bar:

$$\frac{d^4y}{dx^4} + k^2 \frac{d^2y}{dx^2} = k^2 \frac{n^2\pi^2}{L^2} a_n \sin \frac{n\pi x}{L} \tag{2.2.11}$$

To find a particular solution, we take

$$y_p = \sum_{n=1}^{\infty} A_n \sin \frac{n\pi x}{L} \tag{2.2.12}$$

When this is substituted into (2.2.11), the coefficients $A_n$ are found to be

$$A_n = \frac{-k^2 a_n}{k^2 - n^2\pi^2/L^2} = \frac{-Pa_n}{P - P_n} \tag{2.2.13}$$

The complete solution is then

$$y = A \sin kx + B \cos kx + Cx + D - \sum_{n=1}^{\infty} \frac{Pa_n}{P - P_n} \sin \frac{n\pi x}{L} \tag{2.2.14}$$

Since P, and hence k, is arbitrary, application of the boundary conditions (2.2.7) gives A = B = C = D = 0 and the general solution is simply

$$y = -\sum_{n=1}^{\infty} \frac{Pa_n}{P - P_n} \sin \frac{n\pi x}{L} \tag{2.2.15}$$

From this solution we see that the deflection becomes arbitrarily large as P approaches the critical loads $P_n$ given by (2.2.8). However, the dynamic solution given in subsequent sections shows that the motion is unstable for *any* load greater than the lowest critical load $P_1$, which, from (2.2.8), is given by

$$P_1 = \frac{\pi^2 EI}{L^2} \tag{2.2.16}$$

In the neighborhood of $P = P_1$, the first term dominates the deflection. Neglecting the higher terms, the midspan deflection for $P < P_1$ is given approximately by

$$\delta = y\,(L/2) \approx \frac{-Pa_1}{P - P_1} \tag{2.2.17}$$

Figure 2.2a gives a plot of deflection δ from (2.2.17) versus end load P. On the basis of this formula, Southwell[2] suggested that the critical load $P_1$ could be

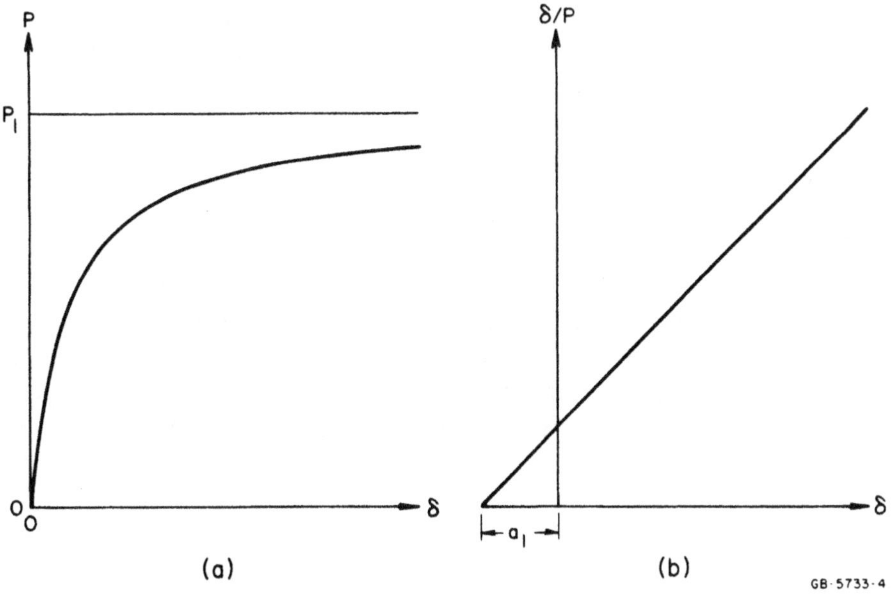


FIGURE 2.2 FORCE-DEFLECTION CURVE AND SOUTHWELL PLOT FOR SMALL DEFLECTION ELASTIC BUCKLING

extracted from test data by plotting $\delta/P$ versus $\delta$. In this form, (2.2.17) becomes

$$\frac{\delta}{P} = \frac{1}{P_1}(\delta + a_1) \tag{2.2.18}$$

which gives the straight line in Figure 2.2b. The inverse of the slope gives the critical load $P_1$ and the $\delta$ intercept gives the coefficient $a_1$ as shown.

If the bar is treated as initially perfectly straight but subjected to an eccentrically placed load, the Southwell procedure can still be used to determine the critical load. Consider, for example, that the load is displaced from the centroidal axis by an amount $\epsilon$, equal at both ends. This can be treated as a bar having an initial displacement given by

$$\begin{aligned} y_o(x) &= \epsilon \qquad x \neq 0, L \\ &= 0 \qquad x = 0, L \end{aligned} \tag{2.2.19}$$

Substituting this displacement into (2.2.10), the coefficient of the first term in its Fourier expansion is

$$a_1 = \frac{4\epsilon}{\pi} \tag{2.2.20}$$

Thus, for P in the neighborhood of $P_1$ the Southwell plot is as described previously, and the $\delta$ intercept is now $4\epsilon/\pi$. If the bar is considered to have both an initial shape and some eccentricity, (2.2.18) becomes

$$\frac{\delta}{P} = \frac{1}{P_1}\left[\delta + \left(a_1 + \frac{4\epsilon}{\pi}\right)\right] \tag{2.2.21}$$

For real columns, in which both $a_1$ and $\epsilon$ are small and difficult to measure, there is therefore no way of telling in a Southwell plot how much of the deflection is caused by load eccentricity and how much is caused by an initial deflection. In experiments run near the turn of the century,[3–5] it was found that the experimental buckling deflections could be calculated, on the average, by using values of equivalent eccentricity given by

$$\epsilon = 0.06\, r^2/c \tag{2.2.22}$$

where $r^2/c$ is the core radius of the cross section, r being the radius of gyration and c being the distance from the elastic axis to the outermost fiber. For a rectangular bar of depth h, this gives $\epsilon = 0.01$ h. In long columns, it is reasonable to assume that initial imperfections in shape become more important, and these can be expected to depend on the length of the column. On this basis, Salmon[6] found that, although equivalent imperfections from a large collection of experimental results were scattered by an order of magnitude at any given length, both the average amplitude of the imperfections and the range of amplitudes increased in proportion to the length of the bars. For the longer columns, almost all imperfections were in the band

$$0.0001 < \frac{a_1}{L} < 0.001 \tag{2.2.23}$$

Several authors have proposed that imperfections depending on both the core radius and the column length can be expected to be present. They suggest that a conservative estimate for an equivalent deflection including both types of imperfections can be taken as

$$a_1 = 0.1\,r^2/c + \frac{L}{750} \tag{2.2.24}$$

In the dynamic problems discussed in later sections, we will see that the range of normalized imperfections found in static buckling gives reasonably good agreement with values observed in dynamic buckling.

### 2.2.3 Theory of Dynamic Elastic Buckling of a Simply Supported Bar

The static buckling considered in the preceding sections was concerned with the steady load that can be safely carried by a column or bar. If, instead, a load is suddenly applied and then removed, as in hammering a nail, the maximum load can far exceed the static buckling load without inducing objectionably large strains or deflections. On the other hand, oscillatory forces such as from reciprocating or unbalanced machinery, even while producing loads smaller than the static buckling load, can nevertheless produce objectionably large deflections if the frequency of oscillation bears a critical relation to the natural frequency of the column. Both of these problems involve dynamic buckling. As discussed in Chapter 1, the impact of a nail is a *pulse buckling* problem, whereas a column under an oscillatory load is a *vibration buckling* problem. In the remainder of this chapter we will examine several examples of elastic and plastic pulse buckling of bars.

In the pulse problem, loads can be applied with no appreciable buckling right up to and beyond the elastic limit, provided only that they are applied for a short enough time. Because of this feature in the dynamic problem, rather than asking for the maximum load that can be carried, we specify a load and ask for the response. Knowing how the buckling grows with time, we then determine the maximum duration for which the given load can safely be applied. In Section 3.5 this procedure will be applied to more general problems in which the load varies continuously with time.

Consider first a simply supported bar under a compressive load P, uniform throughout its iength as shown in Figure 2.1. The force P may be much larger than the critical Euler load $P_1$, but for the present, the average compressive stress is assumed to be within the elastic limit. To keep the bar from buckling during application of the load P, imagine that it is supported all along its length by lateral constraining blocks.* Then, at time $t = 0$, the blocks are suddenly removed and buckling motion begins. The motion is governed by Equation (2.2.4), repeated here.

$$EI \frac{\partial^4 y}{\partial x^4} + P \frac{\partial^2}{\partial x^2}(y + y_o) + \rho A \frac{\partial^2 y}{\partial t^2} = 0 \tag{2.2.25}$$

After dividing through by EI, it is convenient to introduce the parameters

$$k^2 = \frac{P}{EI}, \quad r^2 = \frac{I}{A}, \quad c^2 = \frac{E}{\rho} \tag{2.2.26}$$

The first two parameters have already appeared in the static problem. The new parameter, appearing because of the dynamic inertia term, is the wave speed of longitudinal stress waves in the bar.[7] When these quantities are used, the equation of motion (2.2.25) becomes

$$\frac{\partial^4 y}{\partial x^4} + k^2 \frac{\partial^2 y}{\partial x^2} + \frac{1}{r^2 c^2} \frac{\partial^2 y}{\partial t^2} = -k^2 \frac{\partial^2 y_o}{\partial x^2} \tag{2.2.27}$$

As in the static problem, the boundary conditions of zero moment and displacement at the ends of the bar give

$$y = \frac{\partial^2 y}{\partial x^2} = 0 \quad \text{at} \quad x = 0 \quad \text{and} \quad x = L \tag{2.2.28}$$

The solution to (2.2.27) subject to boundary conditions (2.2.28), as in the static

---

*In practice, the load is suddenly communicated to the bar by an axial stress wave (or waves). Effects of these waves are small, as will be seen in Section 2.2.8.

problem, can be expressed by a Fourier sine series in x. Thus, we assume a product solution

$$y(x,t) = \sum_{n=1}^{\infty} q_n(\tau) \sin \frac{n\pi x}{L} \tag{2.2.29}$$

The initial displacement $y_o(x)$ is also expressed in series form by

$$y_o(x,t) = \sum_{n=1}^{\infty} A_n \sin \frac{n\pi x}{L} \tag{2.2.30}$$

where the coefficients can be found from

$$A_n = \frac{2}{L} \int_0^L y_o(x) \sin \frac{n\pi x}{L} \, dx \tag{2.2.31}$$

Equations (2.2.29) and (2.2.30) are now substituted into (2.2.27) to give the following equation of motion for the Fourier coefficients $q_n(t)$:

$$\left( \frac{n^4\pi^4}{L^4} - k^2 \frac{n^2\pi^2}{L^2} \right) q_n + \frac{1}{r^2c^2} \ddot{q}_n = k^2 \frac{n^2\pi^2}{L^2} A_n \tag{2.2.32}$$

which, upon rearranging to the more standard form, becomes

$$\ddot{q}_n + \frac{r^2c^2n^2\pi^2}{L^2} \left( \frac{n^2\pi^2}{L^2} - k^2 \right) q_n = r^2k^2c^2 \cdot \frac{n^2\pi^2}{L^2} A_n \tag{2.2.33}$$

One of the principal points of the theory of dynamic buckling to be discussed in this volume appears here. The nature of the solutions to Equation (2.2.33) depends on the sign of the coefficient of $q_n$. If $n\pi/L < k$, this coefficient is negative and the solutions are hyperbolic; if $n\pi/L > k$, this coefficient is positive and the solutions are trigonometric. Thus, if the mode numbers n are large enough, $n > kL/\pi$, the displacements are trigonometric and therefore bounded. However, over the lower range of mode numbers, $n < kL/\pi$, the hyperbolic solutions grow exponentially with time and have the potential of greatly amplifying small initial imperfections. These modes are therefore called the *buckling modes.*

The mode number $n = kL/\pi$, separating the trigonometric and hyperbolic solutions, gives a wavelength corresponding to the wavelength of static buckling under the given load P; no matter how long the duration of load application, if $n > kL/\pi$ the motion remains bounded, while for any $n < kL/\pi$ the motion diverges. To see more clearly this relation to a static problem, recall first from Equation (2.2.29) that the deflection curve of the bar is a sine wave with n half-waves. For $n = kL/\pi$, this curve is given by sin kx. One half-wave of this deflection curve, corresponding to the buckle shape of a simple pinned Euler column, therefore occupies a distance from the left support given by

$$kx_{st} = \pi$$

or

$$x_{st} = \pi/k \qquad (2.2.34)$$

Using the definition $k^2 = P/EI$, this relation gives

$$P = \frac{\pi^2 EI}{x_{st}^2} \qquad (2.2.35)$$

This is identical to Equation (2.2.16) for the static buckling of an Euler column of length $x_{st}$ under the load P.

The dynamic equation also demonstrates the statement made in Section 2.2.2 that loads greater than $P_1 = \pi^2 EI/L^2$ give unstable motion. This follows from the observation already made that the motion is unstable if the coefficient of $q_n$ in (2.2.33) is negative, that is, if

$$\frac{n^2\pi^2}{L^2} - k^2 < 0 \qquad (2.2.36)$$

Since $k^2 = P/EI$ is positive, this quantity is most negative for $n = 1$. Using $n = 1$ in Equation (2.2.36), the left-hand side is negative for all $P > \pi^2 EI/L^2$ and the motion is unstable as previously stated.

For the dynamic problems of present interest here, $P >> \pi^2 EI/L^2$ and many modes are unstable. Thus the mode numbers of the buckling modes are very high and the wavelengths of the buckling are so short that the total length of the bar becomes relatively unimportant. In fact, in experiments to be described later, dynamic buckling is produced by impact at one end of the bar and, because of the finite speed of axial wave propagation, buckling occurs before any signal is received from the opposite end. In this problem the total length of the bar has no significance at all. We should therefore seek a characteristic length other than the length of the bar. Because the nature of the motion changes at the static Euler wavelength $x_{st} = \pi/k$, it is quite natural to use $1/k$ as the characteristic length in the x-direction, along the bar. Similarly, it is natural to normalize lateral deflections with respect to the radius gyration r of the cross section. The ratio between these lengths is a significant parameter and will be denoted by s:

$$s^2 = r^2k^2 = \frac{r^2P}{EI} = \frac{P}{AE} = \epsilon \qquad (2.2.37)$$

Thus the wavelength of the buckling varies inversely with the square root of the strain $\epsilon$ due to the compressive load P. This will be discussed more fully later.

To incorporate these lengths into the equation of motion, we introduce the dimensionless variables

$$w = \frac{y}{r} \quad , \quad \xi = kx = \frac{sx}{r} \quad , \quad \tau = \frac{s^2ct}{r} \qquad (2.2.38)$$

With these variables, Equation (2.2.25) becomes

$$w'''' + w'' + \ddot{w} = w_o'' \qquad (2.2.39)$$

where primes indicate differentiation with respect to $\xi$ and dots indicate differentiation with respect to $\tau$. Boundary conditions (2.2.28) become

$$w = w'' = 0 \quad \text{at} \quad \xi = 0 \quad \text{and} \quad \xi = \ell = \frac{sL}{r} \qquad (2.2.40)$$

and the product form of solution is now expressed by

$$w(\xi, \tau) = \sum_{n=1}^{\infty} g_n(\tau) \sin \frac{n\pi\xi}{\ell} \qquad (2.2.41)$$

Similarly, the initial displacements are

$$w_o(\xi) = \sum_{n=1}^{\infty} a_n \sin \frac{n\pi\xi}{\ell} \qquad (2.2.42)$$

where

$$a_n = \frac{2}{\ell} \int_0^{\ell} w_o(\xi) \sin \frac{n\pi\xi}{\ell} \, d\xi \qquad (2.2.43)$$

A wave number $\eta$ is introduced by

$$\eta = \frac{n\pi}{\ell} \qquad (2.2.44)$$

and finally (2.2.41) and (2.2.42) are substituted into (2.2.39) to give the equations of motion for the Fourier coefficients $g_n(\tau)$:

$$\ddot{g}_n + \eta^2(\eta^2 - 1)\, g_n = \eta^2 a_n \tag{2.2.45}$$

This equation corresponds to (2.2.33); in the new notation, the transition from hyperbolic to trigonometric solutions occurs at $\eta = 1$.

The general solution to (2.2.45) is

$$\begin{aligned} g_n(\tau) &= C_n \cosh p_n\tau + D_n \sinh p_n\tau - \frac{a_n}{1-\eta^2} \qquad \text{for } \eta < 1 \\ g_n(\tau) &= C_n \cos p_n\tau + D_n \sin p_n\tau - \frac{a_n}{1-\eta^2} \qquad \text{for } \eta > 1 \end{aligned} \tag{2.2.46}$$

where

$$p_n = \eta |1 - \eta^2|^{1/2}$$

These equations are substituted into (2.2.41) to obtain the general solution for the lateral displacement.

$$\begin{aligned} w(\xi, \tau) = \sum_{n=1}^{N} \left( C_n \cosh p_n\tau + D_n \sinh p_n\tau - \frac{a_n}{1-\eta^2} \right) \sin \frac{n\pi\xi}{\ell} \\ + \sum_{n=N+1}^{\infty} \left( C_n \cos p_n\tau + D_n \sin p_n\tau - \frac{a_n}{1-\eta^2} \right) \sin \frac{n\pi\xi}{\ell} \end{aligned} \tag{2.2.47}$$

where N is the largest integer for which $\eta < 1$.

The bar is assumed to be initially at rest. Also, recall that w is measured from the initial displacement $w_o$, so the initial conditions are

$$w(\xi, 0) = \dot{w}(\xi, 0) = 0 \tag{2.2.48}$$

Applying these to (2.2.47) yields $D_n = 0$ and $C_n = a_n/(1 - \eta^2)$. The final solution is then

$$w(\xi, \tau) = \sum_{n=1}^{\infty} \frac{a_n}{1-\eta^2} \left[ \begin{matrix} \cos \\ \cosh \end{matrix}\; p_n\tau - 1 \right] \sin \frac{n\pi\xi}{\ell} \tag{2.2.49}$$

in which the hyperbolic form is taken for $\eta < 1$ and the trigonometric form for $\eta > 1$.

### 2.2.4 Amplification Functions

Equation (2.2.49) shows quantitatively the exponential growth of the buckling terms. The ratio between the Fourier coefficients $a_n$ of the initial displacement and the coefficients $g_n(\tau)$ in the buckling bar will be called the *amplification function* and in this problem is given by

$$G_n(\tau) = \frac{g_n(\tau)}{a_n} = \frac{1}{1-\eta^2}\left[\begin{matrix}\cos\\ \cosh\end{matrix}\; p_n\tau - 1\right] \qquad (2.2.50)$$

A plot of this function, treating $\eta$ as a continuous variable, is given in Figure 2.3 for several values of dimensionless time $\tau$.

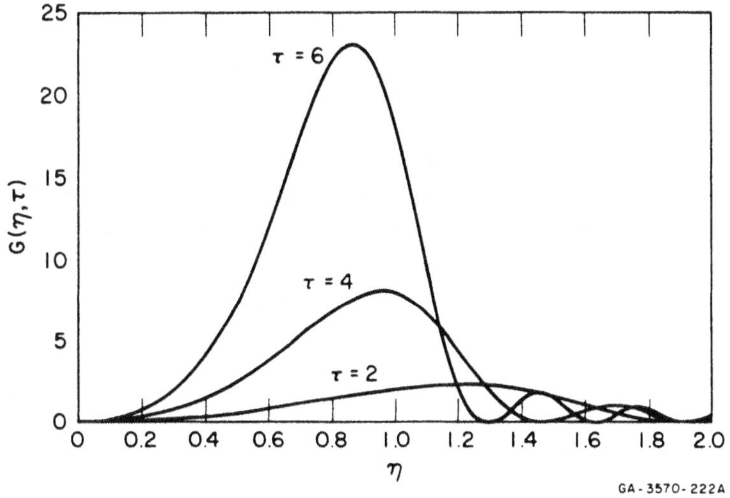


FIGURE 2.3 AMPLIFICATION FUNCTION

It is apparent that, as time increases, a narrow band of wavelengths is amplified having wave numbers centered at somewhat less than $\eta = 1$. To find the wave number of the most amplified mode for late times, we differentiate (2.2.50) for $\eta < 1$.

$$\frac{dG_n}{d\eta^2} = \frac{(1-2\eta^2)}{2\eta^2(1-\eta^2)^2}\, p_n\tau \sinh p_n\tau + \frac{1}{(1-\eta^2)^2}(\cosh p_n\tau - 1) \qquad (2.2.51)$$

Setting this to zero yields

$$1 - \frac{1}{2\eta^2} = \frac{1}{p_n\tau} \cdot \frac{\cosh\, p_n\tau - 1}{\sinh\, p_n\tau} \tag{2.2.52}$$

For times large enough that significant amplification has occurred, $\cosh\, p_n\tau - 1 \approx \sinh\, p_n\tau$ and (2.2.52) is approximated by

$$\eta_{cr}^2 = \frac{1}{2} \cdot \frac{p_n\tau}{p_n\tau - 1} \tag{2.2.53}$$

To a lesser approximation, for large $\tau$ such that $p_n\tau >> 1$, the wave number of the most amplified mode is therefore

$$\eta_{cr} \approx \frac{1}{\sqrt{2}} = 0.707 \tag{2.2.54}$$

Using this to obtain an estimate for $p_{cr} = \eta_{cr}(1 - \eta_{cr}^2)^{1/2} \approx 1/2$, a better estimate for $\eta_{cr}$, from (2.2.53), is

$$\eta_{cr} \approx \frac{1}{\sqrt{2}} \cdot \sqrt{\frac{\tau}{\tau - 2}} \tag{2.2.55}$$

For example, at $\tau = 6$, Equation (2.2.55) gives $\eta_{cr} = 0.866$, which is about 22% larger than the value in (2.2.54). At $\tau = 10$, the estimate in (2.2.54) is only about 12% low. Thus, for practical purposes, the wave number of the most amplified mode can be taken as $\eta_p = 1/\sqrt{2}$. This will be called the "preferred" mode of buckling. The corresponding wavelength is found from

$$\eta_p \xi_p = 2\pi \quad , \quad \text{or} \quad \xi_p \equiv \lambda_p = 2\pi\sqrt{2} \tag{2.2.56}$$

In dimensional units, from (2.2.38), this length is

$$x_p = \frac{r}{s}\lambda_p = \frac{2\pi\sqrt{2}}{\sqrt{\epsilon}}\, r = 8.88\, r/\sqrt{\epsilon} \tag{2.2.57}$$

A graph of the maximum amplification plotted against $\tau$ is given in Figure 2.4. Beyond $\tau = 4$, growth is very rapid; at $\tau = 12$ initial imperfections are amplified by more than 400. These results suggest that a bar under very high compression will buckle into wavelengths near $8.88\, r/\sqrt{\epsilon}$ at nondimensional times between 4 and 12. Better estimates for critical buckling times are given in succeeding sections.

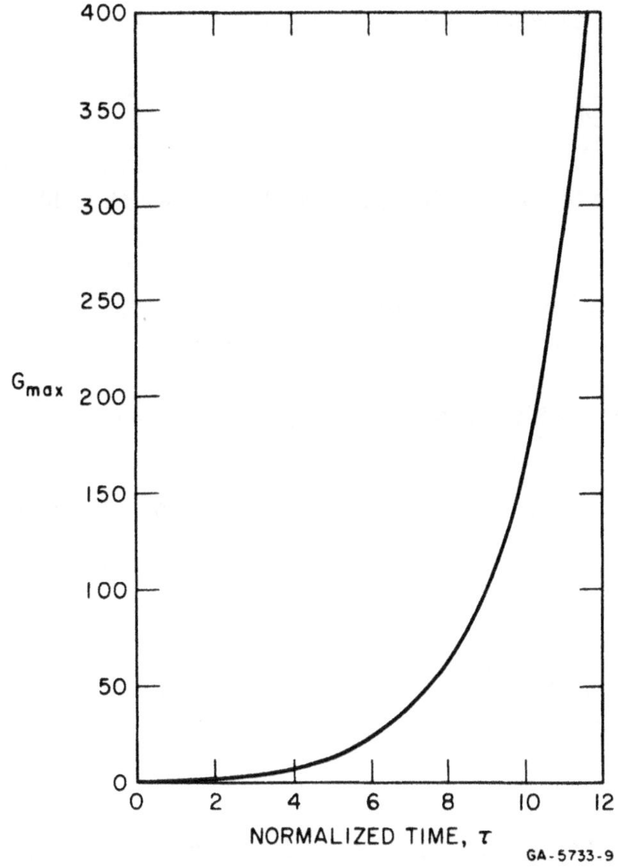


FIGURE 2.4 MAXIMUM AMPLIFICATION VERSUS TIME

### 2.2.5 Dynamic Elastic Buckling Under Eccentric Load

As an example, consider a bar eccentrically loaded as in Figure 2.5. For this problem, the initial shape of the bar is taken as

$$\begin{aligned} w_o(\xi) &= \delta/r \qquad & \xi \neq 0, \\ w_o(\xi) &= 0 \qquad & \xi = 0, \end{aligned} \tag{2.2.58}$$

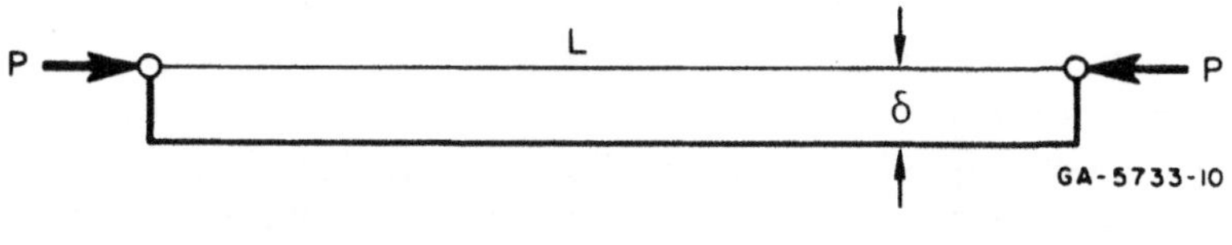


FIGURE 2.5 ECCENTRICALLY LOADED BAR

This initial shape is expanded into the Fourier sine series

$$w_o(\xi) = \sum_{n=1}^{\infty} a_n \sin \frac{n\pi\xi}{\ell} \tag{2.2.59}$$

The coefficients are found by using formula (2.2.43), which yields

$$\begin{aligned} a_n &= \frac{4\delta}{n\pi r} \qquad && \text{n odd} \\ a_n &= 0 && \text{n even} \end{aligned} \tag{2.2.60}$$

From (2.2.49), the buckled shape is given by

$$w(\xi, \tau) = \sum_{n=1,3...}^{\infty} \frac{4\delta}{n\pi r} \cdot \frac{1}{1 - \eta^2} \left[ \begin{matrix} \cos \\ \cosh \end{matrix} \; p_n\tau - 1 \right] \sin \frac{n\pi\xi}{\ell} \tag{2.2.61}$$

To evaluate this sum, recall that

$$\eta = \frac{n\pi}{\ell} \quad ; \quad \text{and} \quad \text{for n odd,} \quad \Delta\eta = \frac{2\pi}{\ell} \tag{2.2.62}$$

Then

$$\frac{4\delta}{n\pi r} = \frac{4\delta}{\eta \ell r} = \frac{4\delta}{r\eta} \cdot \frac{1}{2\pi} \cdot \frac{2\pi}{\ell} = \frac{2\delta}{\pi r \eta} \Delta\eta \tag{2.2.63}$$

and (2.2.61) can be written

$$w(\xi, \tau) = \frac{2\delta}{\pi r} \sum_{n=1,3...}^{\infty} \frac{1}{\eta(1 - \eta^2)} \left[ \begin{matrix} \cos \\ \cosh \end{matrix} \; p_n\tau - 1 \right] \sin \eta\xi \; \Delta\eta \tag{2.2.64}$$

If we assume that the bar is very long compared with the wavelengths of the buckling, $\Delta\eta \rightarrow d\eta$ and $\eta$ can be treated as a continuous variable. The sum (2.2.64) can then be replaced by the integral*

$$w(\xi, \tau) = \frac{2\delta}{\pi r} \int_0^{\infty} \frac{1}{\eta(1 - \eta^2)} \left[ \begin{matrix} \cos \\ \cosh \end{matrix} \; p_n\tau - 1 \right] \sin \eta\xi \, d\eta \tag{2.2.65}$$

---

*This is merely a plausible argument, but the result is correct, as can be confirmed by using a Fourier integral representation from the start. Converting from a sum to an integral here can be done because the function multiplying $\sin \eta\xi$ in the integrand dies off for large $\eta$ such that there is no difficulty with $\sin \eta\xi$ oscillating in the interval $\Delta\eta = 2\pi/\ell$. For a more rigorous discussion see Reference 8.

A plot of the function

$$f(\eta,\tau) = \frac{1}{\eta(1-\eta^2)}\left[\begin{matrix}\cos\\ \cosh\end{matrix}\; p_n\tau - 1\right] \qquad (2.2.66)$$

in the integrand is given in Figure 2.6 for $\tau = 6$. To obtain an approximate analytical expression for the integral in (2.2.65), we replace this curve by the triangle of height A in Figure 2.6, where $A(\tau) = f(1/\sqrt{2}, \tau)$. Then

$$w(\xi,\tau) \approx \frac{2\delta}{\pi r}\int_0^1 A(\tau)\,\eta \sin \eta\xi d\eta = \frac{2\delta A(\tau)}{\pi r\xi^2}(\sin\eta\xi - \eta\xi \cos \eta\xi)\Big|_0^1$$

$$= \frac{2\delta A(\tau)}{\pi r\xi^2}(\sin\xi - \xi \cos \xi) \qquad (2.2.67)$$

where

$$A(\tau) = \frac{1}{\frac{1}{\sqrt{2}}\left(1-\frac{1}{2}\right)}[\cosh \tau/2 - 1] \qquad (2.2.68)$$

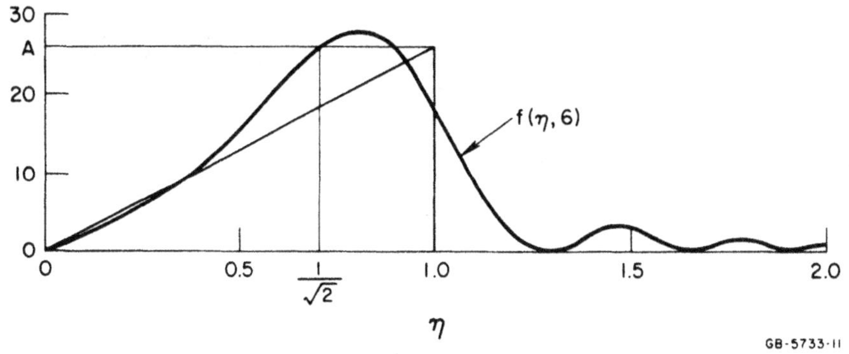


FIGURE 2.6 FOURIER COEFFICIENTS (transform) OF BUCKLED SHAPE

The function

$$W(\xi) = \frac{1}{\xi^2}(\sin \xi - \xi \cos \xi) \qquad (2.2.69)$$

which gives the approximate shape of the buckling bar, is plotted in Figure 2.7. The wavelengths between peaks are slightly greater than $2\pi$ near the support and approach $2\pi$ away from the support.

This discussion gives an estimate for the buckled shape of a bar under idealized eccentric thrust and also shows how the amplitude of the buckled form grows with time. Specification of a criterion for failure by dynamic buckling, however, depends on the particular structural problem at hand. For example, if the bar is a push rod used to measure rapid displacements, large deflections within the elastic limit could constitute failure. If a rod is used as a hammer, however, large displacements are probably not objectionable so long as the motion remains elastic and the rod returns to its initial shape.

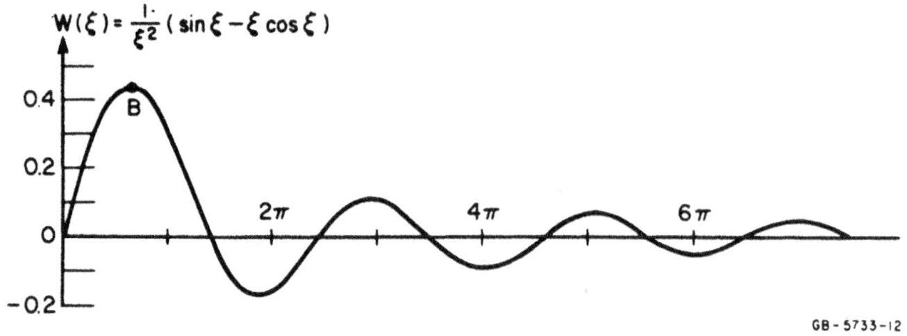


FIGURE 2.7 APPROXIMATE BUCKLED SHAPE OF BAR UNDER SUDDENLY APPLIED ECCENTRIC LOAD

To give a concrete example, let us calculate the duration of load application required to produce a combined bending-compression stress equal to the yield stress. The maximum bending stress occurs at point B in Figure 2.7 where the curvature $W'' = -0.235$ and is a maximum. In general, the compressive bending stress in the inner fiber, for a rectangular bar of height h, is

$$\sigma_b = \frac{M\frac{h}{2}}{I} = \frac{Eh}{2}\frac{\partial^2 y}{\partial x^2} = \frac{Eh}{2}\frac{s^2}{r^2} r w'' = \sqrt{3}\, Es^2 w'' \qquad (2.2.70)$$

Using (2.2.67) with $W'' = -0.235$ and the time variation from (2.2.68), the bending stress at B is

$$\sigma_b = \sqrt{3}\, Es^2 \frac{2\delta A(\tau)}{\pi r}(-0.235) = -0.732\frac{\delta}{r}\sigma_c[\cosh(\frac{\tau}{2}) - 1] \quad (2.2.71)$$

where $\sigma_c$ is the compressive impact stress.

The threshold of buckling is defined by the total stress $\sigma_b + \sigma_c$ reaching the yield stress $\sigma_y$. With $\sigma_b$ from (2.2.71), this condition gives the following relation between the compressive stress $\sigma_c$ and the time $\tau_{cr}$ at which first yield occurs:

$$\left(\frac{\sigma_c}{\sigma_y}\right)^{-1} = 1 + 0.732\frac{\delta}{r}\left[\cosh\left(\frac{\tau_{cr}}{2}\right) - 1\right] \tag{2.2.72}$$

A graph of $\tau_{cr}$ versus $\sigma_c/\sigma_y$ from (2.2.72) is given in Figure 2.8 for several values of eccentricity $\delta$, with $\delta$ expressed in terms of depth h of a rectangular bar for later comparison with experiment. The values chosen range over an order of magnitude, from $\delta = 0.00316$ h to $\delta = 0.0316$ h. The mid-value $\delta = 0.01$ h is a representative value found from static experiments, as given by Equation (2.2.22). We shall see that the dynamic buckling experiments described in Section 2.2.10 suggest that the static data do indeed give equivalent imperfections in the appropriate range for the dynamic problem.

Also given in Figure 2.8 is a curve of the amplification $G_p$ [from (2.2.50) with $\eta = 1/\sqrt{2}$] required to produce first yield for an eccentricity $\delta = 0.01$ h. Similar curves for $\delta = 0.00316$ h and $\delta = 0.0316$ h are omitted for clarity. This curve shows that, for small values of impact stress, the amplification must be very large to produce yield because the bending contribution must be larger and because the wavelength of the buckling is longer. Under these conditions, depending on the practical application, larger buckling deformations may constitute buckling before the yield stress is reached, thus placing an upper limit on $\tau_{cr}$. However, with the yield definition of buckling here, $\tau_{cr}$ approaches infinity (as does the length of the bar) as P approaches zero.

At the other end of the curves, as the impact stress approaches the yield stress, the amplification required to produce first yield is quite small (less than 10 for $\sigma_c/\sigma_y = 0.9$). Also, in a real material the yield stress is not sharply defined and, more important, the tangent modulus begins to fall rapidly as the material yields so the elastic modulus in the present buckling formulation is inappropriate. Thus, application of the curves in Figure 2.8 has little meaning for real materials beyond about $\sigma_c/\sigma_y = 0.9$. Buckling in this range of loads is considered in Section 2.3.

To obtain a physical interpretation of the curves, we observe that in physical units dimensionless time $\tau$ corresponds to the impulse of the applied load. Thus, from the definition of $\tau$ in Equation (2.2.38), this impulse is

$$Pt = \frac{AEr}{c}\tau \tag{2.2.73}$$

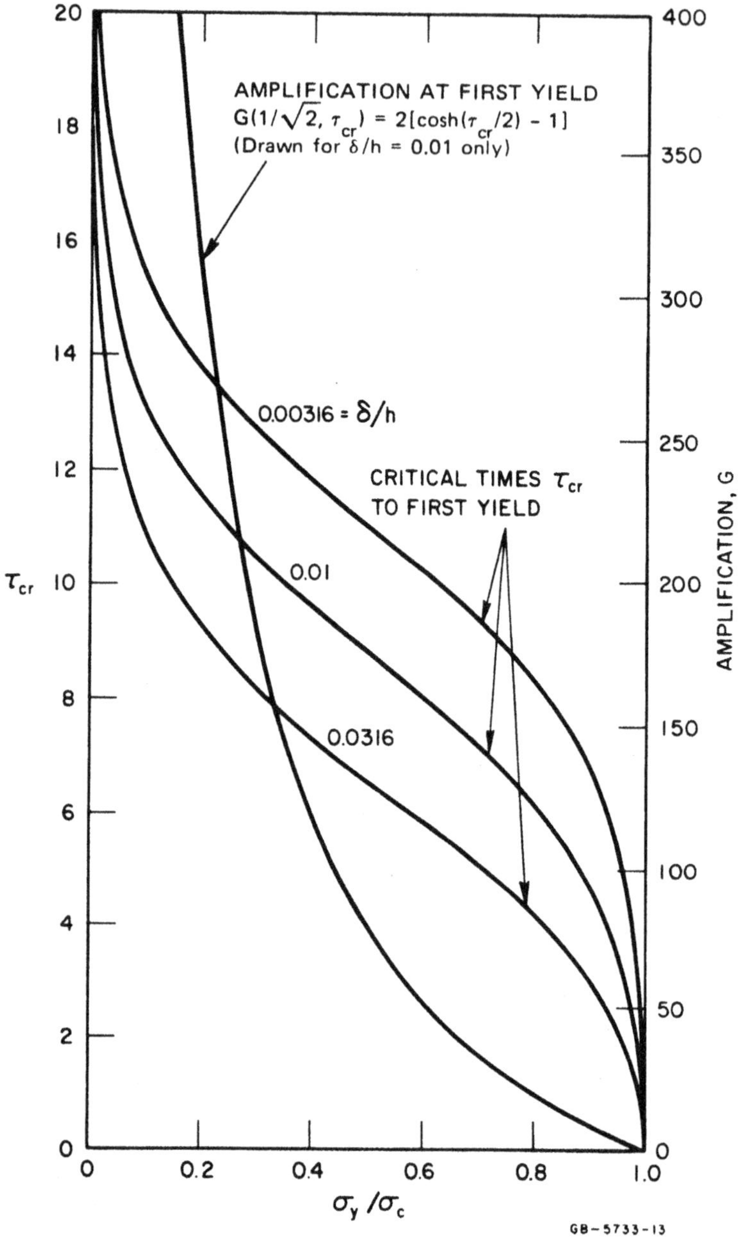


FIGURE 2.8 CRITICAL BUCKLING TIMES TO FIRST YIELD FOR BAR UNDER ECCENTRIC LOAD

and the critical impulse that causes first yield from buckling is

$$I_{cr} = \frac{AEr}{c} \tau_{cr} \tag{2.2.74}$$

Also, the applied load can be expressed by

$$P = A\sigma_c = A\sigma_y\left(\frac{\sigma_c}{\sigma_y}\right) \tag{2.2.75}$$

Thus the curves in Figure 2.8 can be interpreted as giving the combinations of load amplitude P and load impulse I that produce threshold buckling. Load points above the curves give more severe buckling, while load points below the curves give no permanent buckling deformations. We shall see in Section 3.5 that amplitude-impulse curves of this type can be applied to more complex structures, such as a cylindrical shell under lateral pressure.

### 2.2.6 Dynamic Elastic Buckling with Random Imperfections

Another form of imperfection, more uniquely concerned with the dynamic problem, is suggested by experiments to be described later in which a large collection of rubber strips were buckled over a wide range of dynamic thrusts. It was found that the strips buckled into wavelengths that varied randomly at each thrust, with a mean and standard deviation both inversely proportional to the square root of the thrust as suggested by Equation (2.2.57). These results are consistent with the assumption that random imperfections in the strips are amplified by the buckling motion. Thus the resulting buckled form, although still random, has statistics determined by the buckling amplification function given by Equation (2.2.50) and in Figure 2.3.

Several methods of representing a random function have been described by Rice[9] in the study of filtering electrical noise. In the electrical problem, the function represents the variation of current with time, $I = I(t)$. In the buckling problem here, the random function represents the variation of lateral displacement with distance along the bar, $w = w(\xi)$. Thus there is an analogy between the two problems, with electrical current being associated with mechanical displacement, and time in the electrical problem being associated with axial position in the mechanical problem.

In the electrical problem, a noise signal $I_o(t)$, having Fourier components $a_n(\omega_n)$, is fed into a filter having an attenuation characteristic $F(\omega_n)$. The output signal is $I(t)$, having Fourier components $A_n(\omega_n) = F(\omega_n)\, a_n(\omega_n)$. In the mechanical problem, the "input" is the initial displacement $w_o(\xi)$, having Fourier components $a_n(\eta)$, and the "output" is the buckled form $w(\xi)$, having Fourier components $g_n(\eta) = G(\eta, \tau)\, a_n(\eta)$. Because the mechanical problem contains one added variable, time $\tau$, the amplification characteristic also depends on time as indicated by $G_n(\tau)$ in Equation (2.2.50), which is denoted here by $G(\eta, \tau)$. However, at each instant the analogy is quite close. The only difference is that in the electrical problem the process is *stationary*; that is, the currents continue indefinitely in time and the statistics are taken to be independent of time.

In the buckling problem, the boundary conditions at the ends of the bar must be met so the statistics depend also on the position $\xi$, the variable analogous to time. If the buckle wavelengths are very short compared with the length of the bar, however, one would expect that some distance from the end of the bar the effect of $\xi$ diminishes and the assumption of stationary white noise would be acceptable. With this assumption the two problems are completely analogous and all the theory available for the electrical problem can be used here.

It is not necessary to assume that the random imperfections are stationary; this assumption merely makes the mathematics simpler. Before this is done, consider a random form of imperfection that does satisfy the boundary conditions of simple supports at $\xi = 0$ and $\xi = \ell$. These imperfections are given by

$$w_o(\xi) = \sum_{n=1}^{N} a_n \sin \eta\xi \tag{2.2.76}$$

in which N will be specified later. The coefficients $a_n$ are random normal, having mean value zero and standard deviation $\sigma(\eta)$. The normal or Gaussian probability distribution is shown in Figure 2.9. If it is further assumed that $\sigma$ is constant over all wave numbers of interest, then Equation (2.2.76) is called (nonstationary) *white noise*. For $w_o(\xi)$ to remain bounded, $\sigma$ must ultimately die off for large $\eta$. Since

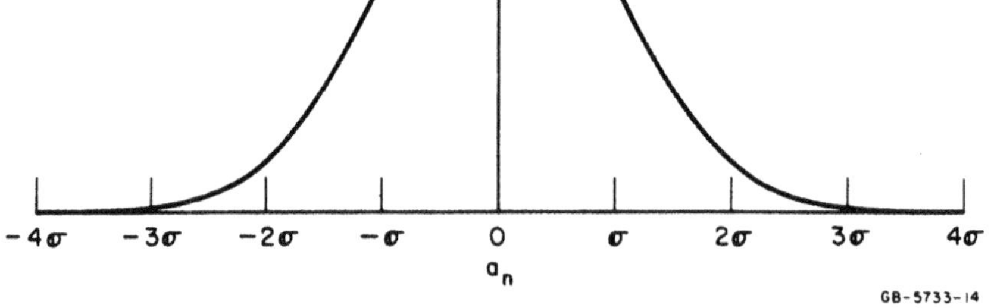


FIGURE 2.9 ASSUMED NORMAL DISTRIBUTION OF FOURIER COEFFICIENTS OF INITIAL IMPERFECTIONS

our central concern is in the buckled shape $w(\xi)$ after the Fourier coefficients have been amplified by $G(\eta, \tau)$, and Figure 2.3 shows that for $\eta > 2$ the amplification is very small, harmonics with $\eta > 2$ can safely be neglected. Thus, in the initial deflections given by (2.2.76), we merely specify that $\sigma(\eta)$ dies off in some unspecified manner for $\eta > 2$ and is constant for $0 < \eta \leqslant 2$. This is the usual assumption justifying the use of white noise as a filter input.

Since the concept of white noise can be applied only when associated with a process passing a finite band of wave numbers, we must defer any examples of random functions until after the amplification function with its inherent cutoff has been applied to give the buckled shapes. This function, repeated from Equation (2.2.50), is

$$G(\eta, \tau) = \frac{1}{1 - \eta^2} \left[ \begin{matrix} \cos \\ \cosh \end{matrix} \; p(\eta)\tau - 1 \right] \tag{2.2.77}$$

where

$$p(\eta) = \eta \, |1 - \eta^2|^{1/2}$$

and the hyperbolic form is taken for $\eta < 1$. The buckled form is given by

$$w(\xi, \tau) = \sum_{n=1}^{N} a_n G(\eta, \tau) \sin \eta\xi \tag{2.2.78}$$

where N is the largest value of n for which $\eta < 2$.

With a cutoff characteristic now applied, we can give examples of the functions characteristic of buckling from random imperfections. Figure 2.10 gives two examples of buckled forms calculated from Equation (2.2.50) using a length $\ell = 50\pi$, which is 25 complete Euler lengths and very long compared with the highly amplified wavelength $\lambda_p = 2\pi\sqrt{2}$ corresponding to $\eta = 1/\sqrt{2}$. With this choice for $\ell$, $n = 100$. The procedure was to select 100 random numbers from a population having a Gaussian distribution as in Figure 2.9, with $\sigma = 1$. These were then used as the coefficients $a_n$ in Equation (2.2.77) and the summation was taken over 100 modes, corresponding to $0 < \eta \leqslant 2$. Higher harmonics would have had a negligible effect as already mentioned because of the rapid decrease of $G(\eta, \tau)$ with $\eta$ for $\eta > 2$.

In each example in Figure 2.10 (i.e., for each set of 100 random coefficients), the buckled shape is plotted at $\tau = 4$ and $\tau = 6$. In both examples, there are more crests (waves) at $\tau = 4$ than at $\tau = 6$ because of the shift in the peak of the amplification function in Figure 2.3 from $\eta \approx 1$ at $\tau = 4$ to $\eta \approx 0.8$ at $\tau = 6$. At still later times, little further change in the number of crests would be expected because, as discussed in Section 2.2.4, the point of maximum amplification does not shift below $\eta = 1/\sqrt{2} \approx 0.707$.

Another feature exhibited in these examples is typical of buckled forms from white noise: although they consist of a random assemblage of harmonics, they exhibit a surprisingly regular pattern of waves. The average wavelength of this pattern depends, of course, on the region of amplification defined by the amplification function. In fact, an amplification function that is square in shape, constant for $\eta < 2$ and zero for $\eta > 2$, would give a wave pattern similar to those shown in Figure 2.10. This is exactly the waveform of the imperfection $w_o(\xi)$ corresponding to the computational procedure used in generating the curves in Figure 2.10, but it is *not* the waveform of the "actual" imperfection, whose Fourier components do not cut off abruptly at $\eta = 2$. This is the reason that numerical examples had to be deferred to the discussion of buckled shapes; any specification of a cutoff wavenumber already implies *filtered* noise.

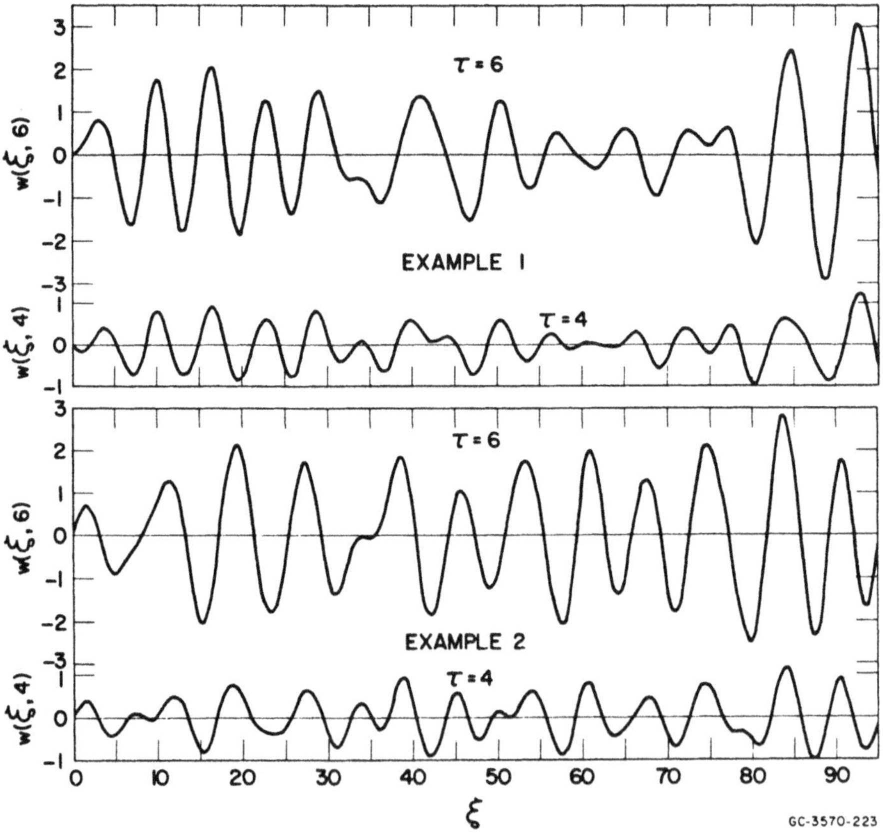


FIGURE 2.10 TWO EXAMPLES OF BUCKLED FORMS FROM RANDOM IMPERFECTIONS

The only way of quantitatively describing buckled shapes such as in Figure 2.10 is to give statistics of the features of interest. The most easily measured quantity in experiments is the buckled wavelengths, so statistics of wavelengths will be calculated for later comparison with experiment. Direct calculation of these statistics is beyond the means of currently available analysis except for a special case to be given later. Instead, the statistics are calculated by the Monte Carlo method; a large sample of random buckled forms is generated numerically by the procedure just described, and the resulting data are plotted directly in the form of a probability distribution (histogram) for the feature of interest.

To determine the distribution of wavelengths, we calculated 65 random buckled shapes as in Figure 2.10, each with a different set of 100 random values for $a_n$. Wavelengths in each buckled shape were then measured for $\tau = 6$, and the histogram in Figure 2.11a was prepared. The wavelengths were measured between alternate zero crossings for the first three waves from the support $\xi = 0$, not counting the support as a crossing. Separate histograms were also prepared for the first, second, and third waves individually. No significant differences were found, indicating that the end support does not seriously affect the wavelengths even a small distance from the support.

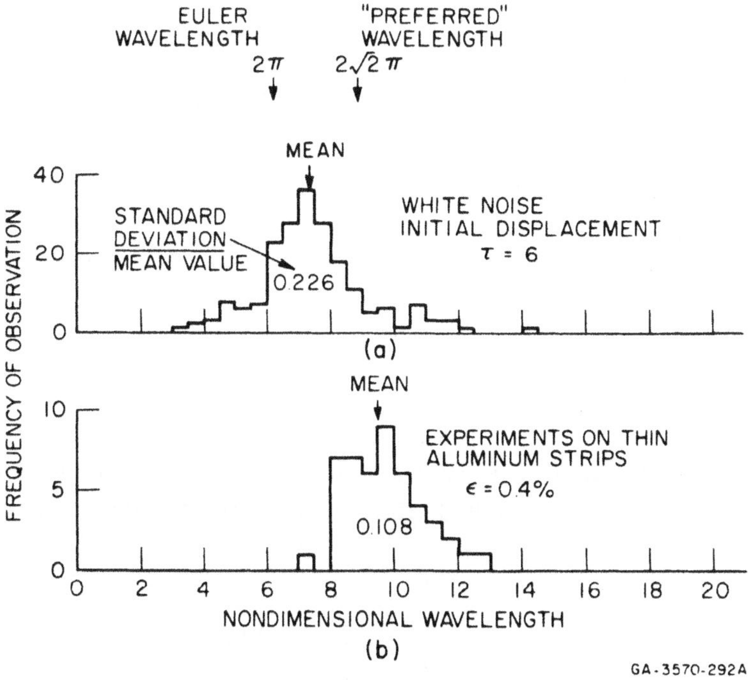


FIGURE 2.11 THEORETICAL AND EXPERIMENTAL HISTOGRAMS OF BUCKLED WAVELENGTHS

Many more computations would have to be added before this would approximate the probability distribution, but the main features of the distribution are apparent. The mean wavelength is $\lambda_m = 7.4$, which lies between the Euler wavelength $\lambda_e = 2\pi = 6.28$ and the "preferred" wavelength $\lambda_p \approx 2\pi\sqrt{2} = 8.88$, as shown. The standard deviation of the wavelength is $\sigma_\lambda = 1.7$ and the ratio of standard deviation to mean wavelength is $\sigma_\lambda/\lambda_m = 0.23$.

Figure 2.11b gives a histogram prepared from experiments on about 50 aluminum strips buckled under axial impact as described in Section 2.2.7: The mean value of the buckled wavelengths is somewhat larger than in the theoretical histogram ($\lambda_m = 9.5$ compared with $\lambda_m = 7.4$ in Figure 2.11a), and the spread in wavelengths is somewhat smaller. The narrower spread possibly results because part of the initial imperfection was in the form of an eccentric impact, which tends to produce a fixed wavelength as described in Section 2.2.5. However, the general features of the observed distribution are adequately represented by the white noise theory. More extensive experimental examples are given in Section 2.2.9.

An analytical expression for the mean wavelength directly in terms of the amplification function $G(\eta, \tau)$ can be given if it is assumed that the buckling displacements are *stationary*, i.e., if the end conditions are neglected as discussed earlier. With this assumption, the initial imperfections can be represented by *stationary white noise* as follows:

$$w_o(\xi) = \sum_{n=1}^{N} a_n \sin(\eta\xi + \phi_n) \tag{2.2.79}$$

This form is similar to Equation (2.2.76) except that here the Fourier components are added in random phase, with the phase angles $\phi_n$ uniformly distributed (with equal probability) in the interval $0 \leqslant \phi_n \leqslant 2\pi$. The buckled displacements are then

$$w_o(\xi) = \sum_{n=1}^{N} a_n G(\eta, \tau) \sin(\eta\xi + \phi_n) \tag{2.2.80}$$

With the standard deviation of $a_n$ constant, it is reasonably simple to demonstrate[9] that the mean wavelength between alternate zero crossings in the buckled form is

$$\lambda_m(\tau) = 2\pi \left[ \frac{\int_0^\infty G^2(\eta, \tau)\, d\eta}{\int_0^\infty \eta^2 G^2(\eta, \tau)\, d\eta} \right]^{1/2} \tag{2.2.81}$$

No analytical expression has yet been found for the standard deviation of wavelengths, even with the stationary process assumption (Slepian[10] discusses the status of this perennial problem in information theory).

For the complicated $G(\eta, \tau)$ in Equation (2.2.77), no closed form expressions for the integrals in Equation (2.2.81) were found. Instead, the integrals were evaluated numerically over the region $0 < \eta < 2$ of significant amplification for a sequence of values of $\tau$. The resulting mean wavelengths are plotted against $\tau$ in Figure 2.12. The mean wavelength increases monotonically with $\tau$, but in the region $\tau > 6$ of significant amplification (see Figure 2.4) the increase is very small. At $\tau = 6$, Figure 2.12 gives $\lambda_m = 7.4$, which is the same result found in Figure 2.11 for buckles satisfying the pinned end conditions. Also plotted is the wavelength corresponding to the most amplified mode, given approximately by Equation (2.2.55) for large $\tau$. The mean and most amplified wavelengths are very close together and have very nearly the same variation with $\tau$. For large $\tau$, both approach the preferred wavelength $\lambda_p = 2\pi\sqrt{2}$.

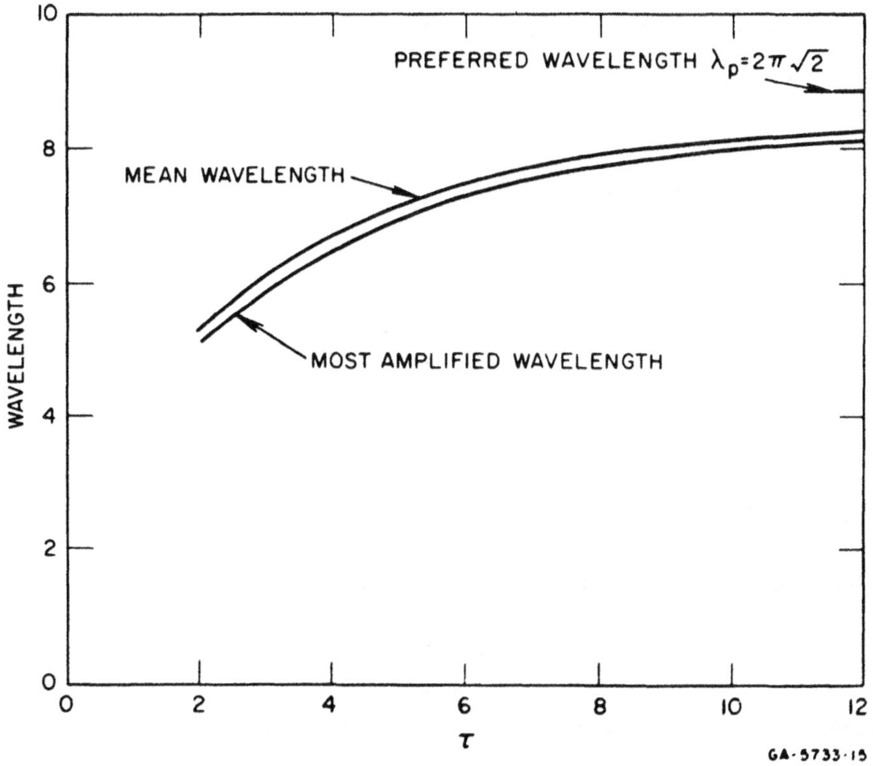


FIGURE 2.12 MEAN AND MOST AMPLIFIED WAVELENGTHS VERSUS TIME

These results suggest that, if it is reasonable to assume that random imperfections are present in a bar as described, then the bar will buckle over the entire compressed length, and the wavelength of the buckles will be reasonably well characterized by the preferred wavelength $\lambda_p = 2\pi\sqrt{2}$. To calculate a threshold of buckling, one can make the simplifying assumption that the motion consists of response in only the preferred wavelength, with an assumed single equivalent imperfection at this wavelength. This will now be done.

As in static buckling, imperfections can be divided into two types, one type having amplitudes proportional to the thickness of the bar and the other having amplitudes proportional to the wavelength of the buckling. In the following, both types will be considered and it will be shown that the resulting critical times $\tau_{cr}$ for buckling do not depend strongly on which type is assumed.

We treat first imperfections having amplitudes proportional to the buckle wavelength $\lambda_p$ and denote the coefficient of this Fourier component by $A_p$, in physical units. Thus we assume

$$A_p = \beta L_p \tag{2.2.82}$$

where $L_p$ is the preferred half-wavelength (the buckled shape of an Euler column) under the applied load P, corresponding to a half-wavelength $\lambda_p/2$ in dimensionaless units. In dimensionless form these quantities, using (2.2.38), are expressed by

$$a_p = \frac{A_p}{r} \; , \quad \frac{\lambda_p}{2} = \frac{sL_p}{r} \; , \quad \lambda_p = 2\pi\sqrt{2} \tag{2.2.83}$$

and the imperfection is now given by

$$a_p = \frac{\beta}{s} \cdot \frac{\lambda_p}{2} = \frac{\pi\sqrt{2}\,\beta}{s} \tag{2.2.84}$$

The criterion for buckling is taken as in Section 2.2.5 on eccentric impact; a critical time $\tau_{cr}$ is determined such that the bending stress plus the direct stress due to P reach the yield stress.

The bending stress, from Equation (2.2.70), is

$$\sigma_b = \sqrt{3}\; Es^2 w'' \tag{2.2.85}$$

The idealized buckled shape is simply a sine wave, given from Equations (2.2.77) and (2.2.78) as

$$w(\xi, \tau) = \frac{a_p}{1 - \eta^2} \left[\cosh p(\eta_p)\tau - 1\right] \sin \eta_p \xi \tag{2.2.86}$$

with $\eta_p = 1/\sqrt{2}$. Differentiating (2.2.86) and substituting the result into (2.2.85) gives the peak bending stress, at $\sin \eta_p \xi = 1$, as

$$\sigma_b = \sqrt{3}\ Es^2 \cdot a_p \left[\cosh \frac{\tau}{2} - 1\right] \tag{2.2.87}$$

which, using $a_p$ from (2.2.84), becomes

$$\sigma_b = \pi \sqrt{6}\ \beta Es \left[\cosh \frac{\tau}{2} - 1\right] \tag{2.2.88}$$

Finally, we use $s^2 = \sigma_c/E$ and the buckling criterion $\sigma_b + \sigma_c = \sigma_y$ to obtain

$$\frac{1 - \sigma_c/\sigma_y}{\sqrt{\sigma_c/\sigma_y}} = \pi\beta \sqrt{\frac{6}{\epsilon_y}} \left[\cosh \frac{\tau_{cr}}{2} - 1\right] \tag{2.2.89}$$

This equation is the counterpart of Equation (2.2.72) for buckling from eccentric impact. An essential difference is that here the critical curves for buckling depend not only on the imperfection amplitude $\beta$ but also on the yield strain $\epsilon_y$. This results from taking the imperfections proportional to the buckle wavelengths.

Curves of $\tau_{cr}$ versus $\sigma_c/\sigma_y$ from Equation (2.2.89) are given in Figure 2.13 for $\epsilon_y = 0.005$, a representative value for engineering metals. Values of $\beta$ are taken from 0.0001 to 0.001, corresponding to the range of imperfection amplitudes observed in static buckling as given in Equation (2.2.23). The curves are quite similar to those in Figure 2.8 for eccentric impact except that the critical times $\tau_{cr}$ change more slowly with $\sigma_c/\sigma_y$ (i.e., the curves are more nearly horizontal for intermediate values of $\sigma_c/\sigma_y$). Also, $\tau_{cr}$ does not shoot up to very large values until $\sigma_c/\sigma_y$ is very small. These observations can be made by comparing the solid curves (imperfections proportional to wavelength) with the dashed curve (which has the same functional form as in the curves for eccentric impact).

Critical buckling times for imperfections proportional to the depth of the bar are found in essentially the same way. The equivalent imperfection amplitude in the preferred mode is then given by

$$A_p = \gamma r \tag{2.2.90}$$

Using this in place of Equation (2.2.82) and applying the same procedure as for imperfections proportional to wavelength, the expression for $\tau_{cr}$ becomes

$$\left(\frac{\sigma_c}{\sigma_y}\right)^{-1} = 1 + \sqrt{3}\ \gamma \left[\cosh \frac{\tau_{cr}}{2} - 1\right] \tag{2.2.91}$$

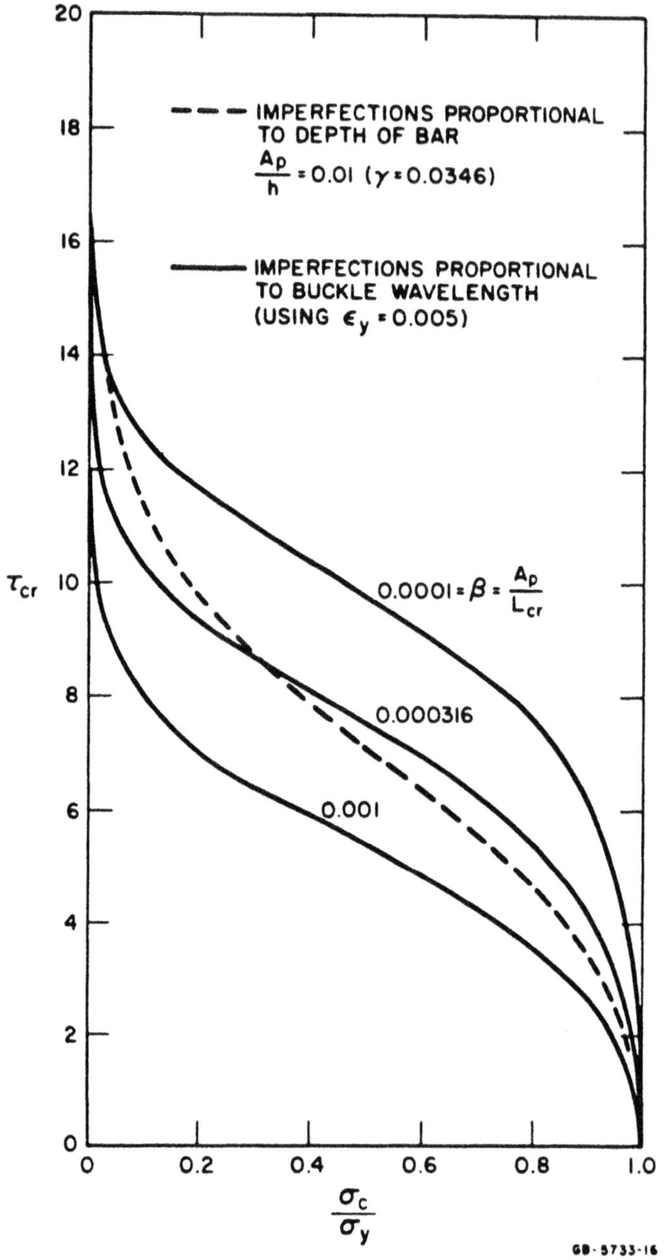


FIGURE 2.13 CRITICAL TIMES TO FIRST YIELD FOR BUCKLING IN "PREFERRED" MODE

This is exactly the same functional form as found for eccentric impact, with the constant 0.732 $\delta/r$ replaced by $\sqrt{3}\gamma = \sqrt{3}A_p/r$. Again, $\tau_{cr}$ depends only on $\sigma_c/\sigma_y$ and not on the magnitude of the yield strain $\epsilon_y$.

As for imperfections proportional to wavelength, we take as estimates for $\gamma$ the values found appropriate in static buckling. For a rectangular bar of depth h, the static empirical formula (2.2.24) gives the conservatively large value $\gamma = 0.1\ r/(h/2) = 0.058$. In Figure 2.13 the dashed curve is a plot of Equation (2.2.91) for a somewhat smaller value ($\gamma = 0.0346$, corresponding to $A_p/h = 0.01$) to give an intermediate value for comparison with the solid curves. This comparison shows that the values of $\tau_{cr}$ calculated for either type of imperfections (with representative values for both taken from static buckling) give very nearly the same result. More important, as we shall see in the next sections, these curves compare favorably with observed thresholds of dynamic buckling.

### 2.2.7 Framing Camera Observations of Dynamic Elastic Buckling

In practice, the most directly applicable physical problem for the preceding theory is the impact of a long bar against a massive object. We consider that the bar is originally stress-free and moving toward the object with velocity V as shown in Figure 2.14a. Since to a good approximation the object can be considered a rigid wall, on impact the left end of the bar immediately comes to rest. Adjacent particles to the right come to rest later as a stress wave of magnitude $\sigma$ propagates to the right at the bar sound velocity c. When the stress wave has passed a distance $x_\sigma$ into the bar, the impulse applied by the end load at the rigid wall must be equal to the initial momentum of the length $x_\sigma$ brought to rest by the stress wave. This

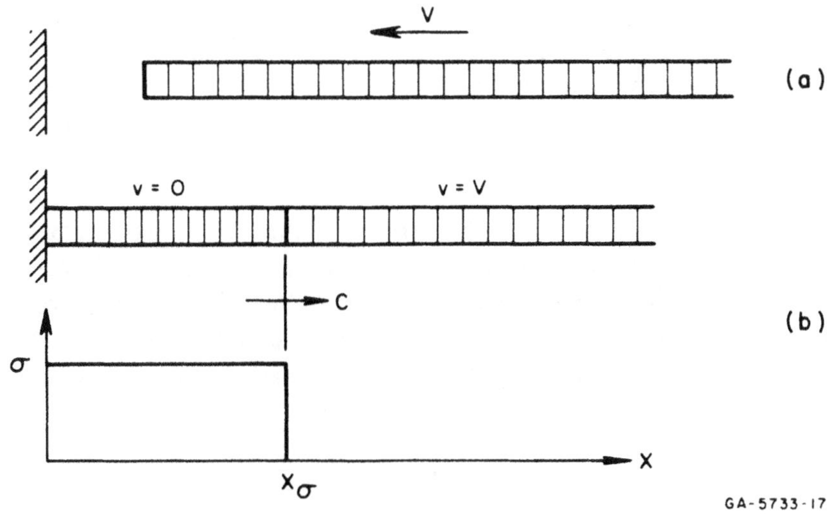


FIGURE 2.14 AXIAL STRESS WAVE IN A BAR IMPACTING A RIGID WALL

condition is expressed by

$$\sigma A \cdot \frac{x_\sigma}{c} = \rho A x_\sigma \cdot V$$

or

$$\sigma = \rho c V \tag{2.2.92}$$

This situation is conveniently produced experimentally by using a tensile testing machine.[11] The initial velocity V is produced by first pulling the bar to a tensile stress $\sigma$. Before the tension is applied, a notch is filed in the bar near the upper jaw with its depth adjusted so that fracture occurs at the notch when the stress in the remainder of the bar is near the desired stress $\sigma$. After fracture, a (compressive) relief wave travels down the bar at velocity c, leaving the bar stress-free behind the wave and traveling at velocity $V = \sigma/\rho c$ by the same argument just made for axial impact. When the wave arrives at the lower jaw it reflects, again as a compressive wave. Since the bar is completely stress-free and traveling at velocity V at the instant of this reflection, Equation (2.2.92) can be used, giving a compressive stress equal to the initial tensile stress $\sigma$. In reality the stress rises to this value in a finite time comparable to the time for stress waves to cross the bar and communicate the notch fracture to the full cross section.

An example[12] of a strip buckled by this procedure is given in Figure 2.15. The strip is made of aluminum alloy 6061-T6 with a 0.5 by 0.0125-inch (12.7 by 0.32 mm) cross section and a length of 30 inches (726 mm) between notch and lower jaw. The photographs show only a few inches of the strip just above the lower jaw. The magnitude of the compressive wave was approximately 40,000 psi (276 MPa), between 10% and 20% below the yield stress. It was photographed by an ultrahigh-speed framing camera at a framing rate giving 6 microseconds between frames.

In the figure, at 18 $\mu$s after the arrival of the compressive wave, the strip appears straight, but careful measurements show that it is slightly buckled even at this early time. At 24 $\mu$s the deflection is perceptible in the printed reproduction here, and at later times the developing buckles are clearly visible. All the buckles remain nearly fixed in position and merely grow in amplitude, just as in the idealized eccentric impact example. The lowermost buckle continues to grow throughout the time shown, but the upper buckles oscillate beyond 70 $\mu$s because the very large deflection of the lower buckles reduces the thrust by allowing the remainder of the bar to move toward the jaw. The rapidity of the buckling is demonstrated by the lateral velocity of the crest of the lowermost wave, measured to be 75 ft/sec (23 m/s). The wavelength of the lower buckle is about 0.47 inch

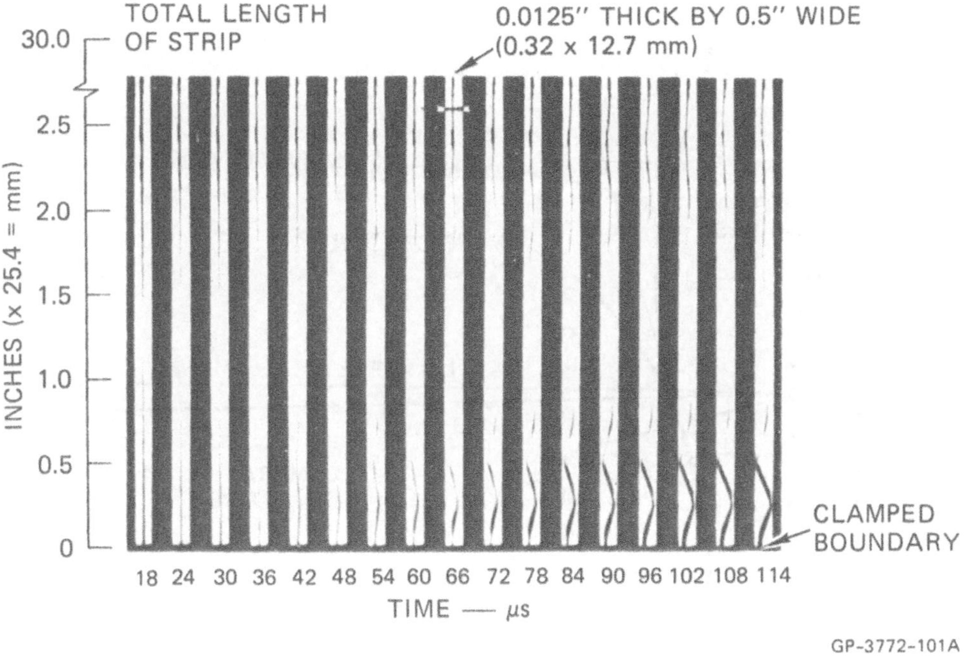


FIGURE 2.15 ALUMINUM STRIP BUCKLING WITHIN A 40,000-psi (276 MPa) AXIAL STRESS WAVE

(Time measured from compressive reflection at lower jaw)

(11.9 mm), very close to the value of 0.50 inch (12.7 mm) calculated for the preferred wavelength $\lambda_p$ from the theory.

### 2.2.8 Streak Camera Observations--Effects of the Moving Stress Wave

The theory, of course, is not strictly applicable to the impact problem because it assumes that the thrust is uniform throughout the length of the bar. In impact, the thrust is applied by the moving axial stress wave, and at each instant only the distance enveloped by the wave is under compression. To observe possible effects of this moving wave, and also to observe early exponential buckling growth as predicted by the theory, we used another experimental arrangement[13] to amplify recordings of the tiny early motion. Instead of the buckling being observed directly in an edge-on view as in Figure 2.15, the strip was polished on one side and the reflected image of a series of light sources was viewed with a streak camera as shown in Figure 2.16. The shift in position of the light source is proportional to the product of the small change in slope of the strip at the point in which the image forms and the distance between the light source and the strip. With this method,

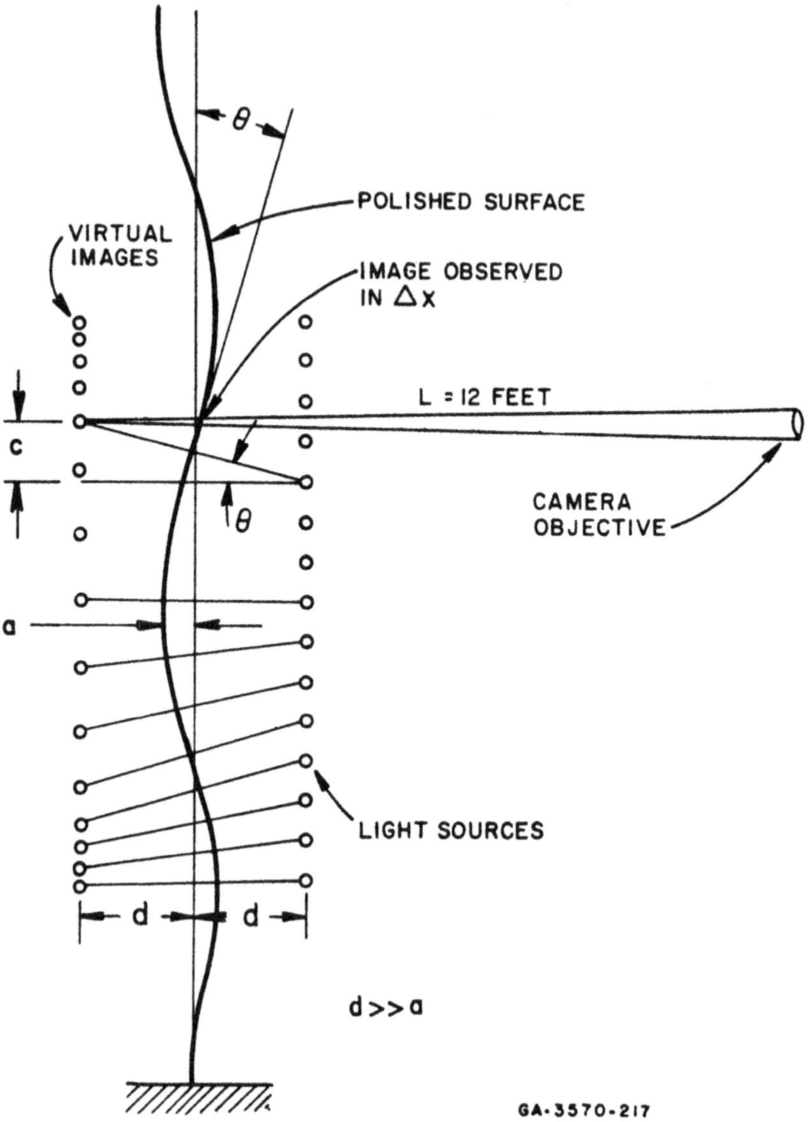


FIGURE 2.16 OPTICAL LEVER METHOD OF OBSERVING BUCKLING SLOPE

deflections of the order of 50 millionths of an inch (1.3 μm) were easily resolved, and the exponential growth was observed.

A plot of peak displacement versus time [assuming the buckle was a simple sine wave at the observed 0.65-inch (16.5 mm) wavelength] is shown in Figure 2.17 for one such experiment. The magnitude of the stress wave in this experiment was approximately 30,000 psi (207 MPa), and the cross section of the aluminum strip was 0.50 by 0.0116 inch (12.7 by 0.29 mm). The experimental points are

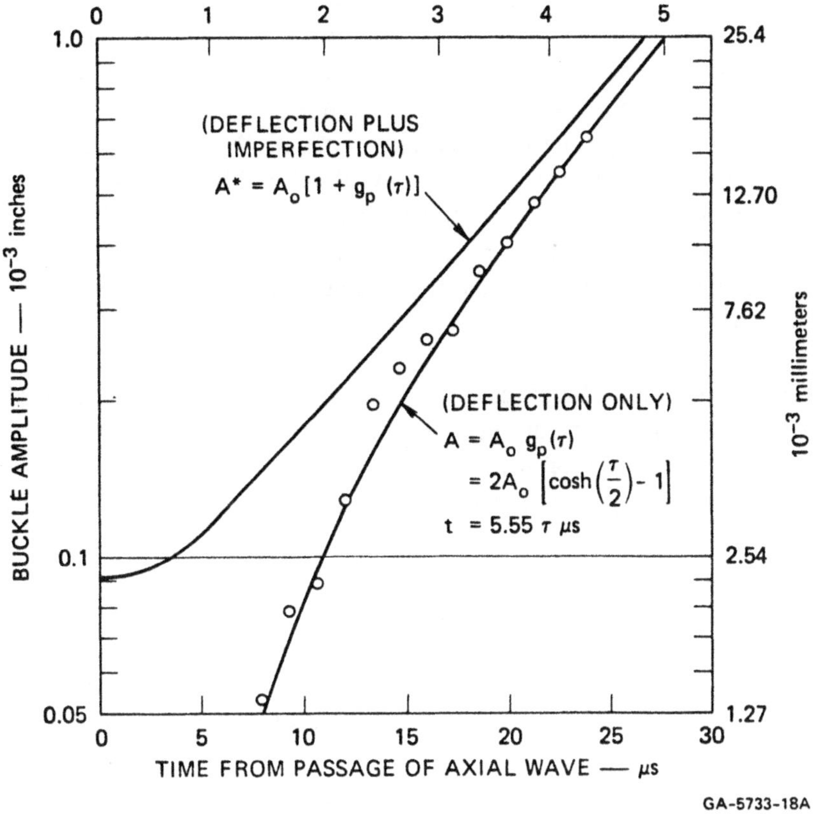


FIGURE 2.17 EXPERIMENTAL (POINTS, FOR DEFLECTION ONLY) AND THEORETICAL (CURVES) BUCKLE AMPLITUDE VERSUS TIME (MATCHED AT 22 μs)

peak displacements A(t) measured from the initial (idealized) displacement $A_o$. The lower smooth curve passing through these points is a theoretical curve calculated under the assumption that the growth is adequately represented by the preferred mode. Taking $\eta_p = 1/\sqrt{2}$ in Equation (2.2.50), the amplitude of this mode is

$$A(\tau) = 2A_o[\cosh(\tau/2) - 1] \tag{2.2.93}$$

Using $\epsilon = 0.003$, $c = 0.20$ inch/$\mu$s (5 mm/$\mu$s), and $r = 0.0116/\sqrt{12}$ inch (0.085 mm) in Equation (2.2.38) gives $\tau = 0.18$ t, with t in $\mu$s. The Fourier coefficient $A_o$ of the equivalent initial imperfection was adjusted to $9.1 \times 10^{-5}$ inch (2.3 $\mu$m) to fit the experimental data as shown. The upper curve is the calculated total amplitude $A_o + A(\tau)$.

This experiment demonstrates that the observed buckling consists of exponential growth that can be calculated quite adequately by the simple theory. The simple uniform thrust theory is adequate, even though the thrust is applied by a moving stress wave, because the stress wave has moved a large distance along the bar before significant buckling displacements appear. For example, in Figure 2.17, the peak amplitude of the buckling is only about 0.001 inch [0.03 mm, giving a bending stress of 4600 psi (32 MPa), well within the elastic limit] at 30 $\mu$s after passage of the axial stress wave. At 30 $\mu$s the stress wave has propagated about 6 inches (152 mm) along the bar, about 10 times the observed wavelength of 0.65 inch (16.5 mm).

However, the high magnification of the optical lever did reveal that the axial impact produced very high frequency bending vibrations superimposed on the buckling motion.* On the original streak camera record, an oscillation was observed having a period of 3.1 $\mu$s (320 kHz) and a peak-to-peak amplitude of about $5 \times 10^{-6}$ inch (0.13 $\mu$m). The oscillations appeared to be a wave train propagating along the bar from the impact at the lower jaw at a phase velocity of 0.075 inch/$\mu$s (1.9 mm/$\mu$s), giving a wavelength of (0.075)(3.1) = 0.23 inch (5.8 mm). These oscillations had little effect on the buckling, apparently because of this short wavelength and because their period was so short compared with the buckling motion (3.1 $\mu$s corresponds to $\Delta\tau = 0.55$). Thus we can conclude that effects dependent on the moving axial stress front had a negligible effect on the buckling.

The argument concerning the distance the axial stress wave has traveled during the buckling motion can be stated analytically. From the theory, we have seen that whether we assume the imperfections are local in nature, as in eccentric impact, or consist of a general random form of imperfections, the wavelength of the buckles is always quite close to the wavelength $\lambda_p = 2\pi\sqrt{2}$ of the preferred mode. Also, the magnification of the buckling motion depends only on $\tau$, all other essential parameters having been included in its definition. It seems reasonable to assume that effects of the axial stress wave will be small as long as significant magnification takes place only after the axial wave has passed several buckle wavelengths along the bar. Without specifying a numerical value, we assume that the buckled form is unalterably determined (e.g., the buckled deformations are much larger than the initial imperfections) at a critical time $\tau_{cr}$. Using the definition $\tau$ in Equation (2.2.38) gives for the corresponding real time

$$t_{cr} = \frac{r}{s^2 c} \tau_{cr} \tag{2.2.94}$$

---

*These were observed on all three experiments performed.

Real time t can be expressed in terms of the number N of preferred wavelengths $L_p$ through which the axial stress wave passes at velocity c, giving

$$t = \frac{L_p N}{c} = \frac{2\pi\sqrt{2}r}{cs} N \qquad (2.2.95)$$

Putting this into expression (2.2.94) for critical time and using the definition of s in Equation (2.2.38) gives

$$N_{cr} = \frac{\tau_{cr}}{2\pi\sqrt{2}} \cdot \frac{1}{\sqrt{\epsilon}} \qquad (2.2.96)$$

This suggests that the reasonableness of neglecting axial wave effects depends only on the compressive strain of the axial thrust. In metals this strain is very small within the elastic limit and, as we have observed, elastic buckling is adequately represented by the constant thrust theory.

### 2.2.9 Experiments on Rubber Strips--Statistical Observations

Since Equation (2.2.96) suggests that axial wave front effects, if any, would be more pronounced at large compressive strains, confidence in the theory would be enhanced for metals if it could be demonstrated experimentally that the theory is acceptable in a material that can withstand large elastic compressive strains. Pure gum rubber is such a material, and experiments have been performed using this material to strains up to about 15%.[13]

The apparatus for these experiments, shown in Figure 2.18, is very simple and can be used for classroom demonstrations. A strip of pure gum rubber 0.0375

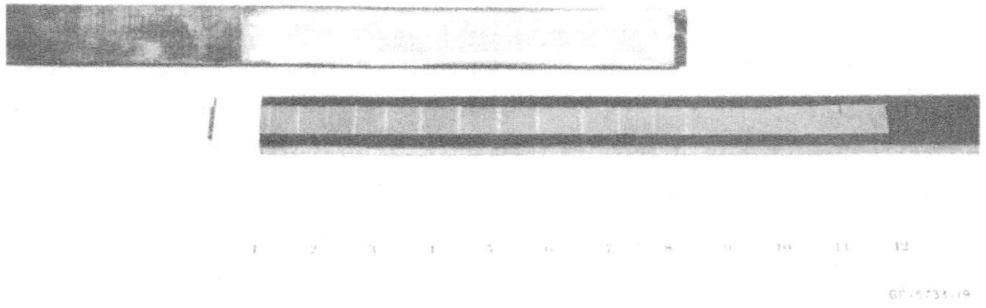

FIGURE 2.18 APPARATUS AND TYPICAL RECORD FOR BUCKLING RUBBER STRIPS

× 0.50 inch (0.95 × 12.7 mm) in section and about 1 foot (0.31 m) long was looped over one end of a rigid support bar and secured by means of masking tape as shown, with extra layers of tape wound above and below the rubber strip so that its end was separated from the support bar and the cover bar. The cover bar is shown above this assembly in the photograph. A strip of emery cloth has been glued to it and saturated with chalk dust.

In an experiment, the free end of the strip was held between thumb and forefinger, the cover bar placed over the strip, chalked side down and not touching the strip, and then the strip was stretched to a specified strain and released. The wrinkled strip impacted the chalk bar with sufficient velocity that a well-defined line was left on the strip at the crest of each wave, as shown. The positions of these lines were easily measured to an accuracy of 0.01 inch (0.25 mm).

To examine the applicability of the random noise assumption for imperfections, in addition to the applicability of the constant thrust theory, many experiments were performed so that statistical distributions could be prepared. Figure 2.19 gives histograms of the measured wavelengths for several values of initial tensile elongation. These data were taken from tests on 18 strips, each tested at all the strains, from smallest to largest strain in order to minimize any perturbations caused by the wrinkling of a previous test. Buckling at a strain greater than 25% is rather violent and leaves the strip with a definite bias toward the corresponding wavelength. The number of waves observed in each test varied from 2 to 3 at 3% strain up to 12 at 16% strain. The same strip tested repeatedly at the same strain gave an almost identical wave pattern each time, consistent with our mathematical model in which the imperfections are assumed random but fixed for any given bar. Data from only the first test at each strain were used for the histograms. Each histogram has a total of 65 observations so they can be compared directly.

It is significant that the general shape of all the histograms is the same and that the ratio between the standard deviation and mean value is nearly constant over the entire range of strains, as shown in Figure 2.19. This demonstrates that the statistics are inherent in the buckling process and are not the result of errors in measurement. It also indicates that the strips had no preferred wavelength characteristic of a manufacturing process. If these distributions are compared with the distribution in Figure 2.11, calculated assuming that initial imperfections can be represented by white noise, we see that the white noise assumption gives a very good description of the observed buckling.

To compare the observed wavelengths with the uniform thrust theory, we must take into account the large strains involved. Only the final compressive strain resulting from the initial tensile strain is needed, so the corrections can be obtained without reference to the details of large strain-wave propagation. It is sufficient to

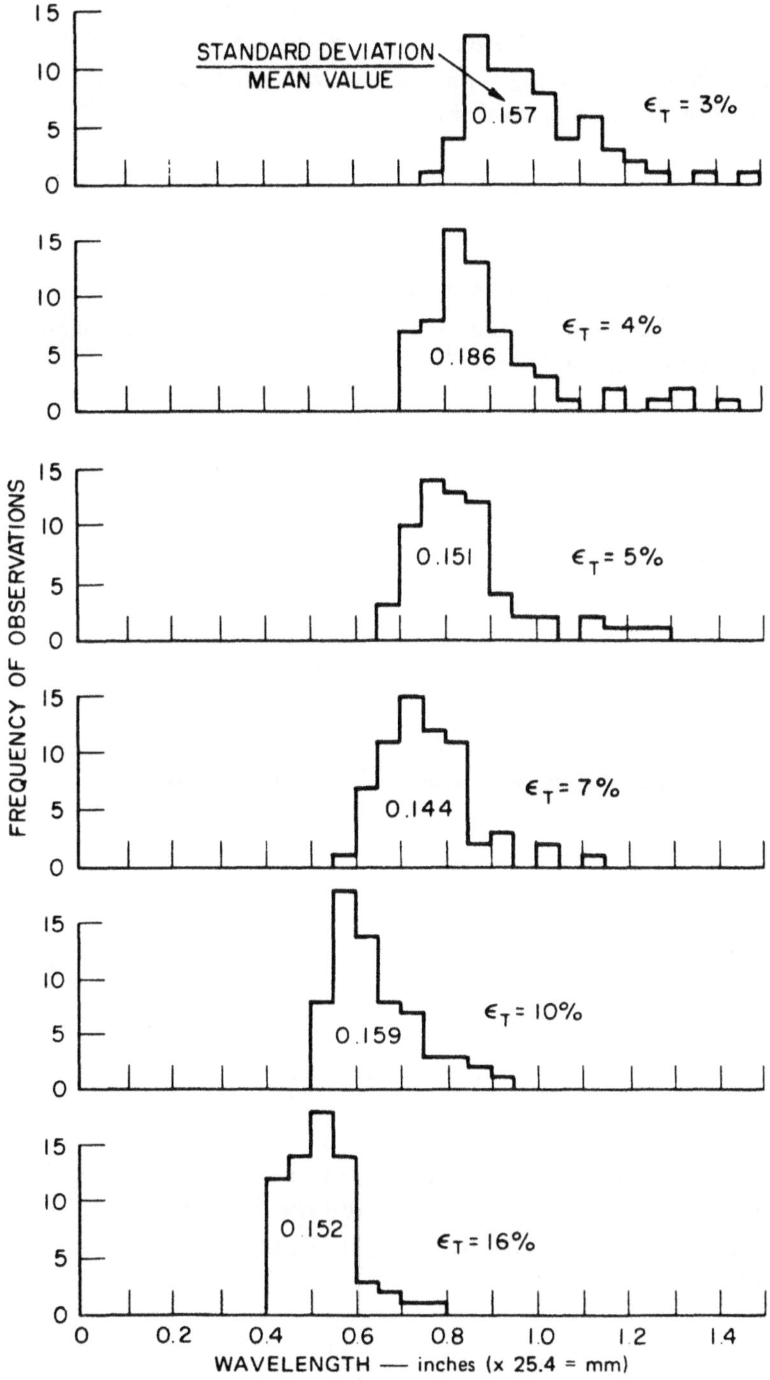


FIGURE 2.19 HISTOGRAMS OF OBSERVED BUCKLE WAVELENGTHS FOR SEVERAL INITIAL TENSILE STRAINS

assume that the rubber behaves elastically so that the potential energy stored in compression equals the initial potential energy in tension. Tensile stress-strain tests performed on sample strips showed that true stress was linear with elongation out to at least 100% with a Young's modulus of 285 psi (1.97 MPa). Thus the initial tensile force F in the strip is given by

$$F = E\epsilon \frac{A_o}{1 + \epsilon} \tag{2.2.97}$$

where $A_o$ is the unstressed cross-sectional area of the strip and $\epsilon = (\ell - \ell_o)/\ell_o$ is the elongation. The initial stored energy at uniform tensile elongation $\epsilon_T$ is equal to the work done by the end force F(z),

$$U_T = \int_0^{\ell-\ell_o} F(z)dz = EA_o\ell_o \int_0^{\epsilon_T} \frac{\epsilon d\epsilon}{1 + \epsilon} = EA_o\ell_o \log_e(1 + \epsilon_T) \tag{2.2.98}$$

where z is in the position of the moving end of the strip. Similarly, the compressive energy stored in the strip is

$$U_c = - EA_o\ell_o \log_e(1 - \epsilon_c) \tag{2.2.99}$$

expressed so that the compressive strain $\epsilon_c$ is a positive quantity. When these energies are equated, the compressive strain is found to be simply

$$\epsilon_c = \frac{\epsilon_T}{1 + \epsilon_T} \tag{2.2.100}$$

Further, the increased thickness h from the unstressed thickness $h_o$, assuming the rubber is incompressible, is

$$h = \frac{h_o}{(1 - \epsilon_c)^{1/2}} \tag{2.2.101}$$

The last correction to be made accounts for the wrinkles being formed at axial strain $\epsilon_c$ but measured when the strip has returned to zero strain. The ratio of the observed wavelength $\ell_r$ to the wavelength while under compression is, by the definition of $\epsilon_c$,

$$\frac{\ell_r}{\ell_c} = \frac{1}{1 - \epsilon_c} \tag{2.2.102}$$

The wavelength of the "most amplified" mode in dimensionless coordinate $\xi$ is

$\lambda_p = 2\sqrt{2}\pi$. Using this with Equations (2.2.38) and $r = h/\sqrt{12}$, the wavelength of the most amplified mode while the strip is under compression is

$$\ell_{pc} = \pi \left(\frac{2}{3}\right)^{1/2} \frac{h}{\epsilon_c^{1/2}} \tag{2.2.103}$$

After the strip has relaxed, this preferred length would be elongated according to (2.2.102). Using (2.2.103) in (2.2.102) with (2.2.100) and (2.2.101), the elongated length is given by

$$\ell_{pr} = \pi \left(\frac{2}{3}\right)^{1/2} \frac{(1 + \epsilon_T)^2}{\epsilon_T^{1/2}} h_o \tag{2.2.104}$$

In Figure 2.20 the observed wavelengths of Figure 2.19 are plotted against this preferred length, the circled points giving the mean values and the bars extending one standard deviation above and below the circles. The mean values fall very close to a straight line through the origin, and the ends of the standard deviation bars are also closely bounded by straight lines. These observations suggest that Equation (2.2.104) gives the proper form of variation with strain. However, the ratio between observed and preferred wavelengths (the slope of the line through the circles) is 1.70 here as compared with only 1.07 for the aluminum experiments given in Figure 2.11. This difference is attributed to strain-rate effects in the rubber. If, for example, these effects are lumped into an effective dynamic compressive modulus k times the static tensile modulus, the preceding theory gives a slope of 1.00 for $k = 2$.

Although the foregoing interpretation of the discrepancy between the aluminum and rubber experiments is somewhat speculative, the smooth variation of measured wavelength with strain strongly supports the conclusion that lateral motion immediately behind the axial-stress front has a negligible effect on the wrinkle formation and that a constant-thrust theory can be used with confidence. The main effect of the traveling thrust is that the duration of the thrust decreases as one moves away from the struck end, and this could easily be accounted for by simply assigning a different duration to each wrinkle. This conclusion should also be applicable to more complicated structures, such as cylindrical shells under axial impact. For large deflections, it might also prove necessary to compute a new thrust for each wave, reduced owing to lateral deflections in preceding waves.

### 2.2.10 Buckling Thresholds in Aluminum Strips

To obtain estimates of equivalent imperfections to be used in estimating thresholds of pulse buckling, we ran experiments on thin 6061-T6 aluminum strips

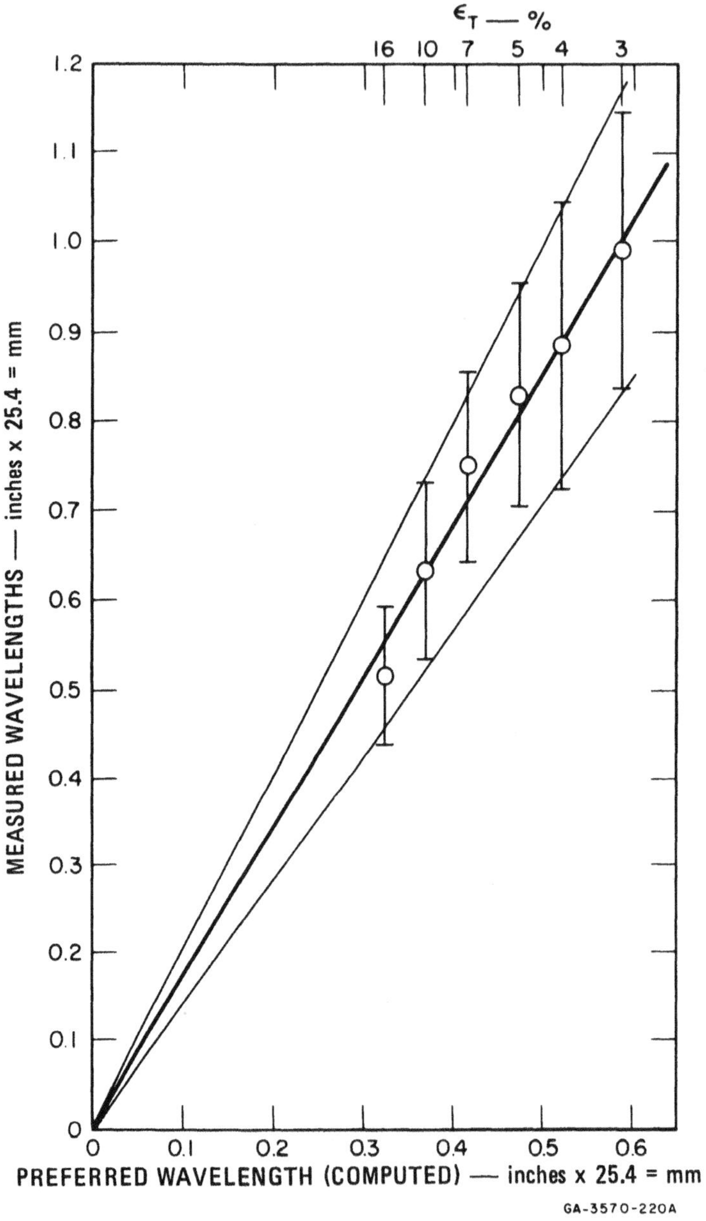


FIGURE 2.20 MEASURED VERSUS THEORETICAL WAVELENGTHS

Bars extend one standard deviation above and below mean value.

using a tensile testing machine as described previously. Tests were run on strips 1/2 and 1/4 inch wide (12.7 and 6.3 mm) and 0.0124 and 0.025 inch thick (0.31 and 0.63 mm).* The initial tensile stress (and reflected compressive stress) was nominally adjusted to 0.4 and 0.7 times the yield stress of 42,000 psi (290 MPa) by appropriately sized fracture notches in the strips. Duration of the thrust at the lower jaw was varied by varying the length L between the notch and lower jaw, the duration being 2L/c. For each combination of strip width, thickness, and compressive stress, tests were run at increasing lengths until plastic buckles appeared. These were observed by sighting down the shiny finish of the strips, a simple procedure with high resolution. The dimensionless time $\tau$, from its definition in Equation (2.2.38), is

$$\tau = \frac{\epsilon_c c}{r} \cdot \frac{2L}{c} = \frac{2\epsilon_c L}{r} \tag{2.2.105}$$

Figure 2.21 gives a plot from tests at many combinations of axial stress and duration, with open points representing tests in which no buckling was observed and solid points tests in which buckling was observed. The upper points (longer duration, buckling) are all solid, and the lower points (shorter duration, no buckling) are nearly all open, as would be expected. At intermediate durations, buckling and no-buckling points are intermingled as a result of the random nature of the imperfections. Also given on the same graph are theoretical curves similar to the dotted curve in Figure 2.13 for assumed imperfections in the preferred mode proportional to strip thickness. The experimental transition band of intermingled points between no buckling and buckling follows the trend of the theoretical curves, with equivalent imperfections in the experiments ranging from about 0.01 to 0.03 times the thickness of the bar.

The most severe buckles generally appeared at the jaw or one plastic hinge from the jaw, as would be expected because of the longer duration of thrust near the jaw and the possibility of eccentric loading (see Figure 2.7). As often as not, however, three or four plastic hinges were observed, suggesting that random imperfections throughout the bar were at least as important as eccentric loading. Buckling a few wavelengths away from the jaw, of course, had to take place in a somewhat shorter time, thus increasing the equivalent imperfections above those implied in Figure 2.21. However, this effect is small because the wavelength of the buckling is small compared with 2L, as discussed in relation to Equation (2.2.96). Thus we can conclude that random imperfections in these tests were equivalent to

---

*Two widths were tested at each thickness to examine the effect of fracture time on buckling. It was found that any possible effects were masked by changes in critical loads caused by random variations in imperfections.

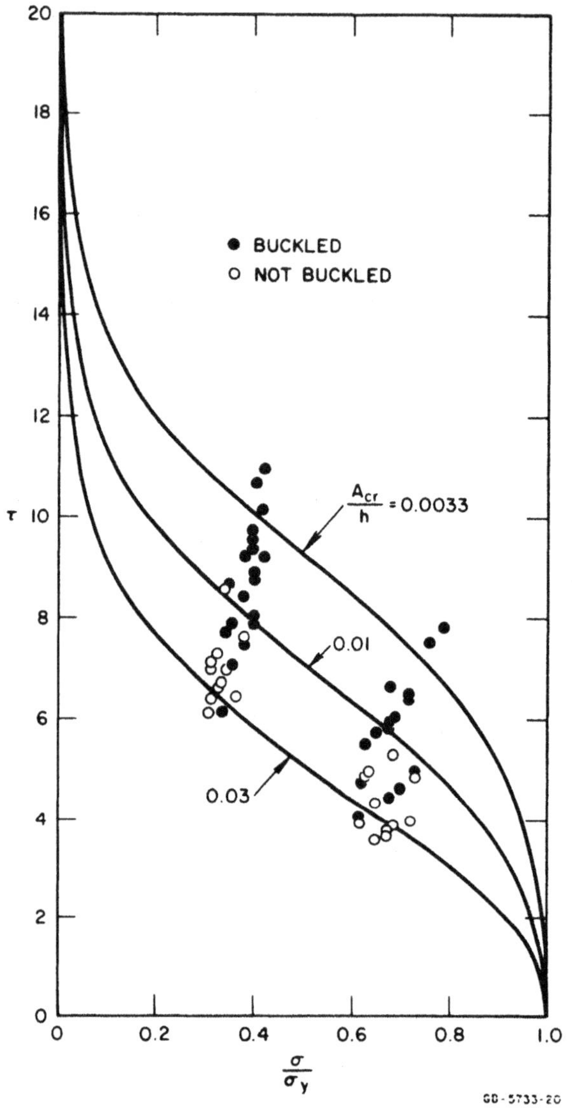


FIGURE 2.21 OBSERVED BUCKLING COMPARED WITH CRITICAL CURVES FOR IMPERFECTIONS IN PREFERRED MODE PROPORTIONAL TO STRIP THICKNESS

single imperfections in the preferred mode of 1% to 3% of the strip thickness. This result can be used to calculate critical pulse loads in a variety of dynamic pulse buckling situations, including buckling of plates and shells. This hypothesis is confirmed by experiments on shells presented in Chapter 3.

## 2.3 DYNAMIC PLASTIC FLOW BUCKLING OF BARS

### 2.3.1 Introduction

In the preceding theory and experiments, the axial stress was much greater than the static Euler buckling stress, but was restricted to be less than the yield stress so that the buckling was elastic before yielding at the crests of the buckles. In thicker (or shorter) bars under axial impact, the time required for significant elastic buckling is long compared with the duration 2L/c of axial compression at the point of impact. In these bars buckling occurs only when the impact stress and strain are well beyond the elastic range. In this section we treat the buckling of bars that takes place during sustained axial plastic flow with final plastic strains from about 1% to 30%, depending on the thickness of the bar.[14]

Figure 2.22 shows several 1/2-inch-diameter (12.7 mm) aluminum-alloy (6061-T6) bars that buckled plastically during axial impact against a heavy steel slab at the velocities indicated. Permanent shortening of the bars is shown by the relative positions of the impacted ends. Thickening extending several inches from the ends is apparent for the higher velocities, and within this thickened length, there is plastic buckling with a distinguishable wavelength, almost uniform, for all specimens. Similar effects are shown in Figure 2.23 for 1/4-inch-diameter (6.3 mm) bars and in Figure 2.24 for 1/2-inch-diameter (12.7 mm) tubes of another aluminum alloy (2024-T3).

Figure 2.25 shows a series of high-speed framing camera photographs of one of the 1/2-inch-diameter (12.7 mm) aluminum (6061-T6) bars during impact against a heavy steel plate. Buckling is first apparent at 11 $\mu$s and is complete at 243 $\mu$s. When the bar strikes the plate, a stress wave traveling to the left is produced in the bar. The associated axial strain, obtained from the thickening of the bar, is shown in Figure 2.26. The lower curve in Figure 2.26, corresponding to the 110 - $\mu$s view of Figure 2.25, indicates axial strains of several percent extending 3 inches from the impact end before buckling. The upper curve indicates that the axial strain at each cross section increases with time during the buckling motion. Axial flow ceases, at each cross section, upon the arrival of the elastic unloading wave from the free end of the bar. Thus the total period during which some part of the bar is in sustained plastic flow is the time required for the elastic wave to propagate from the impact end to the free end and back to the impact end. For the 18-inch-long (0.46 m) aluminum bars described here, this is about 220 $\mu$s.

Since the strain at each cross section continually increases, the buckling is treated as a pertubation of the motion associated with the axial compression. It is supposed that the axial strain rate dominates the extensional strain rate introduced by the bending motion; therefore, no strain-rate reversal occurs until after the

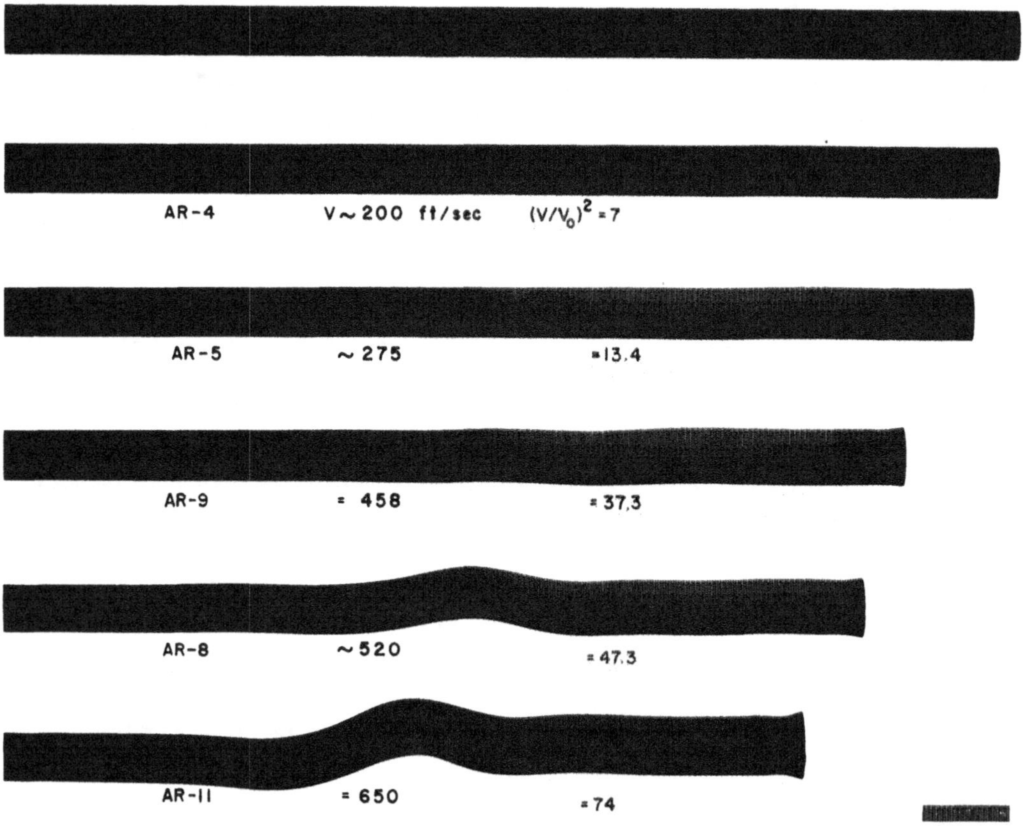


FIGURE 2.22 ALUMINUM (6061-T6) RODS BUCKLED BY LONGITUDINAL IMPACT

Diameter, 0.454 inch (11.5 mm), length, 18 inches (457 mm).
Ratio $(v/v_o)^2$ gives ratio of kinetic energy of rod to elastic strain energy it can absorb. (ft/sec x 0.305 = m/s)

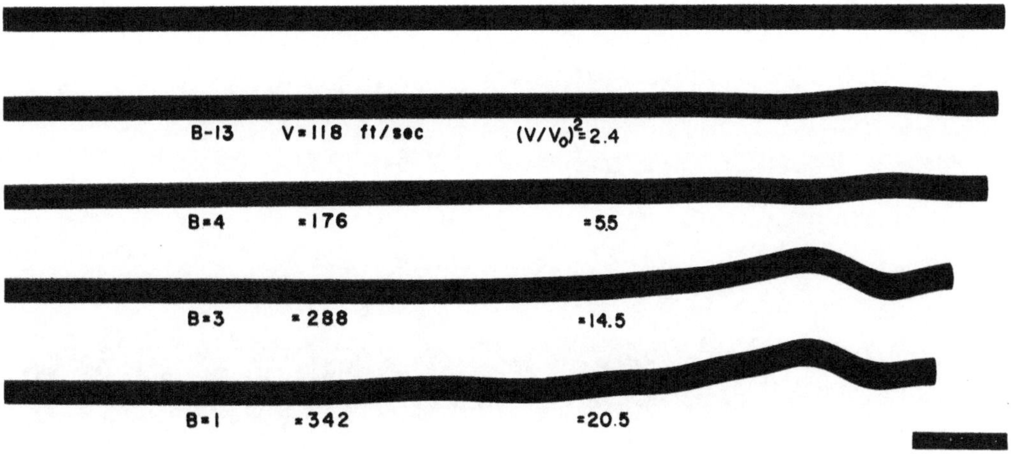


FIGURE 2.23 ALUMINUM (6061-T6) RODS BUCKLED BY LONGITUDINAL IMPACT
Diameter, 0.209 inch (5.3 mm), length, 18 inches (457 mm).
(ft/sec x 0.305 = m/s)

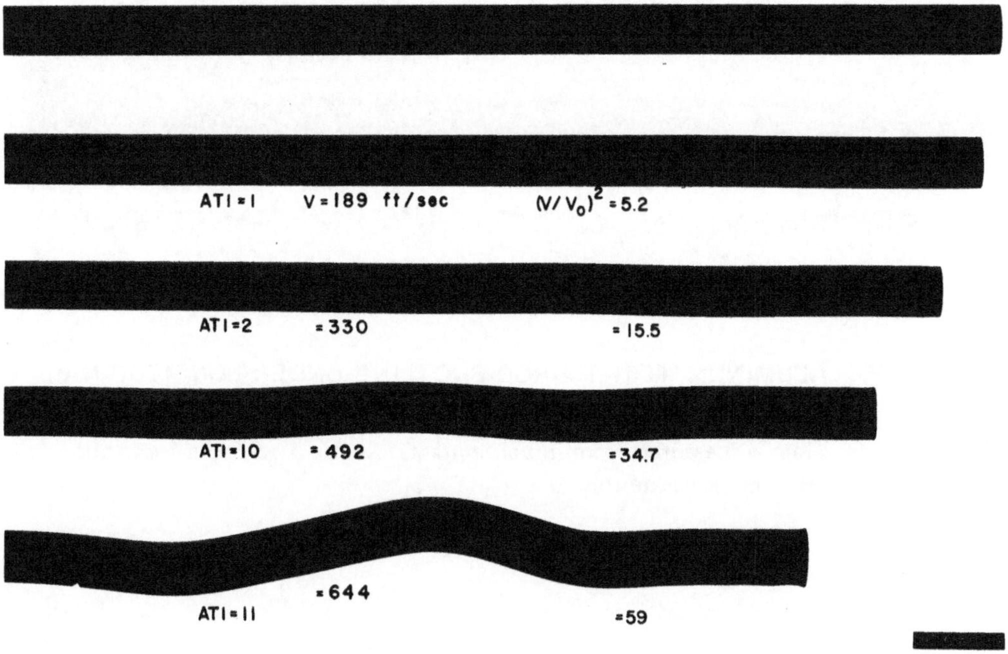


FIGURE 2.24 ALUMINUM (2024-T3) TUBES BUCKLED BY LONGITUDINAL IMPACT
Diameter, 0.454 inch (11.5 mm), wall, 0.075 inch (1.91 mm), length, 18 inches (457 mm). (ft/sec x 0.305 = m/s)

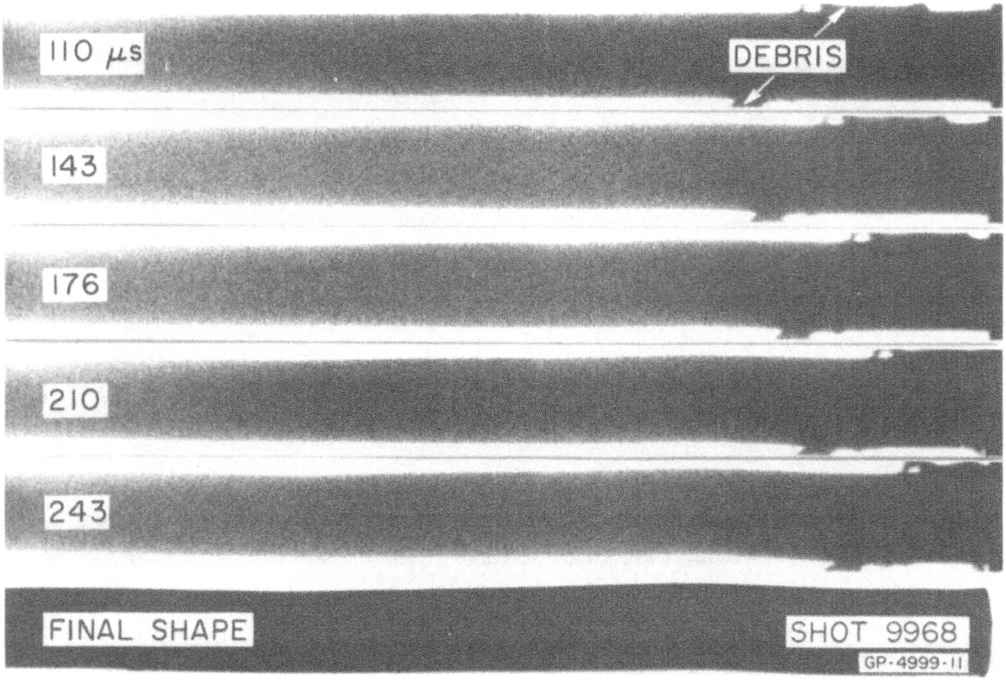


FIGURE 2.25 ALUMINUM (6061-T6) ROD BUCKLING UNDER LONGITUDINAL IMPACT WITH A HEAVY STEEL PLATE AT 420 ft/sec (128 m/s)

Time is measured from initial contact. Debris is wadding from rifle used to accelerate rod.

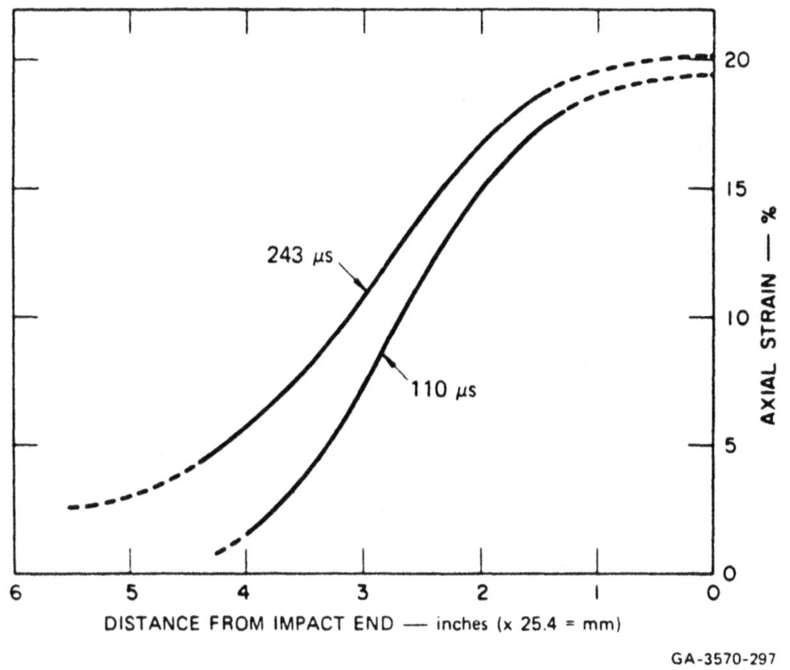


FIGURE 2.26 AXIAL STRAIN VERSUS DISTANCE FROM IMPACT END FOR ROD IN FIGURE 2.25

buckling is well developed. In this period, bending occurs relatively easily since the bending stiffness, being proportional to the tangent modulus in the plastic region of the stress-strain curve, is about 100 times less than the elastic bending stiffness. When strain-rate reversal supervenes, or the elastic unloading wave returns from the free end of the bar, the bending stiffness increases, becoming partly or wholly governed by the elastic modulus. To keep the theory simple, in the following analysis, linear strain-hardening, corresponding to a constant strain-hardening modulus, is assumed. Furthermore, since, as indicated in Figure 2.25, no perceptible buckling occurs while the wave front traverses a wavelength, the wave front is disregarded and the bar is taken to buckle under uniform axial strain.

### 2.3.2 Differential Equation of Motion

We consider a uniform bar, as shown in Figure 2.27, which is suddenly subjected to a thrust P, exceeding the yield value. The thrust is assumed to prevail throughout the bar and throughout the motion. Figure 2.28 shows the stress-strain curve taken for the material. Since axial flow is maintained, P must increase steadily; however, the strain-hardening is slight so the increase is small, and P will be taken as constant.

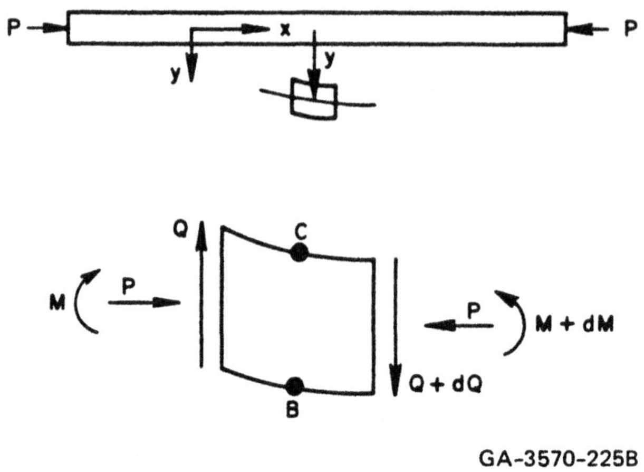


FIGURE 2.27 NOTATION

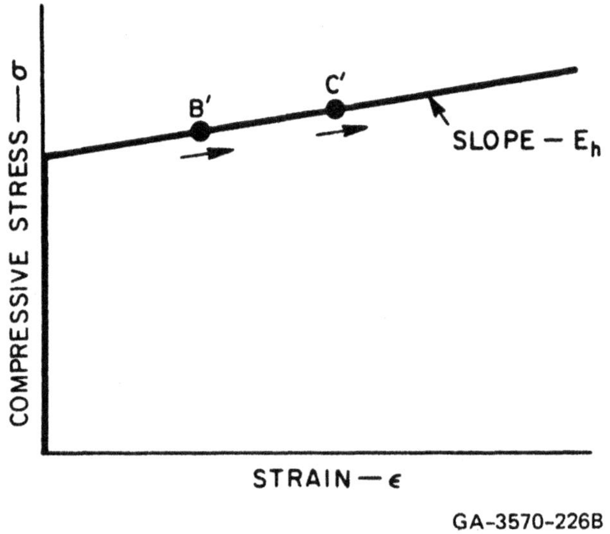


FIGURE 2.28 IDEALIZED STRESS-STRAIN DIAGRAM

Flexural motion may occur if the bar is not quite straight or if transverse initial velocities exist. As mentioned previously, the flexure is treated as relatively small in the sense that the tensile strain rate it contributes at a point such as B in Figure 2.27 is never enough to counteract the compressive strain rate of the axial flow. Thus the corresponding point B′ in Figure 2.28 continues to move steadily upward, but it lags behind the point C′, corresponding to point C in Figure 2.27, where the flexure contributes a compressive strain rate.

The instantaneous difference in strains and stresses at B and at C implies a bending moment M, related to the curvature through the strain-hardening modulus $E_h$. The appropriate relation for a slender bar is of the same form as for elastic behavior and is given by

$$M = -E_h A r^2 \frac{\partial^2 y}{\partial x^2} \tag{2.3.1}$$

where A is the area of the cross section, and r is the radius of gyration.

From Figure 2.27, we have for translation of the element

$$\frac{\partial Q}{\partial x} = \rho A \frac{\partial^2 y}{\partial t^2} \tag{2.3.2}$$

with $\rho$ for density. Neglecting rotational inertia, we have for rotational equilibrium

$$Q + P \frac{\partial}{\partial x}(y + y_o) = \frac{\partial M}{\partial x} \tag{2.3.3}$$

where $y_o$ is the initial deflection in the unstressed state.

We now change to dimensionless quantities defined, with $P = A\sigma$, by

$$s^2 = \frac{\sigma}{E_h}, \quad w = \frac{y}{r}, \quad w_o = \frac{y_o}{r} \tag{2.3.4}$$

$$\xi = \frac{sx}{r}, \quad \tau = \frac{\sigma}{(E_h P)^{1/2}} \cdot \frac{t}{r}$$

These parallel the notation in Section 2.2 except that $E_h$ replaces E.

In (2.3.1), (2.3.2), and (2.3.3), we treat $E_h$, A, $\rho$, and P as constants. Elimination of M and Q yields the differential equation for the dimensionless deflection w in the form

$$w'''' + w'' + \ddot{w} = -w_o'' \tag{2.3.5}$$

where primes indicate derivatives with respect to $\xi$, and dots indicate derivatives with respect to $\tau$. This is identical to Equation (2.2.39) for elastic buckling except for the difference implied by replacing E by $E_h$.

Thus, this simplified theory for dynamic plastic flow buckling is completely analogous to the theory for elastic buckling and the theoretical results, and many of the arguments on the nature of the buckling, can be used directly from Section 2.2. To avoid cumbersome reference back to corresponding equations in Section 2.2, the main results are derived briefly here. The derivation is made in terms of Fourier integrals rather than Fourier sine series to illustrate this alternative method of solution. Use of the Fourier integral representation, which applies to the infinite interval $0 < \xi < \infty$, makes explicit that buckling does not depend on the length of the bar (because it occurs before any signal is received from the free end). Also, perturbations in initial velocity are considered in addition to imperfections in initial shape. Since the Fourier transform derivation leads to amplification functions identical to those from a Fourier series derivation, either approach can be used as preferred.

### 2.3.3 The Initially Straight Bar

For flexure arising from initial transverse perturbational velocities (as, for instance, from elastic bending waves following a slightly oblique impact), we put $w_o = 0$ in (2.3.5) and consider w in $0 < \xi < \infty$ representable by

$$w(\xi,\tau) = \int_0^\infty g(\eta,\tau) \sin \eta\xi \cdot d\eta \qquad (2.3.6)$$

The initial conditions are

$$w(\xi,0) = 0 \ , \qquad \dot{w}(\xi,0) = v_o \int_0^\infty \beta(\eta) \sin \eta\xi \cdot d\eta \qquad (2.3.7)$$

with

$$\dot{w}(0+,\ 0) = 0$$

where $v_o$ is an amplitude factor. The Fourier integral transform $g(\eta, \tau)$ is found from (2.3.5) to satisfy

$$\ddot{g} - \eta^2(1 - \eta^2)\, g = 0 \qquad (2.3.8)$$

With the initial conditions (2.3.7), the solution of (2.3.8) is

$$g = v_o \cdot \beta(\eta) p^{-1} \begin{Bmatrix} \sinh p\tau \\ \sin p\tau \end{Bmatrix} \qquad (2.3.9)$$

where

$$p = \eta|1 - \eta^2|^{1/2} \tag{2.3.10}$$

and the hyperbolic form is taken in (2.3.9) for $\eta < 1$, the circular form for $\eta > 1$.

The possibility of large growth during the buckling period appears in sinh $p\tau$ for wave numbers $\eta$ close to a preferred value $\eta_p$. The buckling form is then attributed to the dominance of the corresponding components sin $\eta\xi$ in (2.3.6). To distinguish this preferred value, we observe that in (2.3.9) the magnification of $v_o\beta(\eta)$ is expressed by $p^{-1}$ sinh $p\tau$, which may be written as

$$\mu_1 = \tau(p\tau)^{-1} \sinh p\tau \tag{2.3.11}$$

The function $x^{-1}$ sinh x increases monotonically with x, for $x > 0$. Thus for a given $\tau$, (2.3.11) is greatest when p is greatest. From (2.3.10), the value for $\eta$ for which p is a maximum is

$$\eta_p = 1/\sqrt{2} \tag{2.3.12}$$

The corresponding magnification (2.3.11) is

$$\mu_1 \max = 2 \sinh \frac{1}{2}\tau \tag{2.3.13}$$

Figure 2.29a plots the values of the magnification $\mu_1$ in (2.3.11) as functions of the wave number $\eta$, for three selected values of $\tau$. The half-wavelengths $\pi r/s\eta$, corresponding to the values of $\eta$, are given in terms of the radius gyration r. These involve the value $s = 0.50$, which as shown later is appropriate to the material of the bars in Figure 2.22. The times in microseconds correspond, for the bars of Figure 2.22, to each of the selected values of $\tau$. At $t = 52$ $\mu$s, the lowest curve, the magnification is only slightly selective. But at $t = 156$ $\mu$s, which is well within the duration (220 $\mu$s) of sustained plastic flow, it is strongly selective. The peak occurs, corresponding to (2.3.12), for the half-wavelength 8.88 r, which for the bars in Figure 2.22, is 1.01 inch. Comparisons with test results are made in Section 2.3.5.

### 2.3.4 The Nearly Straight Bar

To evaluate the effect of an initial deflection represented by $w_o$ in (2.3.6), we

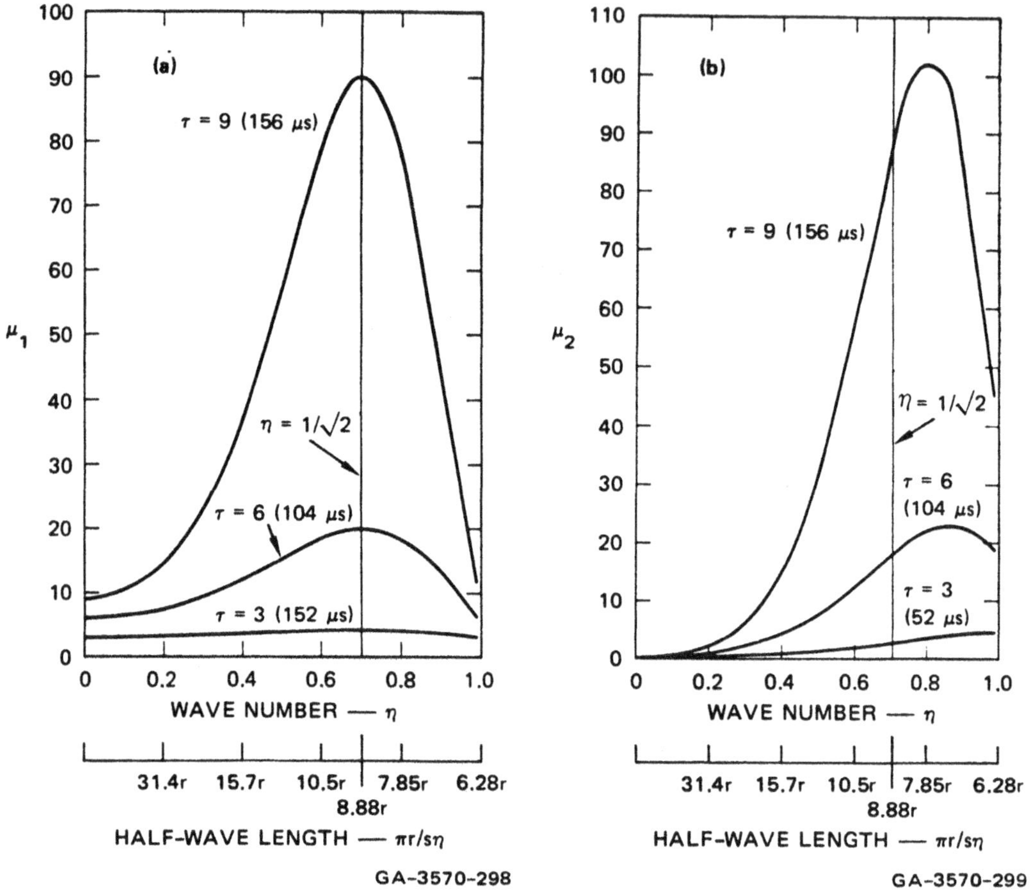


FIGURE 2.29 HYPERBOLIC MAGNIFICATION FACTOR FOR INITIAL VELOCITY [IN (a)] AND DISPLACEMENT [IN (b)] PERTURBATIONS

Times are in μs for rods in Figure 2.22.

represent it in $0 < \xi < \infty$ by

$$w_o(\xi) = \delta_o \int_0^\infty \alpha(\eta) \sin \eta\xi \cdot d\eta \tag{2.3.14}$$

where $\delta_o$ is an amplitude factor. The initial conditions are now

$$w(\xi,0) = 0 \quad , \qquad \dot{w}(\xi,0) = 0 \tag{2.3.15}$$

Again considering a solution of (2.3.5) in the form (2.3.6), we find

$$g = \delta_o\alpha(\eta) \cdot \frac{1}{1 - \eta^2}\left[\begin{matrix}\cosh \\ \cos\end{matrix}\ p\tau - 1\right] \tag{2.3.16}$$

with p given by (2.3.10) and the hyperbolic form being taken for $\eta < 1$.

For $\eta < 1$, the magnification factor multiplying $\delta_o\alpha(\eta)$ in (2.3.16) is

$$\mu_2 = (1 - \eta^2)^{-1}(\cosh p\tau - 1) \tag{2.3.17}$$

This vanishes for $\eta = 0$ and is $\tau^2/2$ for $\eta = 1$. It has a maximum with respect to $\eta$, for given $\tau$, when

$$\frac{1}{2\eta^2} - 1 = -\frac{\cosh p\tau - 1}{p\tau \sinh p\tau} \tag{2.3.18}$$

For conspicuous magnification, $\cosh p\tau$ must be much larger than unity, in which case the right-hand side of (2.3.18) may be replaced by $-1/p\tau$. Since $p\tau$ will then itself be considerably larger than unity, (2.3.18) will yield a value of $\eta^2$ somewhat greater than 1/2. Curves representing $\eta_2$ in (2.3.17) as a function of $\eta$ are shown in Figure 2.29b for three selected values of $\tau$. The upper two curves show the peaks to the right of $\eta = 1/\sqrt{2}$, closer for the higher $\tau$, as expected. The selective magnification resembles that in Figure 2.29a.

### 2.3.5 Comparisons of Theoretical Model and Experimental Results

An indication of the validity of the theoretical model may be obtained by comparing the predicted and observed wavelengths. The peak in the magnification curve of Figure 2.29a, for initial velocity perturbations, occurs at $\eta = 1/\sqrt{2}$ and in Figure 2.29b, for initial displacement perturbations, at a value about 15% higher (for $\tau = 3$). Assuming that there is no dominant peak in the initial perturbations, we take $\eta = 1/\sqrt{2}$ as the theoretical wave number. The corresponding half-wavelength is

$$\lambda = \pi r(2E_h/\sigma)^{1/2} \tag{2.3.19}$$

Values of $\sigma$ and $E_h$ were obtained from the experimental stress-strain data plotted in Figure 2.30. The elastic region is neglected and, in keeping with the assumption of linear strain-hardening, a straight line is passed through the data points for 1% to 3% strain, this being the range in which buckling is initiated (see column 4 of Tables 2.2, 2.3, and 2.4). The corresponding values of $\sigma$ and $E_h$ are listed in Table 2.1, together with the half-wavelengths calculated from (2.3.19).

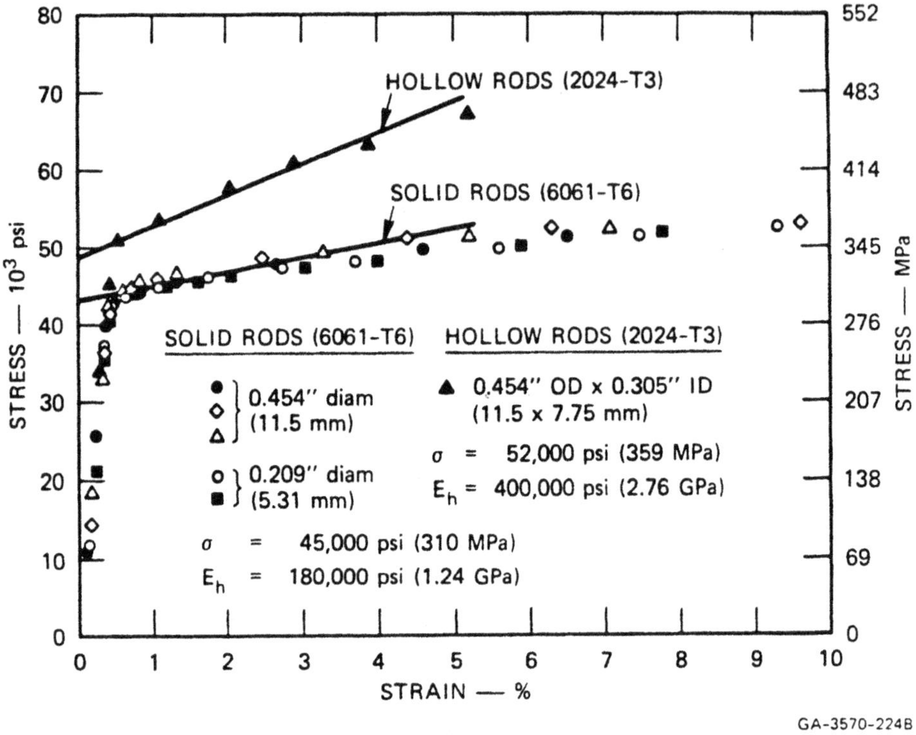


FIGURE 2.30 STRESS-STRAIN DATA

Table 2.1

THEORETICAL HALF-WAVELENGTHS (λ)

| Specimen Type | Material | $\sigma$ | $E_h$ | $s^2 = \sigma/E_h$ | r | λ |
|---|---|---|---|---|---|---|
| **British Units** | | | | | | |
| | | (lb/in.$^2$) | (lb/in.$^2$) | | (inch) | (inch) |
| Rod, 0.454-inch-diameter | 6061-T6 | 45 | 180 | 0.25 | 0.113 | 1.01 |
| Rod, 0.209-inch-diameter | 6061-T6 | 45 | 180 | 0.25 | 0.052 | 0.46 |
| Tubing, 0.454-inch-diameter, 0.075-inch wall | 2024-T3 | 53 | 400 | 0.13 | 0.137 | 1.70 |
| **SI Units** | | | | | | |
| | | (MPa) | (GPa) | | (mm) | (mm) |
| Rod, 11.5-mm-diameter | 6061-T6 | 310 | 1.24 | 0.25 | 2.87 | 25.6 |
| Rod, 5.3-mm-diameter | 6061-T6 | 310 | 1.24 | 0.25 | 1.32 | 11.7 |
| Tube, 11.5-mm-diameter, 1.9-mm wall | 2024-T3 | 366 | 2.80 | 0.25 | 3.48 | 43.2 |

The observed half-wavelengths were determined from plots of the buckled shapes such as that shown in Figure 2.31. The half-wavelengths obtained for several ½-inch-diameter (12.7 mm) aluminum (6061-T6) bars are listed in Table 2.2. The observed half-wavelengths are in reasonable agreement with the theoretical value of 1.01 inch (25 mm). At the higher velocities--for example, for the last bar of Table 2.2--the half-wavelengths are shorter near the impact end. This is attributed to the lower average value of $E_h$ (see Figure 2.30), corresponding to the higher average strain, which prevails near the impact end during the buckling motion.

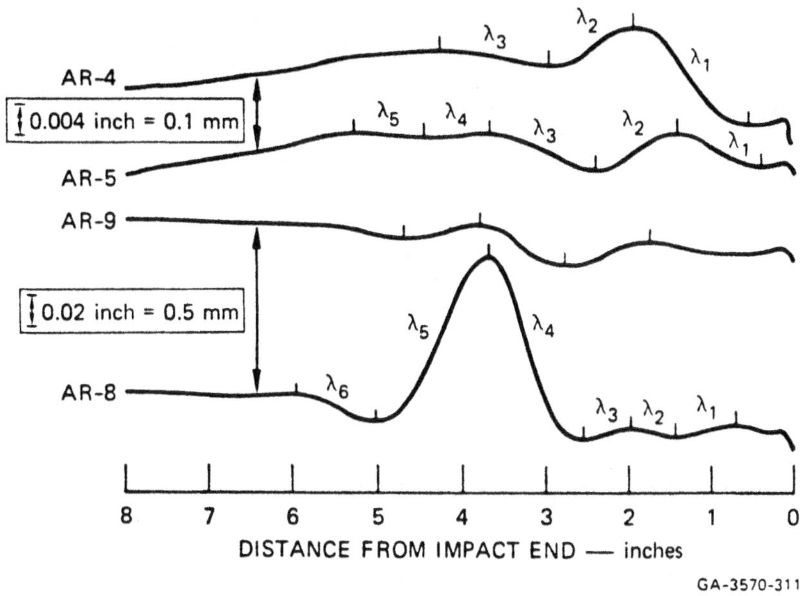


FIGURE 2.31 DISPLACEMENT PLOTS FOR RODS IN FIGURE 2.22

The observed half-wavelengths for the 1/4-inch-diameter (6.3 mm) aluminum (6061-T6) bars are listed in Table 2.3. The observed values are about twice the theoretical value at low velocities and are about 20% greater than the theoretical value at the highest velocity. This difference is also attributed to variations in $E_h$. At the lowest velocity for which buckling occurs, the axial strain at the impact end is about 1%. Thus the average $E_h$ during buckling must be somewhat greater than the value at 1% strain, and from Figure 2.30, this is substantially greater than the indicated slope for solid bars. From (2.3.19), this would result in a longer half-wavelength. At the higher velocities, the strains are higher, and the average value of $E_h$ during the buckling motion is closer to the value used to calculate the theoretical half-wavelength. Thus, at these strains, the observed half-wavelengths are in better agreement with the theoretical value.

Table 2.2

HALF-WAVELENGTHS OF 6061-T6 ALUMINUM BARS OF 0.454-INCH-DIAMETER‡

| Rod | Velocity (ft/sec) | Decrease in Length (inches) | Axial Strain* (%) | Observed Half-Wavelength (inches) $\lambda_1$ | $\lambda_2$ | $\lambda_3$ | $\lambda_4$ | $\lambda_5$ | $\lambda_6$ | $\lambda_7$ | Average |
|---|---|---|---|---|---|---|---|---|---|---|---|
| AR-1 | ~ 70 | 0.11 | 2.8 | 1.30 | — | — | — | — | — | — | 1.30 |
| AR-2 | — | 0.12 | 3.2 | 1.20 | — | — | — | — | — | — | 1.30 |
| AR-3 | ~ 150 | 0.12 | 3.2 | 1.20 | — | — | — | — | — | — | 1.20 |
| AR-4† | ~ 200 | 0.22 | 5.4 | 1.35 | 0.98 | 1.28 | — | — | — | — | 1.20 |
| AR-5† | ~ 275 | 0.46 | 10.6 | 0.98 | 0.97 | 1.25 | 0.76 | 0.82 | — | — | 0.96 |
| AR-6 | ~ 400 | 1.08 | 24 | 0.85 | 1.03 | 0.95 | 0.80 | 0.88 | — | — | 0.90 |
| AR-9† | 458 | 1.12 | 24 | 0.98 | 1.00 | 0.90 | — | — | — | — | 0.96 |
| AR-12 | 461 | 1.20 | 26 | 0.70 | 0.95 | 0.80 | 0.75 | 1.03 | 1.13 | 1.13 | 0.93 |
| AR-13 | 481 | 1.26 | 27 | 0.80 | 0.88 | 0.88 | 0.85 | 1.08 | 1.13 | 1.05 | 0.95 |
| AR-8† | ~ 520 | 1.52 | 34 | 0.68 | 0.52 | 0.57 | 1.10 | 1.33 | 0.94 | — | 0.86 |

* One diameter from impact end. Axial strain is taken to be twice the diametral strain.

† Shown in Figure 2.22.

‡ 1 inch = 25.4 mm, 1 ft/s = 0.305 m/s.

Table 2.3

HALF-WAVELENGTHS OF 6061-T6 ALUMINUM BARS OF 0.209-INCH-DIAMETER‡

| Rod | Velocity (ft/sec) | Decrease in Length (inches) | Axial Strain* (%) | Observed Half-Wavelength (inches) $\lambda_1$ | $\lambda_2$ | $\lambda_3$ | $\lambda_4$ | $\lambda_5$ | $\lambda_6$ | Average |
|---|---|---|---|---|---|---|---|---|---|---|
| B-7 | 79 | 0.01 | 0.96 | — | — | — | — | — | — | — |
| B-10 | 86 | 0.02 | 0.96 | — | — | — | — | — | — | — |
| B-16 | 101 | 0.04 | 1.16 | 0.83 | 1.10 | — | — | — | — | 0.96 |
| B-13† | 118 | 0.07 | 1.7 | 0.93 | 0.80 | — | — | — | — | 0.86 |
| B-5 | 126 | 0.09 | 3.0 | 0.83 | 0.88 | — | — | — | — | 0.85 |
| B-15 | 147 | 0.12 | 3.6 | 0.73 | 0.75 | — | — | — | — | 0.74 |
| B-11 | 164 | 0.14 | 4.4 | 0.53 | 0.68 | — | — | — | — | 0.60 |
| B-19 | 168 | 0.14 | 4.5 | 0.75 | 0.68 | — | — | — | — | 0.71 |
| B-14 | 171 | 0.16 | 4.5 | 0.73 | 0.83 | — | — | — | — | 0.78 |
| B-4† | 176 | 0.17 | 5.1 | 0.77 | 0.73 | — | — | — | — | 0.75 |
| B-3† | 288 | 0.50 | 12.2 | 0.55 | 0.48 | 0.38 | 0.38 | 0.45 | 0.93 | 0.67 |
| B-1† | 342 | 0.67 | 17.0 | 0.55 | 0.73 | 0.95 | 0.53 | 0.48 | 0.75 | 0.55 |

* One diameter from impact end. Axial strain is taken to be twice the diametral strain.

† Shown in Figure 2.23.

‡ 1 inch = 25.4 mm, 1 ft/s = 0.305 m/s.

The half-wavelengths for the ½-inch-diameter (12.7 mm) aluminum (2024-T3) tubes are listed in Table 2.4. The observed values agree favorably with the theoretical value of 1.70 inch (43 mm).

Table 2.4

HALF-WAVELENGTHS OF 2024-T3 ALUMINUM TUBES OF 0.454-INCH-DIAMETER AND 0.075 INCH WALL‡

| Tube | Velocity (ft/sec) | Decrease in Length (inches) | Axial Strain* (%) | Observed Half-Wavelength (inches) $\lambda_1$ | $\lambda_2$ | $\lambda_3$ | $\lambda_4$ | Average |
|---|---|---|---|---|---|---|---|---|
| AT1-1† | 189 | 0.19 | 3.6 | 1.75 | 1.50 | — | — | 1.63 |
| AT1-2† | 330 | 0.59 | 8.2 | 1.35 | 1.53 | 1.30 | — | 1.39 |
| AT1-8 | 462 | 1.04 | 14.8 | 1.40 | 1.38 | 1.30 | — | 1.36 |
| AT1-10† | 492 | 1.25 | 17.6 | 1.28 | 1.80 | 1.60 | 1.85 | 1.63 |
| AT1-9 | 576 | 1.48 | 20 | 0.93 | 1.68 | 2.38 | — | 1.66 |
| AT1-11† | 644 | 1.92 | 26 | 1.75 | 1.75 | 2.48 | 1.65 | 1.90 |
| AT1-14 | 667 | 2.05 | 29 | 1.65 | 1.75 | 2.25 | 1.40 | 1.82 |

* One diameter from impact end. Axial strain is taken to be twice the diametral strain.

† Shown in Figure 2.24.

‡ 1 inch = 25.4 mm, 1 ft/s = 0.305 m/s.

Another check on the theoretical model can be made by comparing the theoretical and experimental buckling times. As indicated in Figure 2.29, a preferred wavelength is apparent at $\tau = 6$ and is well established at $\tau = 9$. The corresponding actual times calculated from the last equation in (2.3.4) are indicated in Table 2.5.

The bar shown in the process of buckling in Figure 2.25 exhibits observable buckling at 110 $\mu$s, and buckling is complete at 243 $\mu$s. This is to be compared with the calculated times indicated in the first line of Table 2.5. The initial buckling motion occurs within the period during which selective magnification is apparent, Figure 2.29a. Also, the buckling motion is complete within the time required for the elastic wave to transverse the bar and return to the impact end. Thus the simple theory gives a buckling time that is in good agreement with the experimental results.

Table 2.5

THEORETICAL BUCKLING TIMES

| Specimen type | Time t ($\mu$s) $\tau = 3$ | $\tau = 6$ | $\tau = 9$ |
|---|---|---|---|
| Rod, 0.454-inch-diameter (11.5 mm) | 52 | 104 | 156 |
| Rod, 0.209-inch-diameter (5.3 mm) | 24 | 48 | 72 |
| Tube, 0.454-inch-diameter (11.5 mm), 0.075-inch wall (1.9 mm) | 81 | 162 | 243 |

The times in Table 2.5 indicate that the small-diameter bars are the most unstable, i.e., buckle fastest. The large-diameter bars are next, and the tubes are the least unstable. This is consistent with the buckled forms in Figures 2.22 and 2.23, which show that, at comparable velocities, the small bars buckle more violently. To a lesser extent, at comparable velocities, the bars in Figure 2.22 show more severe buckling than those in Figure 2.24.

Thus, the simple theoretical model presented here accounts reasonably well for the observed buckling. The predicted time scale and wavelengths are in essential agreement with the observations. The most significant feature yet to be accounted for in the theory is the variation of $E_h$ during the buckling motion. This variation has an important effect and has practical significance in dynamic plastic flow buckling of rings and shells from radial impulses. Theoretical and experimental results for a continuously decreasing hardening modulus are given for these structural elements in Sections 3.4 and 3.5.

## REFERENCES

1. E. Sevin, "On the Elastic Bending of Columns Due to Dynamic Axial Forces Including Effects of Axial Inertia," ASME Trans., **82**, Series E, J. Appl. Mech., **27**, pp. 125-131, 1960.

2. R.V. Southwell, "On the Analysis of Experimental Observations and Problems of Elastic Stability," Proc. Royal Soc., London, Series A, **135**, pp. 601-616, 1932.

3. A. Marston, "Correspondence on the Theory of the Ideal Column," Trans. ASCE, **39**, pp. 108-120, 1898.

4. C. Jensen, "Quebec Bridge Disaster," Engineering, **85**, London, pp. 433-434, 1908.

5. W.E. Lilly, "The Strength of Columns," Trans. ASCE, **76**, pp. 258-274, 1913.

6. E.H. Salmon, "Columns," London, 1921. Salmon's data and a more complete discussion of equivalent imperfections are given in S.P. Timoshenko and J.M. Gere, *Theory of Elastic Stability,* 2nd ed. McGraw-Hill Book Co., New York, 1961.

7. H. Kolsky, *Stress Waves in Solids,* Dover Publications, New York, pp. 41-47, 1963.

8. R.V. Churchill, *Fourier Series and Boundary Value Problems,* McGraw-Hill Book Co., New York, pp. 85-90, 1941.

9. S.O. Rice, "Mathematical Analysis of Random Noise," in N. Wax, *Selected Papers on Noise and Stochastic Processes,* Dover Publications, New York, 1954.

10. D. Slepian, "Contributions to Signal and Noise Theory," IEEE Trans. on Information Theory **IT-9**, pp. 229-233, 1963.

11. G. Gerard and H. Becker, "Column Behavior under Conditions of Impact," J. Aeronaut. Sci. **19**, pp. 58-65, 1952.

12. H.E. Lindberg, "Buckling of a Very Thin Cylindrical Shell Due to an Impulsive Pressure," ASME Trans. **86**, Series E, J. Appl. Mech., **31**, pp. 267-272, June 1964.

13. H.E. Lindberg, "Impact Buckling of a Thin Bar," ASME Trans. **87**, Series E, J. Appl. Mech., **32**, pp. 315-322, 1965.

14. G.R. Abrahamson and J.N. Goodier, "Dynamic Flexural Buckling of Rods Within an Axial Plastic Compression Wave," ASME Trans. **88**, Series E, J. Appl. Mech., **33**, pp. 241-247, June 1966.

# 3. DYNAMIC PULSE BUCKLING OF RINGS AND CYLINDRICAL SHELLS FROM RADIAL LOADS

## 3.1 INTRODUCTION

This chapter presents theoretical and experimental results that can be used to analyze pulse buckling of rings and cylindrical shells subjected to radial pulse loads ranging from an ideal impulse to a step in radial pressure. It is shown that under impulsive and nearly impulsive loads, threshold buckling can be either elastic or plastic, depending on the radius-to-thickness ratio of the ring or shell. This buckling is closely analogous to the dynamic elastic and plastic-flow buckling described for long bars in Chapter 2, and as in the bars, wavelengths of buckling are short compared with the structure dimensions. Under long duration loads, buckling is elastic because of the increased time for buckling to take place; the load need be only slightly higher than the static buckling load.

For end-supported shells, these features of buckling near the two load duration extremes allow a great simplification in the analysis. Because the wavelengths for impulsive buckling are short compared with the shell length, buckling from impulsive loads is assumed to be independent of the axial coordinate. This gives an adequate description of buckling in the main span of the shell, where the buckle amplitudes are greatest and nearly constant along the length. With this simplification, the complexities of plastic-flow buckling can be analyzed with a reasonably simple theory. In buckling under long duration loads, the pressures are much lower and stresses are elastic. Then the end boundary conditions and the variation of buckle amplitude with length are easily included by using elastic shell theory. There is a range of loads of intermediate durations for which neither of these simplifications is appropriate, but extrapolation of results from the impulsive and long-duration load theories gives reasonable estimates of critical loads in this range.

Dynamic plastic flow buckling of rings or long cylindrical shells from ideal radial impulse is presented first, in Section 3.2, because for these structural elements the corresponding theory, in its simplest form, is simpler than for elastic buckling. The simple form of the plastic theory is similar to that given in Section 2.3 for bars. The main difference is that the thrust results from hoop mode compression as the shell is brought to rest from the initial inward radial velocity, rather than from axial impact. As for the bar, bending moments from strain hardening are crucial in the theory. For mathematical simplicity the strain-hardening modulus is taken to be constant.

Dynamic plastic flow buckling with a strain-hardening modulus that decreases continuously with increasing compressional strain is presented in Section 3.4. The variable modulus theory includes both elastic and plastic flow buckling. Sections 3.2 and 3.3 present theoretical analyses and experimental results that demonstrate the phenomena of plastic and elastic buckling and the appropriateness and accuracy of the theory. Section 3.4 then uses the theory to develop formulas for critical radial impulse for buckling over the entire range of elastic and plastic flow buckling. This range corresponds to buckling in very thin to very thick rings and shells; the transition from elastic to plastic flow buckling occurs at a radius-to-thickness ratio of about 200 in engineering metals. Section 3.4 also contains a numerical analysis of post-buckling response beyond strain-rate reversal, showing that the simple analytical theory with no strain-rate reversal gives reasonably good predictions of both critical buckling modes and critical impulses for buckling.

Section 3.5 combines the theoretical developments of the earlier sections to present the complete analysis of simply supported cylindrical shells subjected to exponentially decaying and triangular radial pressure pulses ranging from ideal impulses to step loads.

## 3.2 DYNAMIC PLASTIC FLOW BUCKLING OF RINGS AND LONG CYLINDRICAL SHELLS FROM UNIFORM RADIAL IMPULSE

### 3.2.1 Introduction

The theory for this form of shell buckling is closely analogous to the theory for bar buckling given in Chapter 2. The mathematical development is presented such that the reader need not read Chapter 2 first; however, a reading of Chapter 2 is helpful because many of the buckling features described there carry over to the shell. As for the bar, we describe theoretical buckling motion for both discrete and random imperfections, but this time for imperfections in velocity rather than in shell shape. Discrete imperfections are taken as a small change in velocity from one value to another as one moves around the shell. Wrinkles are found to be localized in the vicinity of the velocity change in a manner similar to the localization of buckling near the end of a bar undergoing eccentric impact. Since experiments and theory for shell buckling were performed before those for the bar, the conceptual development in the original paper[1] is given in its entirety to retain the element of discovery in the original work.

Cylindrical shells subjected to all-round impulsive external pressures of sufficient intensity exhibit a characteristic behavior as indicated in Figure 3.1. The circle represents the original size (outer surface), and the thick black somewhat crumpled circle inside is the end of the shell. It shows a general radius reduction of

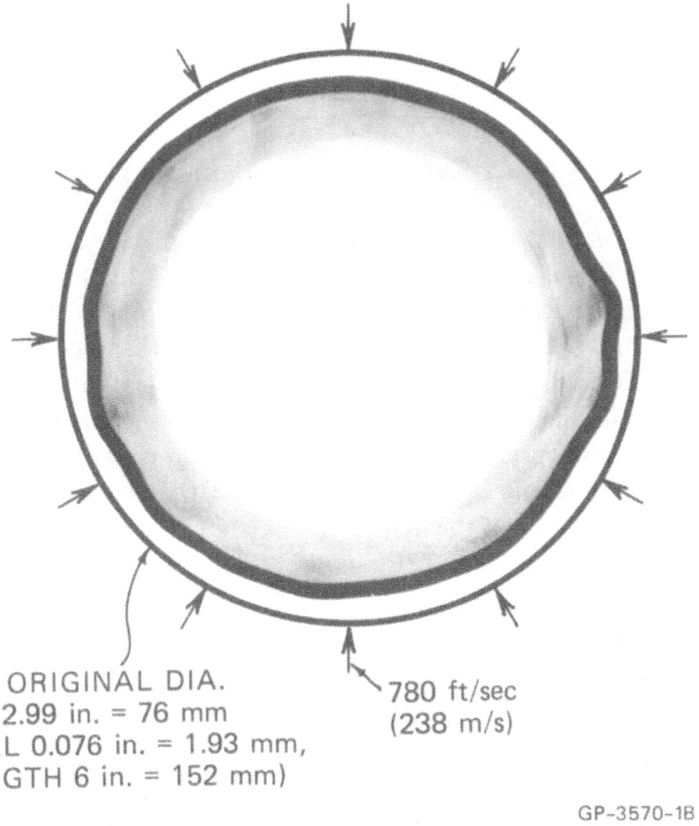


FIGURE 3.1 TYPICAL BUCKLED SHAPE OF CIRCULAR CYLINDRICAL ALUMINUM (2024-T3) SHELL SUBJECTED TO A "UNIFORM" INWARD RADIAL VELOCITY

about 8% and is wrinkled in a manner sufficiently regular to permit a count of the number of crests and troughs. The number of crests so counted is 16. The gray band inside is the interior surface, the white center being the open far end of the shell.

Experimental results of this kind are obtained with a reproducibility indicating a characteristic form of dynamic plastic buckling. This section presents a theory of the formation of the wrinkled shape, based on the dynamics of a metal shell that deforms plastically with strain-hardening (the stress required to maintain flow increases as the flow proceeds). This property is crucial to the theory presented here, a definite buckled shape being predictable only when it is present.* Comparison of such predictions with several experimental results for a range of dimensions and materials is given later. The agreement suggests that the main features of the buckling process are satisfactorily represented in the theory.

### 3.2.2 Postulated Character of the Motion--Dynamic Flow Buckling

We consider a cylindrical shell loaded in such a manner that all elements receive a large initial radial velocity simultaneously. In the absence of imperfections, the material flows into a uniform cylindrical shell of smaller radius (and thicker wall) until the initial kinetic energy has been absorbed in the work of plastic deformation. But with the inevitable small imperfections in the uniformity of initial velocity round the shell and the imperfections in the material properties and the shape, there will be perturbations from this uniform converging motion. If a small part of the circumference begins to lag behind the remainder, its curvature will therefore be increased and the action of the compressive circumferential stress will be to encourage it to lag behind still more.

If the material is perfectly plastic, the steady increase of circumferential compressive strain at all points of the wall thickness implies an unchanging plastic-flow stress, and this means that a perturbational change of curvature induces zero bending moment. Then there is no resistance to the perturbational lag already mentioned, nor to a perturbational acceleration of an element that happens to bend the other way. It becomes evident that the uniform converging motion is simply unstable. In fact, an initial attempt to construct a theory for the perfectly plastic material led to essentially divergent series for any time, and so to the realization that the actual motion depends crucially on strain-hardening. This aspect becomes clear in the analysis to be given, leading to series that begin like divergent series but, in virtue of the character of the later terms, are in fact convergent.

It would complicate matters unduly to include all sorts of imperfections together. Instead, the basic problem is taken as that of a geometrically and materially perfect cylindrical shell, with a slight imperfection in the uniformity of the initial velocity. There is consequently departure from the circular form throughout the motion, but this is regarded as always small. Thus at two points A and B on the outside and inside, respectively, in Figure 3.2, the circumferential strain is

---

*Two other plastic properties that can give a definite buckled shape are associated with biaxial plastic-flow and with an increase in stress with strain rate. These are considered in Chapter 4. The shells described in the present section are long enough that axial flow can be neglected, and strain rate effects are small for the present shell materials.

always supposed to increase during the flow, but at B it increases more rapidly than at A because of the slight curvature perturbation. On the stress-strain curve in Figure 3.2, the representative points A′B′ are both moving to the right, with A′ lagging behind B′. The stress-strain curve is taken as the two straight lines OC and CD, with slopes E (Young's modulus) and $E_h$ (strain-hardening modulus). Because of the difference of circumferential stress between A′ and B′, there is bending moment M in the section AB of the shell, and it is related to the curvature perturbation by the simple proportionality of elastic bar theory, using $E_h$ instead of E. Considering unit axial length regarded as a ring, one has in fact

$$M = E_h I \kappa \quad , \quad \kappa = \frac{1}{a^2}\left(\frac{d^2w}{d\theta^2} + w\right) \quad , \quad I = \frac{1}{12}\, h^3 \tag{3.2.1}$$

in which $\kappa$ is the curvature increase corresponding to a radial inward displacement $w(\theta)$, a is the radius of the shell, and h its wall thickness. Actually, of course, the radius decreases and the thickness increases, but it turns out that the essentials are adequately covered when the initial values are used.

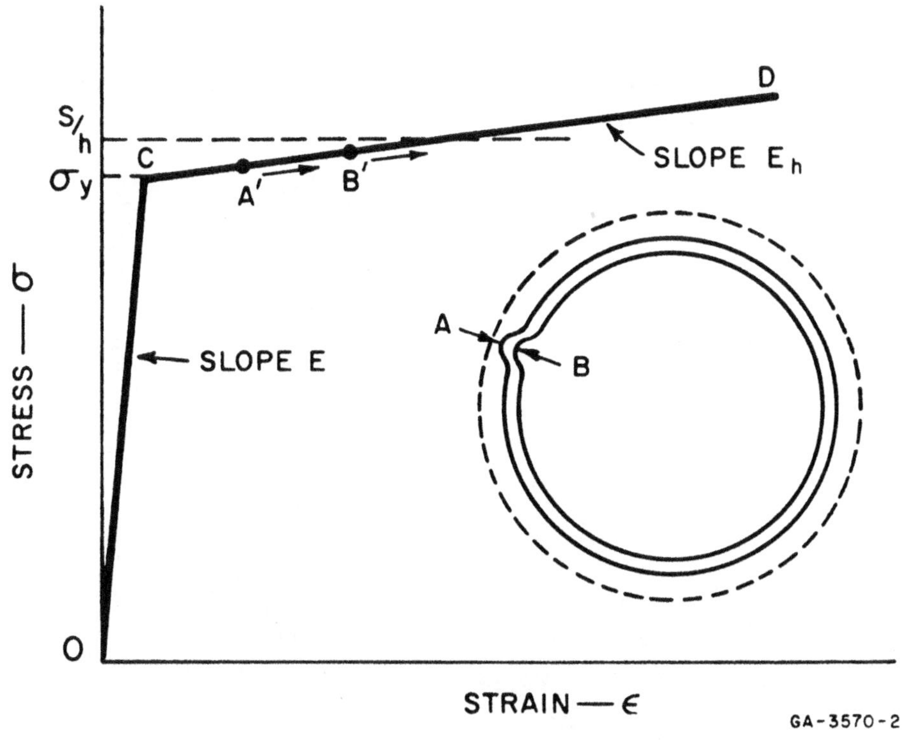


FIGURE 3.2 IDEALIZED STRESS-STRAIN RELATION AND ASSOCIATED NOTATION

As the flow proceeds, the mean circumferential stress $\sigma_m$ is not quite uniform round the "circle," since $w(\theta)$ is not quite uniform. This is neglected and the circumferential force $\sigma_m h$ is treated as uniform. Its magnitude increases as the flow proceeds, in virtue of the strain-hardening. The slope of the line CD, however, is small compared with that of OC for the materials of interest. Accordingly, the circumferential compressive stress is taken as constant, at a mean level such as S/h in Figure 3.2. There is then a constant circumferential force (per unit axial length) throughout the inward motion.

### 3.2.3 Equations of Motion

Neglecting rotary inertia, one obtains from the element of the ring (unit axial length of shell) in Figure 3.3

$$Q = \frac{\partial M}{\partial \lambda} \tag{3.2.2}$$

Q being the shear force and $d\lambda$ the arc element corresponding to $d\theta$, subtending now the angle $d\phi$ at the center of curvature.

The dynamical equation for the radial motion is, with m for mass per unit run of circumference and t for time,

$$\frac{\partial Q}{\partial \lambda} + S\frac{\partial \phi}{\partial \lambda} = -m\frac{\partial^2 w}{\partial t^2} \tag{3.2.3}$$

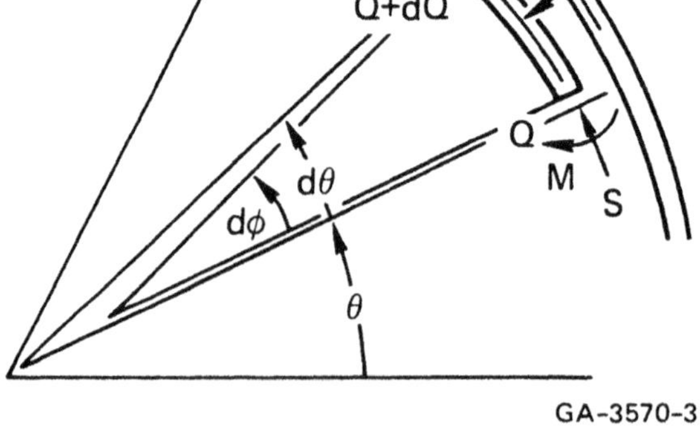


FIGURE 3.3 NOTATION. ELEMENT OF SHELL IN MOTION

and for the curvature

$$\frac{\partial \phi}{\partial \lambda} = \frac{1}{a} + \kappa \tag{3.2.4}$$

Putting Equation (3.2.4) in Equation (3.2.3), eliminating Q by means of Equation (3.2.2) and then M by means of Equation (3.2.1), one finds that w satisfies the equation

$$\frac{E_h I}{a^4}\left(\frac{\partial^4 w}{\partial \theta^4} + \frac{\partial^2 w}{\partial \theta^2}\right) + S\left[\frac{1}{a} + \frac{1}{a^2}\left(\frac{\partial^2 w}{\partial \theta^2} + w\right)\right] = -m\frac{\partial^2 w}{\partial t^2} \tag{3.2.5}$$

It is convenient to change to dimensionless inward radial displacement and time

$$u = \frac{w}{a} \ , \ \tau = \sqrt{\frac{E_h I}{m a^4}} \cdot t = \frac{1}{\sqrt{12}}\sqrt{\frac{E_h}{\rho}} \cdot \frac{h}{a} \cdot \frac{t}{a} \tag{3.2.6}$$

where $\rho$ is density, and to introduce the dimensionless constant

$$s^2 = \frac{Sa^2}{E_h I} = 12\frac{Sa^2}{E_h h^3} = 12\frac{\sigma a^2}{E_h h^2} \tag{3.2.7}$$

where $\sigma = S/h$ is approximately the yield stress in simple compression. Then Equation (3.2.5) can be written in the form

$$\frac{\partial^4 u}{\partial \theta^4} + (1 + s^2)\frac{\partial^2 u}{\partial \theta^2} + s^2 u + \frac{\partial^2 u}{\partial \tau^2} = -s^2 \tag{3.2.8}$$

This is for a ring that initially is perfectly circular. If instead it departs from the circular form by an initial "displacement" $w_i(\theta)$ in the unstressed state, and if one uses $w(\theta,t)$ now for the additional inward radial displacement during the motion, Equations (3.2.1), (3.2.2), and (3.2.3) are unchanged, but Equation (3.2.4) becomes

$$\frac{\partial \phi}{\partial \lambda} = \frac{1}{a} + \frac{1}{a^2}\left(\frac{\partial^2}{\partial \theta^2} + 1\right)(w_i + w) \tag{3.2.9}$$

and retaining Equations (3.2.6) and (3.2.7) with the new meaning of w and u, Equation (3.2.8) changes to

$$\frac{\partial^4 u}{\partial \theta^4} + (1 + s^2)\frac{\partial^2 u}{\partial \theta^2} + s^2 u + \frac{\partial^2 u}{\partial \tau^2} = -s^2\left(1 + u_i + \frac{d^2 u_i}{d\theta^2}\right) \tag{3.2.10}$$

Equation (3.2.10) will be used in Section 3.4 to examine the relative importance of shape and velocity imperfections. In analyzing the present experiments, we are concerned with the effect of velocity perturbations from explosive loading discontinuities. Hence imperfections in initial shape are not considered further here, and Equation (3.2.8) is taken as the basic equation.

### 3.2.4 Perfectly Circular Ring, Almost Uniform Initial Radial Velocity

Since the wrinkling, expressed by the dependence of w on $\theta$, is being regarded as a perturbation of the uniformly converging motion of the perfect ring with perfectly uniform radial displacement at all times, it is convenient to begin with a particular solution, $u_p(\tau)$, of Equation (3.2.8), which is independent of $\theta$. It is a solution of

$$\frac{d^2u}{d\tau^2} + s^2u = -s^2 \tag{3.2.11}$$

and therefore is of the form

$$u_p(\tau) = -1 + A_0 \cos s\tau + B_0 \sin s\tau \tag{3.2.12}$$

If the initial impulsive velocity is $V_0$

$$\left(\frac{\partial w}{\partial t}\right)_{t=0} = V_0 \quad , \quad \text{and} \quad \left(\frac{\partial u}{\partial \tau}\right)_{\tau=0} = V_0\sqrt{\frac{ma^2}{E_hI}}$$

Writing

$$v_0 = V_0\sqrt{\frac{ma^2}{E_hI}} = \sqrt{\frac{12\rho}{E_h}} \cdot \frac{a}{h} \cdot V_0 \tag{3.2.13}$$

and taking $u_p(0) = 0$, Equation (3.2.12) becomes

$$u_p(\tau) = -1 + \cos s\tau + \frac{v_0}{s} \sin s\tau \tag{3.2.14}$$

This is the unperturbed motion. It proceeds until $\dot{u}_p(\tau) = 0$, the dot indicating $\partial/\partial\tau$. Thus the duration is given by the smallest root of

$$\tan s\tau = \frac{v_0}{s} \tag{3.2.15}$$

This will provide an approximation to the duration even when slight wrinkling occurs. It corresponds simply to the absorption of the kinetic energy as plastic work in the shortening of the circumference, and this shortening is not influenced to the first order by the slight wrinkling.

Retaining the initially perfect circular ring, one now considers a slightly non-uniform initial velocity given by the Fourier series

$$\left(\frac{\partial w}{\partial t}\right)_{t=0} = V_0 \left[1 + \sum_{n=2}^{\infty} (\alpha_n \cos n\theta + \beta_n \sin n\theta)\right] \tag{3.2.16}$$

$$\dot{u}(\theta,0) = v_0 \left[1 + \sum_{n=2}^{\infty} (\alpha_n \cos n\theta + \beta_n \sin n\theta)\right] \tag{3.2.17}$$

Terms with $n = 1$ are omitted since, to the present order of accuracy, they do not contribute to deformation of the circular shape. For the appropriate solution of Equation (3.2.8), the particular solution Equation (3.2.14), together with the complementary solution, is taken in the form

$$u = \sum_{n=2}^{\infty} [f_n(\tau) \cos n\theta + g_n(\tau) \sin n\theta] \tag{3.2.18}$$

The homogeneous form of Equation (3.2.8) requires $f_n$ to satisfy

$$\ddot{f}_n + [n^4 - n^2(1 + s^2) + \dot{s}^2] f_n = 0 \tag{3.2.19}$$

and $g_n$ must satisfy the same differential equation.

One of the principal points of the present theory of the dynamic buckling process emerges here. The coefficient of $f_n$ in Equation (3.2.19) is positive, and so $f_n$ and $g_n$ are circular functions of $\tau$, provided n is large enough. But for several of the smaller values of n the coefficient is negative, and $f_n$ and $g_n$ are hyperbolic functions, with the possibility of attaining large values within the duration of the motion. The coefficient of $f_n$ in Equation (3.2.19) can be written as

$$p_n^2 = (n^2 - 1)(n^2 - s^2) \tag{3.2.20}$$

with the properties

$$\begin{aligned} p_n^2 &< 0 \quad \text{if} \quad n < s \\ p_n^2 &> 0 \quad \text{if} \quad n > s \end{aligned} \tag{3.2.21}$$

As an example of the order of magnitude to be expected for s, one may take the following values, which are representative of the aluminum alloy shell of Figure 3.1,

$$E_h = 4 \times 10^5 \text{ psi} = 2.8 \text{ GPa}, \qquad \sigma = 6 \times 10^4 \text{ psi} = 0.41 \text{ GPa}$$

$$a = 1.46 \text{ inch} = 37 \text{ mm}, \qquad h = 0.077 \text{ inch} = 2.0 \text{ mm}, \quad a/h = 18.9 \tag{3.2.22}$$

Then Equation (3.2.7) gives $s = 25.4$. Thus $f_n$ and $g_n$ will be hyperbolic functions for $2 \leqslant n \leqslant 25$, and they will be circular functions for $n \geqslant 26$. In the tests described in a later section, s ranged from 13 to 75.

Writing r for the greatest integer less than s, $p_n^2$ may be redefined conveniently as

$$p_n^2 = (n^2 - 1)(s^2 - n^2) \qquad \text{when } n \leqslant r$$

$$p_n^2 = (n^2 - 1)(n^2 - s^2) \qquad \text{when } n \geqslant r + 1 \tag{3.2.23}$$

The general solution of Equation (3.2.8) can then be put in the form

$$u = -1 + \cos s\tau + \frac{v_0}{s} \sin s\tau$$
$$+ \sum_{n=2}^{n=r} [(A_n \cosh p_n\tau + B_n \sinh p_n\tau)\cos n\theta$$
$$+ (C_n \cosh p_n\tau + D_n \sinh p_n\tau)\sin n\theta]$$
$$+ \sum_{n=r+1}^{\infty} [(A_n \cos p_n\tau + B_n \sin p_n\tau)\cos n\theta$$
$$+ (C_n \cos p_n\tau + D_n \sin p_n\tau)\sin n\theta] \tag{3.2.24}$$

To meet the initial condition $u(\theta,0) = 0$, the constants $A_n$ and $C_n$ must vanish. To meet the initial velocity prescribed as Equation (3.2.17), one must take

$$B_n = \frac{v_0\alpha_n}{p_n}, \quad D_n = \frac{v_0\beta_n}{p_n} \tag{3.2.25}$$

The final solution is then

$$u = -1 + \cos s\tau + \frac{v_0}{s} \sin s\tau$$

$$+ v_0 \sum_{n=2}^{n=r} (\alpha_n \cos n\theta + \beta_n \sin n\theta) \frac{1}{p_n} \sinh p_n\tau$$

$$+ v_0 \sum_{n=r+1}^{\infty} (\alpha_n \cos n\theta + \beta_n \sin n\theta) \frac{1}{p_n} \sin p_n\tau \qquad (3.2.26)$$

In tests aiming at uniform impulsive velocity round the circumference (a uniform thickness of explosive all round), the coefficients $\alpha_n$, $\beta_n$ will all be quite small in comparison with unity. The possibility of observable wrinkling must then come from the hyperbolic functions sinh $p_n\tau$ in the terms in the second line of Equation (3.2.26). It is convenient to refer to these terms as the "buckling terms." If the strain hardening is reduced ($E_h \to 0$), s in Equation (3.2.7) increases without bound and hence r also increases. Ultimately all the terms are of the hyperbolic type and the convergence of the series is lost. Hence a solution is obtainable only for $E_h > 0$.

### 3.2.5 Strain Rate Reversal

The foregoing theory is not valid after strain reversal begins. To determine this limit, we examine the strain implied by Equation (3.2.26).

For points on the outside and inside surfaces, the circumferential compressive strain is, respectively,

$$\left(1 - \frac{h}{2a}\right)\left(\frac{w}{a} - \frac{h}{2a^2}\frac{\partial^2 w}{\partial\theta^2}\right) \quad \text{and} \quad \left(1 + \frac{h}{2a}\right)\left(\frac{w}{a} + \frac{h}{2a^2}\frac{\partial^2 w}{\partial\theta^2}\right) \qquad (3.2.27)$$

Strain reversal begins when one of these reaches a maximum positive value and then decreases. Hence only the second factor in each expression is examined. In dimensionless terms these factors are

$$u(\theta,\tau) \mp \frac{h}{2a} \frac{\partial^2 u(\theta,\tau)}{\partial\theta^2} \qquad (3.2.28)$$

and from Equation (3.2.26) these are representable, until reversal occurs, as

$$u \mp \frac{h}{2a}\frac{\partial^2 u}{\partial\theta^2} = u(\theta,\tau) \mp \frac{h}{2a} v_0 \sum_{n=2}^{\infty} (\alpha_n \cos n\theta$$

$$+ \beta_n \sin n\theta)\ \frac{n^2}{p_n}\ {\sinh \atop \sin}\ p_n\tau \qquad (3.2.29)$$

where,

for $2 \leqslant n < r$, $\sinh p_n\tau$ is used with $p_n^2 = (n^2-1)(s^2-n^2)$

and (3.2.30)

for $n \geqslant r+1$, $\sin p_n\tau$ is used with $p_n^2 = (n^2-1)(n^2-s^2)$

In the calculated results to be given, the strain history was traced by evaluation of Equation (3.2.29) for various times.

### 3.2.6 The Buckling Terms--Representative Numerical Cases

In the result Equation (3.2.26) the buckling terms in the first series consist of velocity perturbation terms magnified by factors $p_n^{-1} \sinh p_n\tau$. Regarding $p_n^{-2} \sinh^2 p_n\tau$, with $p_n^2 = (n^2-1)(s^2-n^2)$, as a function of a continuous variable $n^2$, for $1 \leqslant n \leqslant s$, it approaches the value $\tau^2$ for $n \to 1$, and again for $n \to s$. Its derivative with respect to $n^2$ vanishes where

$$(\tanh p_n\tau - p_n\tau)\ \frac{dp_n}{dn^2} = 0$$

and therefore only where $dp_n/dn^2 = 0$. This occurs where

$$n^2 = n_1^2 = \frac{1}{2}(s^2+1) \qquad (3.2.31)$$

and the corresponding maximum of $p_n^2$ is

$$(p_n^2)_{max} = \frac{1}{4}(s^2-1)^2 \qquad (3.2.32)$$

For the shell specified in Equation (3.2.22) the value of n, obtained from Equation (3.2.31), is very nearly 18, and the value of $(p_n)_{max}$ from Equation (3.2.32) is nearly 324.

A representative value for the initial velocity $V_0$ is 695 ft/s = 212 m/s. The density $\rho$, occurring in Equation (3.2.13), corresponds to 2700 kg/m$^3$ for aluminum. Then the duration of the motion is found from Equation (3.2.15) as $\tau_f$ = 0.02, corresponding to $48 \times 10^{-6}$ s. The final inward displacement (in the absence of perturbations) is 0.22 inch = 5.6 mm. (Values of $p_n^{-1} \sinh p_n\tau$ for $\tau = 0.02$ are shown later in Figure 3.11.)

Since the nonuniform "error" of the intended uniform velocity in the tests is not known, it is not possible to predict from the theory what wrinkled forms to expect. Different tests have different "errors," but the deformed specimens have a common characteristic--several wrinkles of rather short length compared with the circumference. Accordingly calculated results were obtained for several different velocity distributions, the departures from the mean being ±5%. The calculations are later compared with experimental observations.

The first of these, shown in Figure 3.4 (departure from mean only), is the "parabolic distribution"; if the circle is developed, the velocity curve becomes two parabolic arcs.

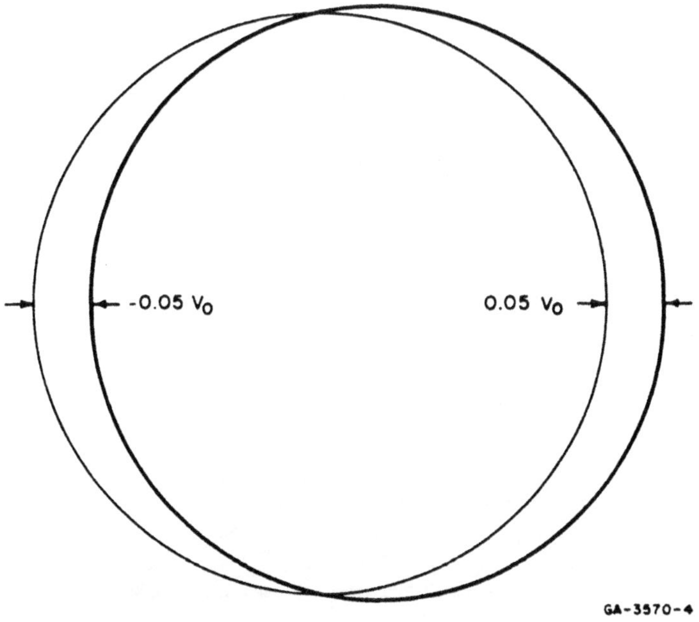


FIGURE 3.4 PARABOLIC VELOCITY PERTURBATION

Curvature is discontinuous at top and bottom crossings.

The series for $\dot{u}(\theta,0)$ takes the form

$$\dot{u}(\theta,0) = v_0 \left[1 - \frac{1}{20} \cdot \frac{32}{\pi^3}\left(\frac{1}{1^3}\cos\theta - \frac{1}{3^3}\cos 3\theta + \frac{1}{5^3}\cos 5\theta - \ldots\right)\right] \tag{3.2.33}$$

and corresponds to Equation (3.2.17) after the unimportant term in cos $\theta$ has been discarded. The manner in which the individual buckling terms in the displacement Equation (3.2.26) grow with time is shown in Figure 3.5, to be read only for the odd integer values of n. Growth is at first slow (lower curves), then rapidly accelerates through the increase of sinh $p_n\tau$. The term having $n = 17$ is the most magnified, i.e., has the greatest value of $p_n^{-1}$ sinh $p_n\tau$, and this term and its neighbors show the most rapid growth. Lower terms in Figure 3.5 have larger ordinates because of their larger values of $\alpha_n$, from Equation (3.2.33). The last buckling term has $n = 25$, the term for $n = 27$ having sin $p_n\tau$ not sinh $p_n\tau$. Its magnitude remains insignificant, the same being true of all further terms. The curve for $\tau = 0.020$ is subject to slight correction for the onset of strain reversal and is included because it suggests rapid development of the wrinkling at the time strain reversal begins. The higher (broken-line) curve for $\tau = 0.022$ emphasizes this suggestion.

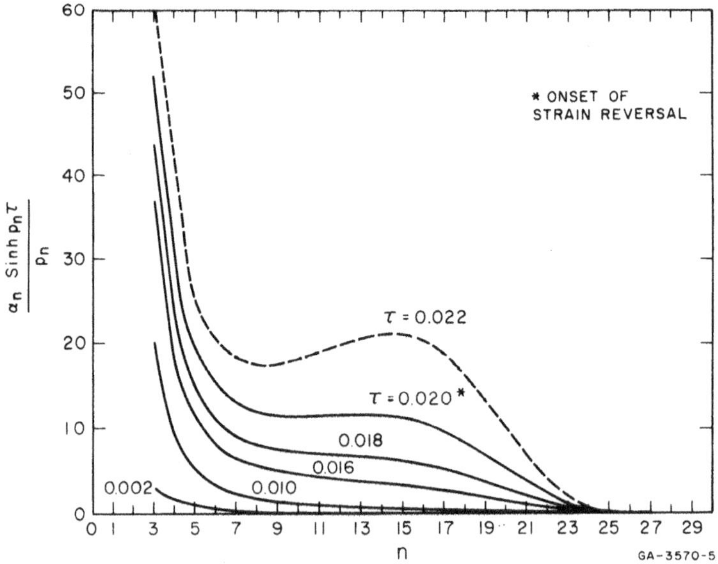


FIGURE 3.5 GROWTH OF FOURIER COEFFICIENTS OF BUCKLING TERMS WITH TIME, FOR THE PARABOLIC VELOCITY PERTURBATION OF FIGURE 3.4

The calculated form of the ring (with radial displacement greatly exaggerated) is shown in Figure 3.6, for $\tau = 0.018$. Figure 3.7 shows the same curve on the developed half-circle (upper half of Figure 3.6), together with the corresponding curves for times $\tau = 0.01$, 0.016, and 0.020. The maximum displacement before strain reversal begins ($\tau = 0.018$) is $u = 10^{-3}$, and corresponds to $w = 1.5$ mils (38 $\mu$m). The angle between the two major crests between 0° and 30° is about 20°.

The second type of velocity variation, shown in Figure 3.8, is called "trapezoidal" from the form it takes in a developed diagram. The transition from $V_0(1 - 0.05)$ on the left to $V_0(1 + 0.05)$ on the right occurs in an arc of 10°. Figure 3.9 shows the calculated wrinkled form for $\tau = 0.008$. The general resemblance to Figure 3.6 is striking. Pronounced wrinkling occurs where the velocity transitions occur in Figure 3.8 at top and bottom.

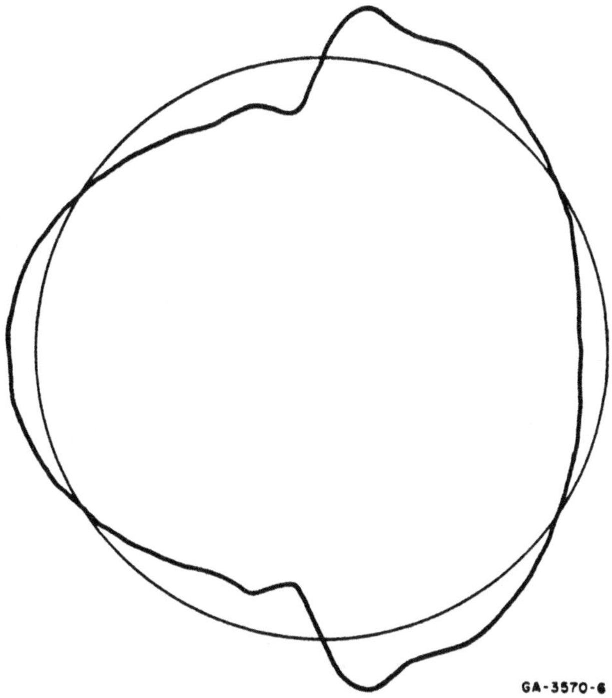


FIGURE 3.6 BUCKLING DISPLACEMENT AT $\tau = 0.018$ FOR THE PARABOLIC VELOCITY PERTURBATION OF FIGURE 3.4

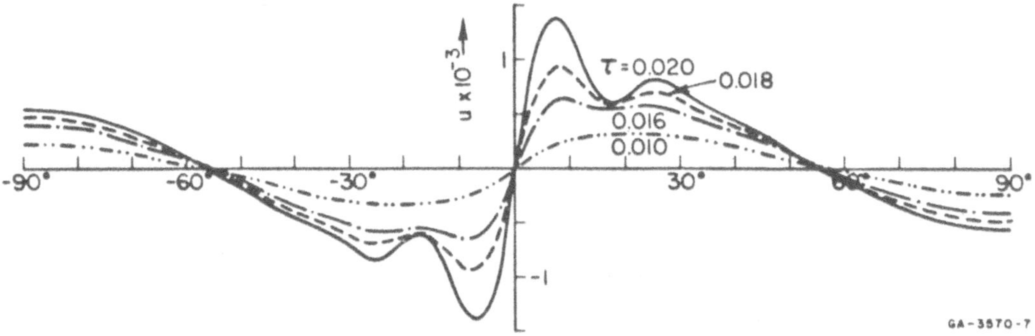


FIGURE 3.7 DEVELOPMENT OF BUCKLING DISPLACEMENT WITH TIME FOR THE PARABOLIC VELOCITY PERTURBATION OF FIGURE 3.4

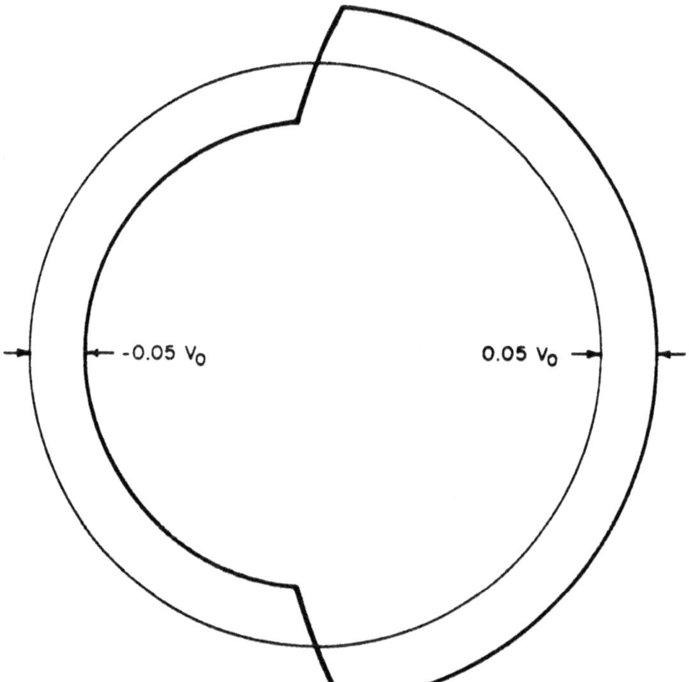


FIGURE 3.8 TRAPEZOIDAL VELOCITY PERTURBATION

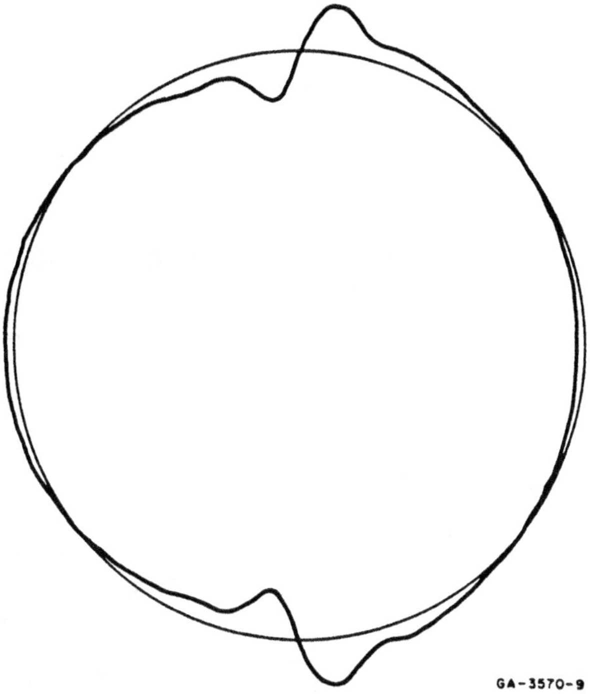


FIGURE 3.9 BUCKLING DISPLACEMENT AT $\tau = 0.008$ FOR THE TRAPEZOIDAL VELOCITY PERTURBATION OF FIGURE 3.8

In these two cases, the parabolic and the trapezoidal, the velocity perturbation had a discretely specified shape. The Fourier coefficients of the initial velocity, $\alpha_n$, diminish with n, and the phase of each harmonic is known. A velocity perturbation in which the coefficients all have the same amplitude but are random in phase is similar to the random perturbation discussed in reference to shape imperfections in Section 2.2.6. Figure 3.10 shows three examples of calculated buckled forms for the shell obtained by taking 25 harmonics ($1 \leqslant n \leqslant 25$), all with the same amplitude for $\alpha_n$. In each, the phasing of these harmonics was chosen at random. Terms in the initial velocity distribution beyond $n = 25$ have a small effect on the buckled form because they are not amplified by the buckling motion (as indicated in Figure 3.11). A more detailed discussion of buckling from random imperfections is given in Section 2.2.6.

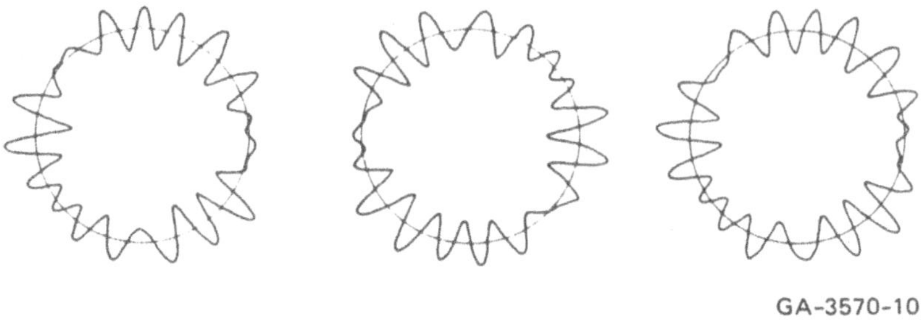


FIGURE 3.10 BUCKLED FORMS FOR "RANDOM PHASE" VELOCITY PERTURBATIONS

Each form is for a different random phasing of the harmonic components of the velocity perturbation.

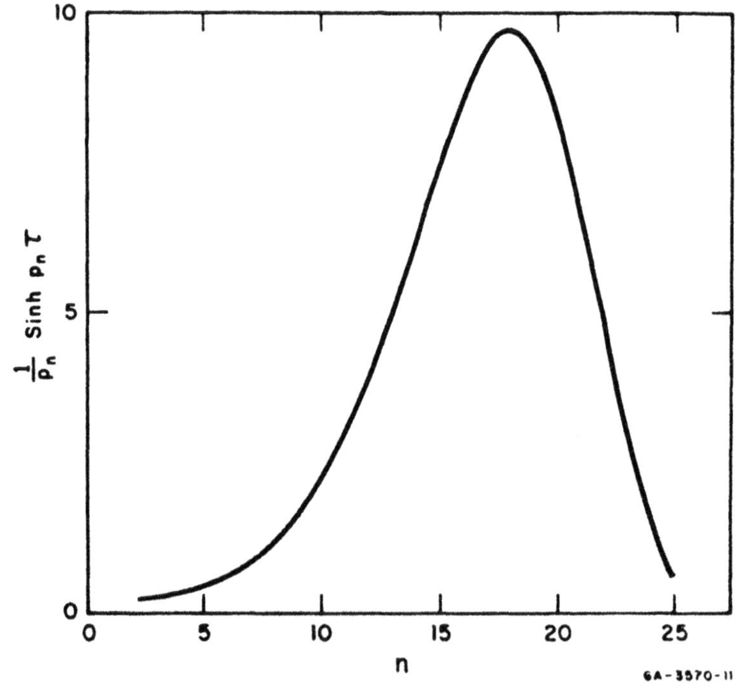


FIGURE 3.11 MAGNIFICATION FACTOR FOR "RANDOM PHASE" EXAMPLES

### 3.2.7 Experimental Technique and Characteristic Results

The buckled cylinder of Figure 3.1 was produced with the experimental arrangement shown in Figure 3.12. Enclosing the cylinder in Figure 3.12 is an attenuator of 1/4-inch-thick (6.3 mm) neoprene foam and a layer of explosive. The attenuator, which is necessary to prevent the cylinder from spalling, extends beyond the end of the cylinder to provide support for the explosive. The detonator starts from an electrical signal and is connected to the main charge by several strips of explosive. Detonation fronts initiated in the main charge by the strips of explosive expand from the initiation points and coalesce into a single ring-shaped detona-

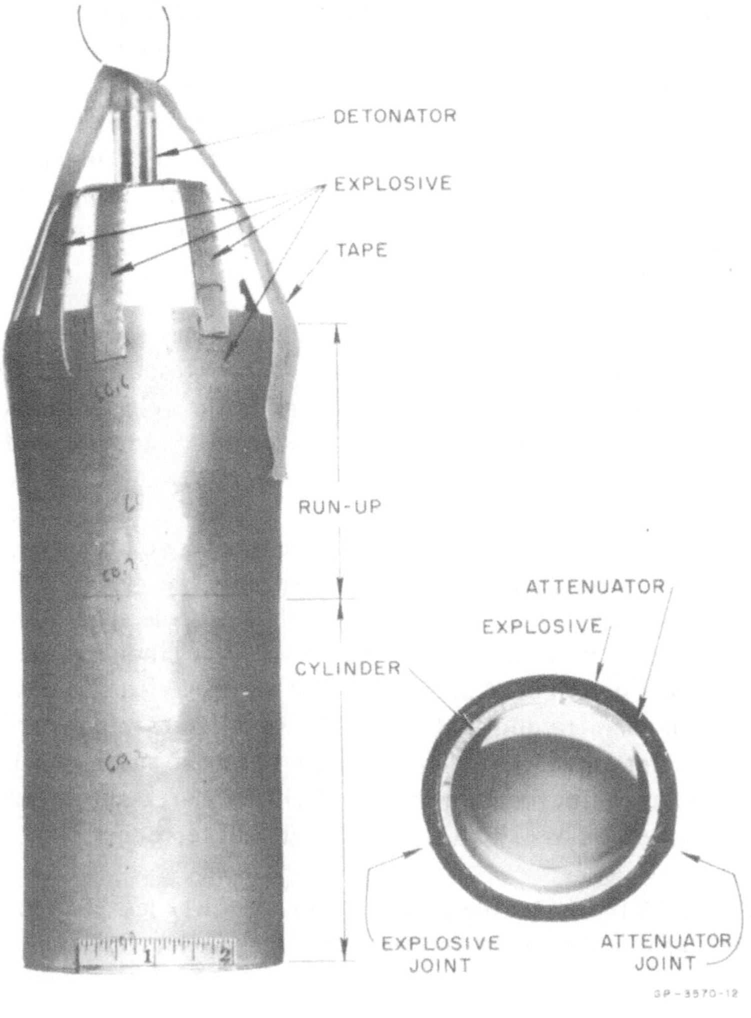


FIGURE 3.12 EXPERIMENTAL ARRANGEMENT

tion front in the "run-up" part of the main charge. The detonation front, which is a few mils ($\approx$ 100 $\mu$m) wide, subsequently sweeps over the cylinder at a rate of 23,300 ft/s (7106 m/s).

A shock wave is induced in the attenuator by the detonation front and is transmitted from the attenuator to the cylinder, imparting to it an impulsive velocity of about 500 ft/s (150 m/s). If this velocity were imparted to the cylinder instantaneously by the shock wave in the cylinder, the cylinder wall would turn through an angle of about 1.5°. Since several reflections may be required before this velocity is achieved, the actual shape of the specimen in the region of the shock front is something between a cylinder and a cone of 3° included angle.

Maximum variations in the thickness of the explosive are from 2% to 10% for charges from 15 to 60 mils thick (0.4 to 1.5 mm), the larger variations going with the thinner charges. The variations in thickness are gradual except across joints. Variations in the thickness of the explosive and the uncertainty in the impulse developed by the explosive per unit thickness combine to give an uncertainty of about 10% in the impulses imparted to the cylinders. The joints in the attenuator and the explosive produce unknown perturbations in the velocity. Since they could not be eliminated, the locations of the joints were recorded and were taken into account in the analysis of the data.

One of the characteristic features of the buckling phenomenon is the reproducibility of the number and the amplitude of the crests, which, according to the theory, depend mainly on the nonuniformities in the velocity distribution. Figure 3.13 shows two cylinders of the same material and the same initial dimensions that were subjected to nominal uniform radial velocities of 650 ft/s (198 m/s). The cylinders were nominally 3 inches (76 mm) in diameter and 6 inches (152 mm) long; the wrinkled shapes of Figure 3.13 represent central cross sections. Except near the joints, the cylinders are similar in shape. The plots around the cylinders show the crests more clearly. The quantity plotted is the square root of the slope; this was chosen to bring out the smaller crests. The limits on the points indicate plus and minus one standard deviation. The number of major crests, obtained by counting all the way around or by determining a representative crest-to-crest distance from a portion of the circumference having a succession of well defined crests, is about 18 for both cylinders. Cylinders of other materials and other radius-to-thickness ratios exhibit a similar degree of reproducibility.

Figure 3.14 illustrates the effect of radius-to-thickness ratio a/h on the buckling phenomenon. As would be expected, the number and amplitude of crests increase with a/h.

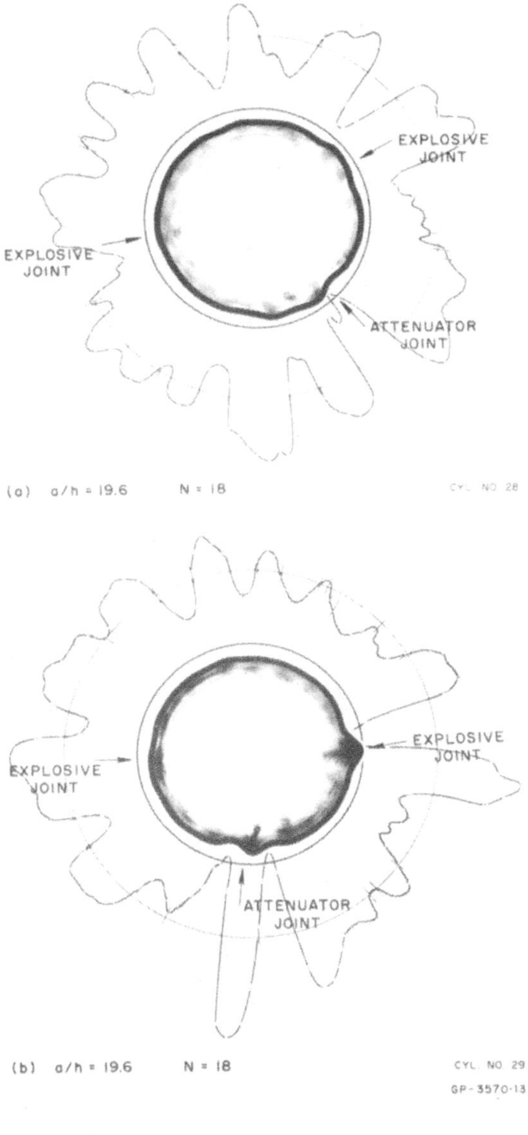


FIGURE 3.13 REPRODUCIBILITY OF BUCKLING OF ALUMINUM (6061-T6) CYLINDERS UNDER "UNIFORM" INITIAL VELOCITY OF 650 ft/sec (198 m/s)

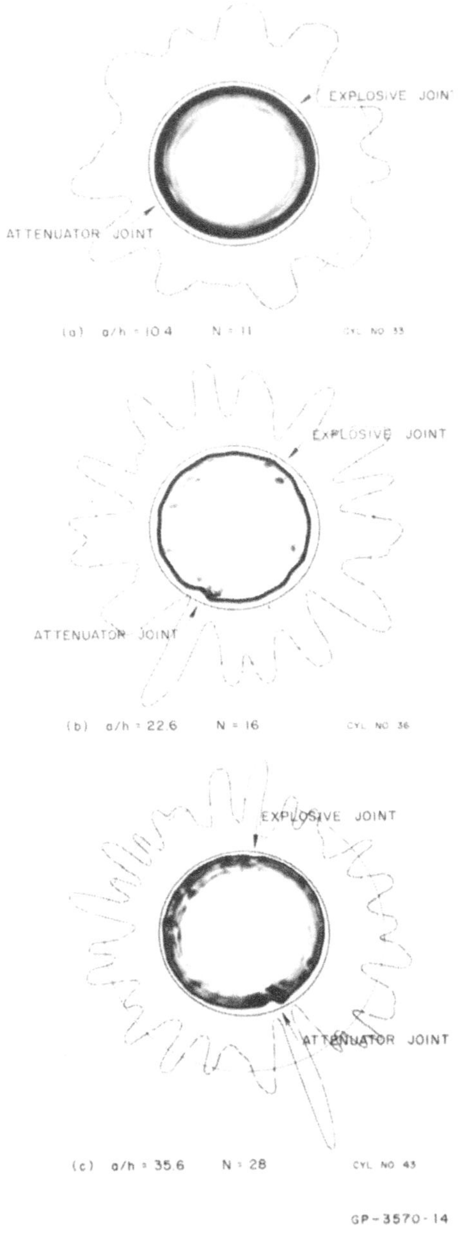


FIGURE 3.14 EFFECT OF a/h ON BUCKLING OF MILD STEEL (1015) CYLINDERS

### 3.2.8 Comparison of Experiment with Theory

A precise comparison of experiment with theory is not possible because the nonuniformities in the experimental velocity distributions, which are required for the theory, are unknown. Instead, the number of crests observed is compared with a theoretical limit that depends on the dimensions and properties of the cylinders and is independent of the nonuniformities, and qualitative comparisons are made of experimental and theoretical amplitudes and permanent strains.

To compare the experimental findings with theory, it is necessary to know the flow stress $\sigma$ and strain-hardening modulus $E_h$ for the cylinder materials. These data were not found in the literature, so tests were undertaken to obtain them. Although it would be desirable to have the data for compressive stresses and high strain rates, expediency required that the tests be made under tensile stresses at ordinary strain rates. Test specimens, consisting of axial strips from actual cylinder stock, were cut to ASTM dimensions and were tested on a conventional tensile testing machine. The results of the tensile tests, corrected to give true strain, are plotted in Figure 3.15.* To the accuracy of the data, the curves do not depart seriously from the simple model of Figure 3.2. More accurate stress-strain curves, required for the variable modulus theory, are presented in Section 3.4. Values of the flow stress $\sigma$ taken at 2% strain and the strain-hardening modulus $E_h$ obtained from the flat parts of the curves are listed in Table 3.1. The materials tested were chosen to give a wide variation in $\sigma$ and $E_h$, and hence in $\sqrt{\sigma/E_h}$. The latter varies from 0.37 for 2024-T3 aluminum to 0.69 for copper.

Table 3.1

STRESS-STRAIN PROPERTIES

| Material | $\sigma$ ($10^3$ psi) | $E_h$ ($10^5$ psi) | $\sqrt{\sigma/E_h}$ |
|---|---|---|---|
| Al 2024-T3 | 56 | 4.1 | 0.37 |
| Al 6061-T6 | 44 | 1.3 | 0.58 |
| Steel 1015 | 102 | 2.5 | 0.64 |
| Brass | 71 | 2.8 | 0.50 |
| Copper | 57 | 1.2 | 0.69 |
| Steel 4130 | 123 | 3.9 | 0.56 |
| Mg AZ31B | 30 | 1.5 | 0.45 |

1 psi = 6.9 kPa.

*Curves for steel 4130 and magnesium AZ31B are included although no buckling experiments were performed with these materials.

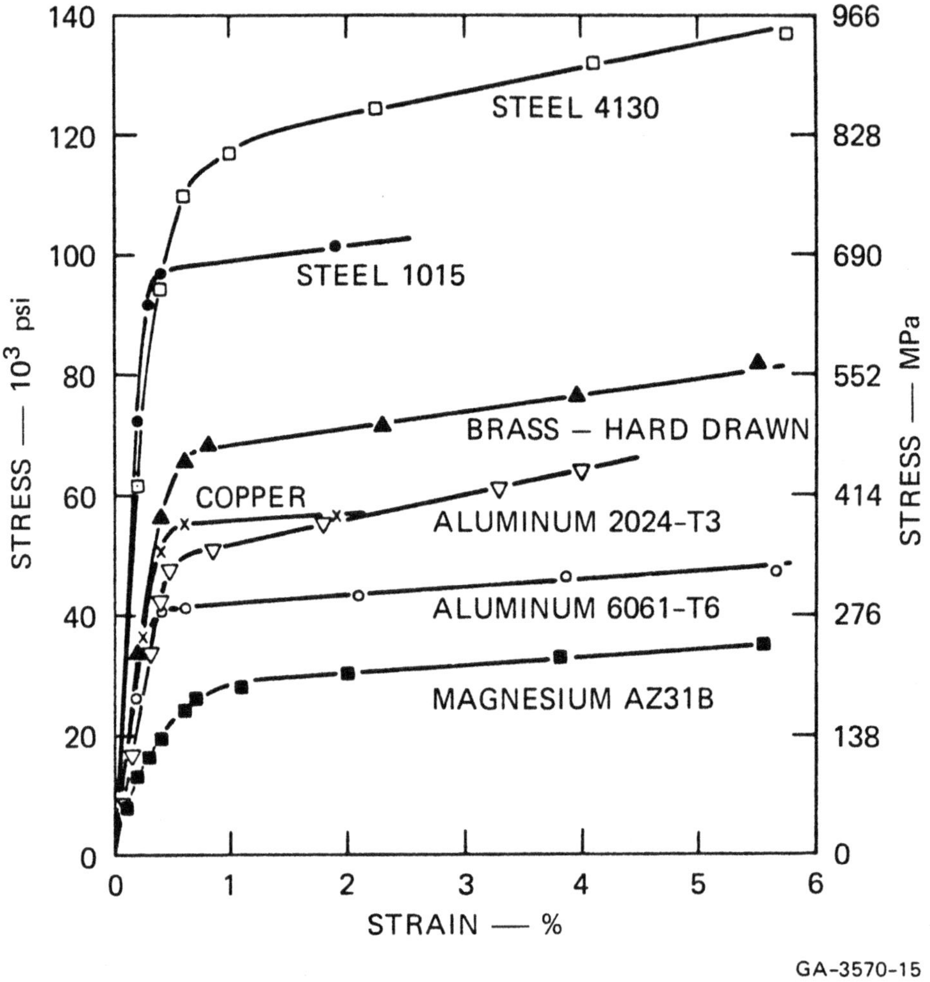


FIGURE 3.15 TRUE STRESS VERSUS STRAIN FOR CYLINDER MATERIALS

In Figure 3.16 the observed number of crests, determined from plots such as shown in Figures 3.13 and 3.14, is plotted against $n_1$, the most magnified component of the buckling terms [see Equation (3.2.26)]. Additional data on the cylinders are given in Table 3.2. The theoretical upper limit on the number of crests is indicated by the line $N = n_1$, since the greatest integer less than s is the last of the buckling terms. The data agree with the theory in that all the points lie below the line.

Table 3.2

NUMBER OF CRESTS

| Cylinder Number | Material | Radius a (inch) | Thickness h (inch) | a/h | Number of Crests N |
|---|---|---|---|---|---|
| 19 | Al 2024-T3 | 1.495 | 0.151 | 9.9 | 12 |
| 16 | | 1.495 | 0.100 | 15.0 | 12 |
| 23 | | 1.495 | 0.077 | 19.4 | 16 |
| 25 | Al 6061-T6 | 1.487 | 0.151 | 9.9 | 9 |
| 26 | | 1.488 | 0.101 | 14.7 | 15 |
| 27 | | 1.488 | 0.101 | 14.7 | 16 |
| 28 | | 1.490 | 0.076 | 19.6 | 18 |
| 29 | | 1.490 | 0.076 | 19.6 | 18 |
| 33 | Steel 1015 | 1.494 | 0.144 | 10.4 | 11 |
| 34 | | 1.495 | 0.089 | 16.8 | 15 |
| 36 | | 1.494 | 0.066 | 22.6 | 16 |
| 43 | | 1.493 | 0.042 | 35.6 | 28 |
| 54 | Brass | 1.495 | 0.113 | 13.2 | 13 |
| 55 | | 1.495 | 0.093 | 16.1 | 16 |
| 57 | | 1.498 | 0.069 | 21.7 | 25 |
| 62 | Copper | 1.429 | 0.146 | 9.8 | 12 |
| 63 | | 1.430 | 0.095 | 15.0 | 14 |

1 inch = 25.4 mm.

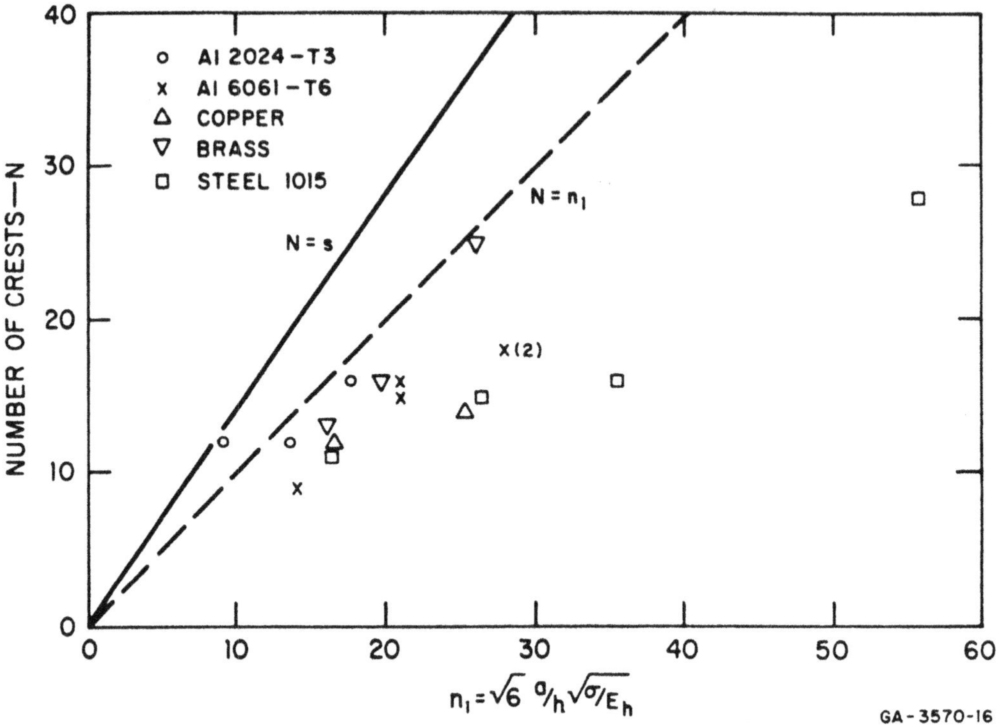


FIGURE 3.16 NUMBER OF CRESTS N VERSUS MOST MAGNIFIED HARMONIC $n_1$

In the vicinity of $n_1 = 10$, the data lie close to the line $N = n_1$. For larger values of $n_1$ the number of crests observed is considerably less than $n_1$. This behavior is to be expected if the Fourier coefficients $\alpha_n$, $\beta_n$, of the initial velocity distributions decrease fairly rapidly with increasing n, as for slowly varying nonuniformities (see the parabolic-perturbation example). As $n_1$ increases, buckling terms of higher order are added to the series of Equation (3.2.26). This could cause an increase in the number of crests, but not so great as the increase in $n_1$ because of the diminishing $\alpha_n$, $\beta_n$. Thus the data of Figure 3.16 are in accord with the theory in this respect also.

As a further check on the theory, we undertook some exploratory experiments in which velocity perturbations of known amplitude and frequency, in addition to the unknown variations, were superimposed on the "uniform" initial velocity. The known perturbations were obtained with strips of metal foil placed adjacent to the explosive. It was found that six strips of copper foil 2 mils (50 $\mu$m) thick and 1/8 inch (3.2 mm) wide spaced uniformly around the cylinders, which produce a local increase in velocity of approximately 20%, changed the number of crests from 16 of 8 mils (0.20 mm) average amplitude (Cylinders No. 26 and 27 of Table 3.2) to 6 of 80 mils (2.0 mm) average amplitude. Thus this enforced nonuniformity is sufficient to overcome the unknown variations and gives an upper limit on the magnitudes of the unknown variations. Since the unknown variations in initial velocity would be expected to consist of several components, the amplitude of the individual components should be less than 10%.

In addition to the number of crests, the amplitudes of the crests provide another readily observable quantity to compare with theory. However, as with the number of crests, a precise comparison is not to be anticipated since the magnitudes of the nonuniformities are unknown. An added complication is the fact that crests of amplitudes greater than a few mils develop strain reversal and hence go beyond the range of the present theory.

In Table 3.3 maximum amplitudes (peak to trough) and average amplitudes of the crests are given for the cylinders of Table 3.2. The maximum amplitudes occur at the explosive or attenuator joints and hence are of only passing interest. Excluding these, average amplitudes range from 2 to 22 mils (0.05 to 0.56 mm).

Table 3.3

AMPLITUDES OF CRESTS

| Material | Cylinder Number | Thickness of Explosive (mils) | Thickness of Cylinder (mils) | Initial Velocity (ft/s) | Maximum* Amplitude (mils) | Average† Amplitude (mils) | Increase in Thickness (%) | Increase in Length (%) |
|---|---|---|---|---|---|---|---|---|
| Al 2024-73 | 19 | 34.7 | 151 | 670 | 35 | 7(6) | 6 | 2.0 |
| | 16 | 20.5 | 100 | 590 | 23 | 7(8) | 15 | 1.2 |
| | 23 | 20.0 | 77 | 780 | 40 | 8(6) | 9 | 1.8 |
| Al 6161-76 | 25 | 27.3 | 151 | 540 | 43 | 11(7) | 5 | – |
| | 26 | 23.5 | 101 | 570 | 74 | 8(11) | 7 | 1.8 |
| | 27 | 22.0 | 101 | 620 | 55 | 8(11) | 6 | 1.9 |
| | 28 | 17.0 | 76 | 650 | 100 | 10(9) | 9 | 1.7 |
| | 29 | 16.4 | 76 | 640 | 202 | 15(7) | 9 | 1.4 |
| Steel 1015 | 33 | 61.0 | 144 | 420 | 30 | 6(10) | 5 | – |
| | 34 | 44.7 | 39 | 500 | 50 | 13(12) | 8 | 2.2 |
| | 36 | 33.0 | 66 | 490 | 94 | 22(12) | 11 | – |
| | 43 | 16.0 | 42 | 380 | 129 | 9(20) | 5 | – |
| Brass | 54 | 24.0 | 113 | 330 | 41 | 9(10) | 4 | 2.1 |
| | 55 | 32.0 | 93 | 320 | 74 | 6(13) | 4 | 2.3 |
| | 57 | 31.6 | 69 | 430 | 270 | 19(9) | 7 | 3.1 |
| Copper | 62 | 45.0 | 146 | 270 | 30 | 8(8) | 5 | 1.9 |
| | 63 | 16.0 | 95 | 150 | 6 | 2(9) | 3 | 0.4 |

* Peak to trough.

† Peak to trough, effects of joints not included.
Number in parenthesis indicates number of crests in average.

1 mil = 0.025 mm, 1 ft/s = 0.305 m/s.

The amplitudes of the crests as a function of time for the parabolic velocity perturbation are shown in Figure 3.7. The maximum amplitude (peak to trough) as strain reversal occurs is approximately 4 mils (0.1 mm, at $\tau = 0.02$). Many disturbances in velocity similar to the parabolic perturbation may be expected around a cylinder in the experiments; hence the calculated maximum amplitude may be compared with the average observed amplitude. The calculation for the parabolic velocity perturbation was made for Cylinder No. 23 (Figure 3.1), listed third in Table 3.3. The observed average amplitude is 8 mils (0.2 mm), and the calculated value is 4 mils (0.1 mm). The difference may be differences in the actual and the postulated velocity perturbations and in the assumed and actual material properties.

The permanent strain produced in the cylinders is given in Table 3.3. The increase in thickness ranges from 3% to 15% and the increase in length from 0.4% to 3.1%. The increase in thickness calculated from Equation (3.26) for an initial velocity of 695 ft/s (212 m/s) is 14%. This is to be compared with 9% observed for Cylinder No. 23. The difference may be attributed to differences in velocity perturbations and material properties.

The observed strains provide a means of checking initial velocities. The energy absorbed (per unit area) in plastic work in increasing the thickness of the shell is $\sigma\epsilon h$, where $\sigma$ is the flow stress, $\epsilon$ is strain, and h is wall thickness. If this is equated to the initial kinetic energy (per unit area), one obtains for the initial velocity

$$V = \sqrt{\frac{2\sigma\epsilon}{\rho}} \qquad (3.2.34)$$

Values of V from Equation (3.2.34), given in Table 3.3, are from 20% to 40% less than the velocities calculated from the explosive thickness. This difference is due partly to neglect of the energy absorbed in plastic bending and possibly to the development of reactions in the shell during loading. The latter effect depends on the ratio of the duration of the load to the response time of the shell. For the aluminum and steel shells of Table 3.3, the response time (quarter period) of the fundamental extensional mode is about 10 $\mu$s. The duration of the load may be of the same order of magnitude. The difference may also be due in part to an increase in yield stress with strain rate.[2]

### 3.2.9 Buckling Threshold

A lower limit on the velocity, or impulse, necessary to produce buckling can be obtained by equating the initial kinetic energy of the shell to the strain energy required to reach the yield stress. For thin shells, the kinetic energy per unit volume of the shell is $\rho V_0^2/2$ and the strain energy per unit volume is $\sigma_y^2/2E$. Equating these yields

$$V_0 = \frac{\sigma_y}{\rho c} \tag{3.2.35}$$

where $c^2 = E/\rho$ is sound speed. The expression for the impulse I (per unit area) corresponding to $V_0$ is

$$I = \frac{\sigma_y h}{c} \tag{3.2.36}$$

Equation (3.2.35) indicates that the threshold velocity depends only on physical properties of the shell and not on its dimensions. The threshold impulse [Equation (3.2.36)] depends in addition on the thickness h of the shell but not on the radius. Table 3.4 gives the threshold velocity and impulse for several metals. Threshold velocities range from 40 to 150 ft/s (12 to 46 m/s). Threshold impulses for h = 5 mm range from 1200 to 4400 $\mu$bar-s (120 to 440 Pa-s).

Table 3.4

BUCKLING THRESHOLDS

| Material | Density ($kg/m^3$) | Sound Speed (m/s) | Yield Stress (MPa) | $V_0$ (m/s) | $I^*$ (Pa-s) |
|---|---|---|---|---|---|
| Al 2024-T3 | 2800 | 5100 | 386 | 27 | 270 |
| Al 6061-T6 | 2800 | 5100 | 304 | 21 | 210 |
| Steel 1015 | 7900 | 5000 | 704 | 18 | 180 |
| Brass | 8500 | 3500 | 490 | 16 | 160 |
| Copper | 8900 | 3500 | 393 | 13 | 130 |
| Magnesium | 1700 | 4700 | 207 | 26 | 260 |
| Titanium | 4500 | 4300 | 897 | 46 | 440 |
| Beryllium | 1800 | 2600 | 276 | 12 | 120 |

* For a shell of surface density 1 $g/cm^2$.

The development of wrinkles of flow buckling depends on the nonuniformities in the initial velocity. For impulsive loads obtained with sheet explosive, it was found (Table 3.3) that velocities of several hundred feet per second were required to produce wrinkles with amplitudes of 5% to 10% of the radius. Hence the thresholds estimated from Equations (3.2.35) and (3.2.36) are quite conservative. In Section 3.4, more accurate estimates are given that allow a small amount of buckling to take place so that reasonable assumed amplitudes of random imperfections lead to just perceptible buckles. In this case, the threshold velocity increases with h at fixed radius. For the relatively thick shells just discussed, the threshold impulse is found to be as much as 10 times the elastic limit impulse given by Equation (3.2.36) (see Figure 3.32).

## 3.3 DYNAMIC ELASTIC BUCKLING OF RINGS AND CYLINDRICAL SHELLS FROM UNIFORM RADIAL IMPULSE

### 3.3.1 Introduction

The preceding section showed that when rings or long cylindrical shells are projected inward by an intense impulsive pressure, a characteristic form of dynamic wrinkling occurs at wavelengths determined by the material tangent modulus in plastic radial flow. In experiments on very thin cylindrical shells[3] (radius-to-thickness ratio 480, compared with 10 to 36 in the preceding section), observed buckle wavelengths were six times those predicted by the plastic flow buckling theory. In these shells it was found that the duration of elastic motion was long enough that elastic buckles formed with amplitudes much larger that any initial imperfections at the plastic wavelengths. The elastic buckling also quickly reduced the average hoop stress so that radial plastic flow, which would have occurred in the absence of any velocity or displacement perturbations, never occurred. As a result, buckling was restricted to the elastic modes. Radial impulse above that for elastic unperturbed hoop motion served only to provide energy eventually absorbed by plastic hinges at the elastic buckle wavelengths.

In this section we present a theoretical analysis of dynamic elastic pulse buckling. Because the motion is elastic, another form of dynamic buckling can also occur, called autoparametric vibration buckling. This is a special form of vibration buckling (described in Section 1.1) in which elastic vibrations in the hoop mode become unstable. At radial impulses well below that required to produce pulse buckling wrinkles as just described, the shell vibrates in and out in the hoop (breathing) mode. Because of inevitable imperfections, flexural modes with natural frequencies near half that of the hoop mode begin to grow in amplitude by extracting energy from the hoop mode. Eventually, nearly all the energy can be transferred to a flexural mode. This buckling motion is also analyzed in this section. It is the only form of vibration buckling analyzed in the monograph. Our purpose is to show the transition between vibration and pulse buckling as the impulsive loading is increased.

### 3.3.2 Theory of Elastic Shell Motion

As in Section 3.2, we consider a cylindrical shell with an inward initial radial velocity imparted to each element uniformly around the circumference. Flexural perturbations are taken in the form of initial imperfections in shape. The relative importance of velocity and shape perturbations is discussed in Section 3.4. To treat shape imperfections, we express the total radial and tangential shell displacements $w^*$ and $v^*$ in the form

$$w^*(\theta, t) = w(\theta, t) + w_i(\theta) \tag{3.3.1}$$

$$v^*(\theta, t) = v(\theta, t) + v_i(\theta) \tag{3.3.2}$$

where $\theta$ is angular coordinate, t is time, $w(\theta, t)$ is radial displacement, measured positive inward from an initial shape imperfection $w_i(\theta)$ in the unstressed shell, and v and $v_i$ are similarly defined for tangential displacements.

The equations of motion are most conveniently derived by the Lagrangian method. This also gives expressions for energy that are used later in discussing energy transfer between hoop and flexural modes. For small displacements, the kinetic energy per unit length of shell is given by

$$T = \frac{1}{2}\rho h a \int_0^{2\pi} \left[\left(\frac{\partial w}{\partial t}\right)^2 + \left(\frac{\partial v}{\partial t}\right)^2\right] d\theta \tag{3.3.3}$$

where $\rho$ is density, h is the wall thickness, and a is the radius of the middle surface. It is convenient to introduce dimensionless quantities

$$\begin{aligned} \zeta &= w/a , & \zeta_i &= w_i/a \\ \psi &= v/a , & \psi_i &= v_i/a \\ \tau &= ct/a \end{aligned} \tag{3.3.4}$$

where

$$c = (E_1/\rho)^{1/2} , \qquad E_1 = E/(1 - \nu^2) \tag{3.3.5}$$

in which E is Young's modulus and $\nu$ is Poisson's ratio. For rings, use E instead of $E_1$. The factor $1/(1 - \nu^2)$ accounts for the zero axial strain constraint in a long shell, as can easily be confirmed by deducing the relation between hoop stress and hoop strain with zero axial strain. The result is $\sigma = E_1\epsilon$. A more complete discussion of elastic shell equations is given in Section 3.5. With dots indicating differentiation with respect to $\tau$, the kinetic energy is expressed by

$$T = \frac{1}{2} E_1 ha \int_0^{2\pi} (\dot{\zeta}^2 + \dot{\psi}^2)\, d\theta \qquad (3.3.6)$$

To find the strain energy, define z as the radial position of a shell fiber from the middle surface. When an element of shell material $(a + z)\, d\theta dz$, per unit length of shell, is brought to stress $\sigma$ from an initially unstressed state, its strain energy is

$$dU = \frac{1}{2} \sigma\epsilon \cdot (a + z)\, d\theta dz \qquad (3.3.7)$$

Substitution of $\sigma = E_1\epsilon$ into Equation (3.3.7) and integration over the shell material volume gives for the total strain energy

$$U = \frac{E_1 a}{2} \int_0^{2\pi} \int_{-h/2}^{h/2} \epsilon^2\, dz d\theta \qquad (3.3.8)$$

in which it has been assumed that the shell is thin enough to neglect z/a in comparison with 1. With the thin shell approximation and with plane sections across the shell thickness assumed to remain plane during flexure, the strain at each fiber is

$$\epsilon = \epsilon_m + z\kappa \qquad (3.3.9)$$

where $\epsilon_m$ is the middle surface strain and $\kappa$ is its change in curvature. Substitution of Equation (3.3.9) into Equation (3.3.8) and integration over the thickness gives

$$U = \frac{1}{2} E_1 ha \int_0^{2\pi} \left[\epsilon_m^2 + a^2\alpha^2\kappa^2\right] d\theta \qquad (3.3.10)$$

where $\alpha^2 = h^2/12a^2$. With the shell assumed thin, $\alpha^2 << 1$ so higher-order terms multiplied by $\alpha^2$ are dropped; we retain for $\kappa$ only terms linear in the displacements, as in simple curved-bar theory. Because of linear superposition, the initial imperfections are therefore not involved in the curvature change, so $\kappa$ is given simply by

$$\kappa = \frac{1}{a^2}\left[\frac{\partial^2 w}{\partial\theta^2} + w\right] = \frac{1}{a}(\zeta'' + \zeta) \qquad (3.3.11)$$

For the middle surface strain, however, terms up to second order for strain and to third order for energy are retained, which gives the least modification to the linear theory that can introduce interaction of circumferential force and flexure. For use in the transition from vibration buckling to pulse buckling, terms to fourth order for energy are also retained. Otherwise, the coupled equations are unbounded for large values of the stability parameter p, as discussed later. The nonlinear terms also introduce the initial imperfections into the strain and energy expressions.

An expression for the middle surface strain is found from the differential displacements shown in Figure 3.17. Points on the undeformed middle surface for an initially perfect cylinder have polar coordinates a, $\theta$. Points on the deformed middle surface have polar coordinates r, $\phi$. The radial displacement from an initially perfect cylindrical shape is therefore

$$w = a - r \tag{3.3.12}$$

and the angular displacement is

$$\psi = \phi - \theta \tag{3.3.13}$$

For the small displacements to be considered here, the angular displacement is the same as the dimensionless tangential displacement defined earlier, so the same

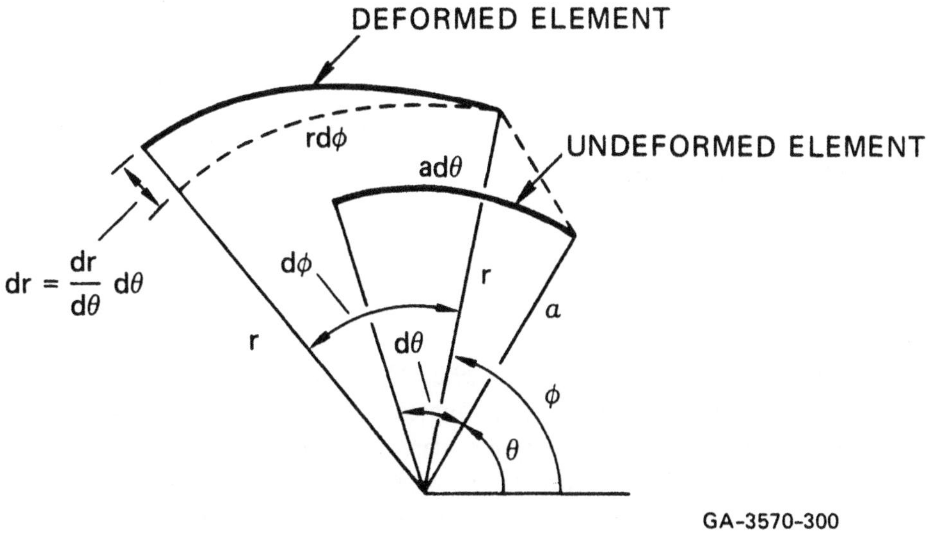


FIGURE 3.17 DIFFERENTIAL DISPLACEMENTS OF MIDDLE SURFACE

symbol has been used for both. Differention of Equations (3.3.12) and (3.3.13) gives

$$\frac{dw}{d\theta} = -\frac{dr}{d\theta}, \qquad \frac{d\phi}{d\theta} = 1 + \frac{d\psi}{d\theta} \tag{3.3.14}$$

The middle surface strain is the change in length of the element in Figure 3.17 divided by the initial length,

$$\bar{\epsilon}_m = \frac{1}{a d\theta}\left\{\left[\left(\frac{dr}{d\theta}\cdot d\theta\right)^2 + \left(r\frac{d\phi}{d\theta}\cdot d\theta\right)^2\right]^{1/2} - a d\theta\right\}$$

$$= \frac{1}{a}(r'^2 + r^2\phi'^2)^{1/2} - 1$$

where primes indicate differentiation with respect to $\theta$ and the bar is used to denote the initially perfect shape. This is a familiar expression from differential geometry. With Equations (3.3.12) and (3.3.13), it becomes

$$\bar{\epsilon}_m = \frac{1}{a}\left[w'^2 + (a - w)^2(1 + \psi'^2)\right]^{1/2} - 1$$

or

$$\bar{\epsilon}_m = \left[\zeta'^2 + (1 - \zeta)^2(1 + \psi'^2)\right]^{1/2} - 1 \tag{3.3.15}$$

in which $\zeta = w/a$ is used from Equation (3.3.4). Upon expansion to terms of second order, we obtain finally

$$\bar{\epsilon}_m = \psi' - \zeta - \zeta\psi' + \frac{1}{2}\zeta'^2 \tag{3.3.16}$$

This is the middle surface strain when the unstressed initial shape is perfectly circular so that its points have coordinates $a, \theta$. The strain in going from an unstressed state that is not perfectly circular, described by initial shape $\zeta_i$, $\psi_i$, to a deformed shape $\zeta^*$, $\psi^*$, is found by using either Equation (3.3.15) or Equation (3.3.16) to calculate the change in strain between these two shapes. (More precisely, the strain is the change in length between these states divided by the unstressed length, but elastic strains are very small so, as in our use of engineering strain, the change in length basis can be neglected.) Thus, for the shell with initial shape imperfections, the middle surface strain is

$$\epsilon_m = \bar{\epsilon}_m^* - \bar{\epsilon}_{mi}$$

$$= \left[ (\psi' + \psi'_i) - (\zeta + \zeta_i) - (\zeta + \zeta_i)(\psi' + \psi'_i) + \frac{1}{2}(\zeta' + \zeta'_i)^2 \right]$$

$$- \left[ \psi'_i - \zeta_i - \zeta_i \psi'_i + \frac{1}{2} \zeta'^2_i \right]$$

which simplifies to

$$e_m = \psi' - \zeta - \zeta\psi' + \frac{1}{2}\zeta'^2 + \zeta'\zeta'_i - \zeta_i\psi' - \zeta\psi'_i \qquad (3.3.17)$$

With terms of fourth order involving the small imperfections neglected, substitution of Equations (3.3.11) and (3.3.17) into Equation (3.3.10) gives for the strain energy

$$U = \frac{1}{2} E_1 ha \int_0^{2\pi} \Big[ (\psi' - \zeta)^2 + (\psi' - \zeta)$$

$$\times (\zeta'^2 - 2\zeta\psi' + \underline{2\zeta'\zeta'_i - 2\zeta_i\psi' - 2\zeta\psi'_i}) \qquad (3.3.18)$$

$$+ \underline{\underline{(\zeta\psi' - \frac{1}{2}\zeta'^2)^2}} + \alpha^2 (\zeta'' + \zeta) \Big] \, d\theta$$

Terms that account for ring imperfections are underlined by a single line, and terms of fourth order in the displacements are underlined by a double line. It will be shown later that the fourth-order terms can be neglected for small initial velocities, but for larger velocities the theoretical coupled breathing and flexural motion becomes unbounded without these terms.

The displacements are now expanded into Fourier series, whose coefficients are to be taken as generalized coordinates.

$$\zeta = a_o(\tau) + \sum_{n=1}^{\infty} [a_n(\tau) \cos n\theta + b_n(\tau) \sin n\theta] \qquad (3.3.19)$$

$$\psi = \sum_{n=1}^{\infty} [c_n(\tau) \cos n\theta + d_n(\tau) \sin n\theta] \qquad (3.3.20)$$

Since flexural wavelengths of interest are long compared with the wall thickness, it is reasonable to neglect membrane stresses due to flexure so that we can impose the inextensionality* condition $\psi' = \zeta - a_o$. Then $c_n = -b_n/n$ and $d_n = a_n/n$, and we omit terms for $n = 1$ because they represent a rigid-body translation for the inextensional model. Similarly, for the imperfections we take

$$\zeta_i = \sum_{n=2}^{\infty} [\delta_n \cos n\theta + \gamma_n \sin n\theta] \tag{3.3.21}$$

and $\psi'_i = \zeta_i$. With these modifications, substitution of Equations (3.3.19)-(3.3.21) into Equations (3.3.6) and (3.3.18) gives

$$T = \pi E h a \left[\dot{a}_o^2 + \frac{1}{2}\sum_{n=2}^{\infty}\left(\frac{n^2+1}{n^2}\right)\left(\dot{a}_n^2 + \dot{b}_n^2\right)\right] \tag{3.3.22}$$

$$U = \pi E h a \left\{a_o^2 + \frac{1}{2}\sum_{n=2}^{\infty}\left[(n^2-1)^2\alpha^2 - (n^2-2)\,a_o\right]\right.$$

$$\times\left(a_n^2 + b_n^2\right) - \frac{1}{2}\sum_{n=2}^{\infty}\frac{3}{8}\left(n^4 - \frac{4}{3}n^2 + 4\right)\left(a_n^4 + b_n^4\right)$$

$$\left. - \sum_{n=2}^{\infty}(n^2-2)(\delta_n a_n + \gamma_n b_n)\,a_o\right\} \tag{3.3.23}$$

In performing the integration in Equation (3.3.18), all second- and third-order terms other than those given in Equation (3.3.23) vanish identically because of the orthogonality of the trigonometric functions. The only third-order coupling is between the flexural modes and the breathing mode. The fourth-order terms in Equation (3.3.18), however, lead to extensive coupling between the flexural modes. This coupling is neglected in Equation (3.3.23). The only fourth-order terms retained are the uncoupled terms $a_n^4$ and $b_n^4$. These are sufficient to suppress the unbounded flexural growth that occurs with all fourth-order terms neglected. In the examples that follow we consider motion only in sequences of single modes. The question of whether further fourth-order terms are needed for accurate representation of several coupled flexural modes has not been addressed.

---

*Demonstration that response of the extensional modes can be reasonably neglected in investigating stability of the breathing mode is the subject of the first half of a paper by Goodier and McIvor.[4]

The equations of motion are found by substituting these energy expressions into Lagrange's equation

$$\frac{d}{dt}\left(\frac{\partial T}{\partial \dot{x}_i}\right) - \frac{\partial T}{\partial x_i} + \frac{\partial U}{\partial x_i} = 0 \qquad (3.3.24)$$

With $a_o$ as $x_i$, the breathing mode equation is found to be

$$\ddot{a}_o + a_o - \frac{1}{4}\sum_{n=2}^{\infty}(n^2 - 2)\left(a_n^2 + b_n^2\right) = 0 \qquad (3.3.25)$$

With $a_n$ as $x_i$, the flexural equations are found to be

$$\ddot{a}_n + \frac{n^2}{n^2+1}\left[(n^2-1)^2\alpha^2 - (n^2-2)\,a_o + \frac{3}{8}\left(n^4 - \frac{4}{3}n^2 + 4\right)a_n^2\right]a_n = \frac{n^2(n^2-2)}{n^2+1}\,\delta_n a_o \qquad (3.3.26)$$

An identical set of equations results for $b_n$.

Because only a few closely spaced "critical" modes dominate the flexural response, the complexity of the study that follows is greatly reduced by considering the response of only a sequence of single flexural modes, to find the critical flexural mode in which stress is most amplified. Thus there is no need to consider the spatial phase of the buckling mode, and the $b_n$ terms can be dropped. With these modifications, Equations (3.3.25) and (3.3.26) become

$$\ddot{a}_o + a_o - \frac{1}{4}(n^2 - 2)\,a_n^2 = 0 \qquad (3.3.27)$$

$$\ddot{a}_n + \left(\omega_n^2 - f_n a_o\right)a_n + \frac{3}{8}n^2 f_n g_n a_n^3 = \sqrt{12}\,\alpha f_n \bar{\delta}_n a_o \qquad (3.3.28)$$

where

$$\omega_n^2 = \frac{n^2(n^2-1)^2\alpha^2}{n^2+1}$$

$$f_n = \frac{n^2(n^2-2)}{n^2+1}$$

$$g_n = \left(n^4 - \frac{4}{3}n^2 + 4\right)/\,n^2(n^2-2)$$

and $\bar{\delta}_n = (a/h)\delta_n$ is the initial imperfection expressed as a fraction of the wall thickness. The quantity $g_n$ is introduced because for thin shells buckling is always in modes with $n^2 >> 1$ so that $g_n \rightarrow 1$ and $f_n \rightarrow n^2$, which simplifies Equation (3.3.28).

### 3.3.3 Initial Growth of the Flexural Modes--The Stability Parameter

Consider first that a small initial velocity is imparted to the shell wall, uniform around its circumference. The distinction between small and larger initial velocities will be made in the next section. For the present, focus also only on the initial buckling motion, before it extracts a substantial fraction of the initial hoop mode energy. Then the $a_n^2$ terms in Equation (3.3.27) can be neglected and the hoop motion is simply

$$a_o = \frac{v_o}{c} \sin \tau \qquad (3.3.29)$$

where $v_o$ is the initial radial velocity and c is the plane strain wave velocity defined in Equation (3.3.4). With this, Equation (3.3.28) reduces to the Mathieu form

$$\ddot{a}_n + (\Omega_n - \mu_n \sin \tau)\, a_n = \sqrt{12}\,\alpha \bar{\delta}_n \mu_n \sin \tau \qquad (3.3.30)$$

where

$$\Omega_n = \omega_n^2 = \alpha^2 n^2 (n^2 - 1)^2 (n^2 + 1)^{-1}$$

$$\mu_n = f_n v_o / c = n^2 (n^2 - 2)(n^2 + 1)^{-1} v_o c^{-1} \qquad (3.3.31)$$

and the term with $a_n^3$ has been dropped for the present because of its smallness. Extensive study of the Mathieu equation[5] has shown that $a_n(\tau)$ grows if the point $\Omega_n$, $\mu_n$ is in an unstable region of the Mathieu stability diagram. Such regions appear unshaded in Figure 3.18.

For a particular case, we should now mark off the points given by Equation (3.3.31) for the integer values of n. Figure 3.19 shows a set of such points on a limited region of the Mathieu diagram near the origin.

As n increases we have

$$\Omega_n \rightarrow n^4 \alpha^2 \;, \qquad \mu_n \rightarrow n^2 v_o / c$$

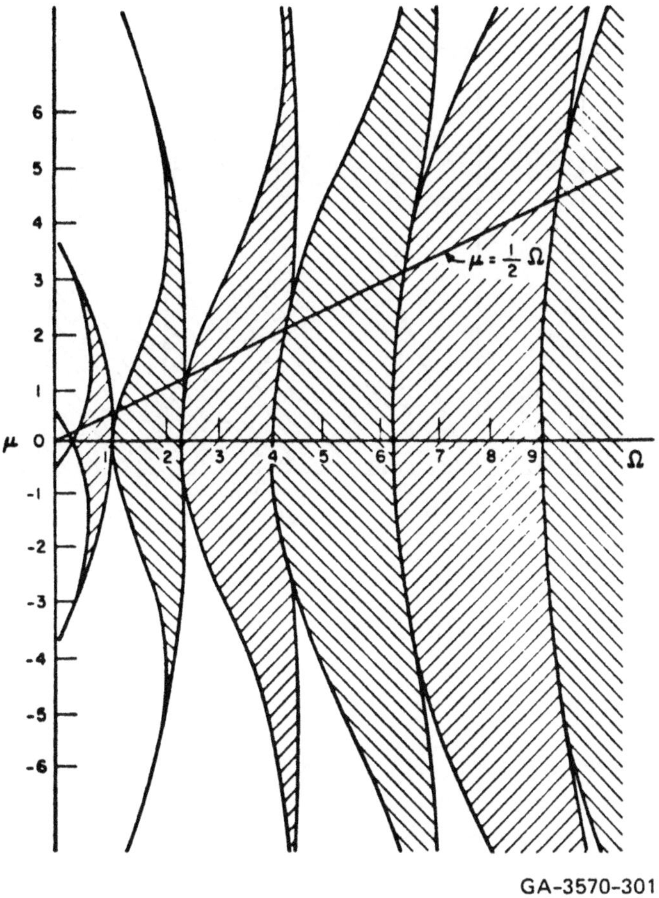


FIGURE 3.18 MATHIEU STABILITY DIAGRAM

and the points tend toward a parabola given by

$$\mu_n = p \, \Omega_n^{1/2} \tag{3.3.32}$$

where*

$$p = v_o / c\alpha \tag{3.3.33}$$

The parabola lies mainly in the stable region in Figure 3.18 for small values of $v_o$ and large values of $\alpha$ (thicker shells). For these cases the higher points on the parabola have little chance of falling between shaded regions. The unstable regions soon become very narrow. A very small change could remove the instability. Small damping also removes it.[5]

---

*This use of the symbol p should not be confused with the coefficients $p_n$ in the equations of motion.

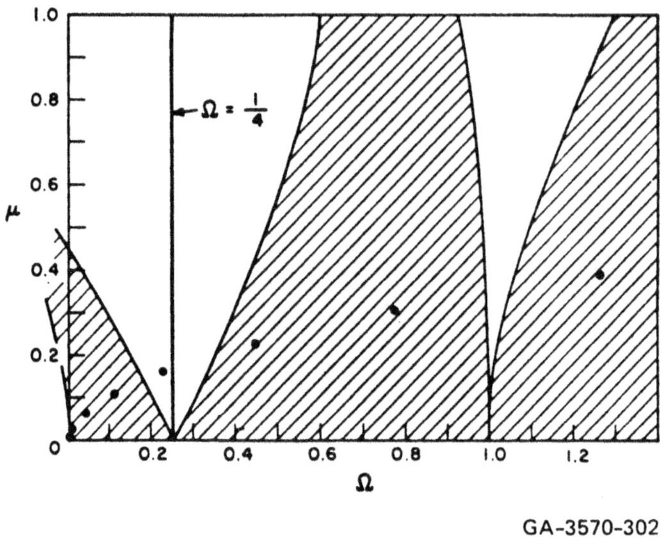


FIGURE 3.19 MATHIEU STABILITY DIAGRAM IN REGION OF $\Omega = 1/4$

For these cases attention is restricted to the first unstable region containing the line $\Omega = 1/4$. It is evident that there will usually be at least one point in this region. We have $\Omega_n = 1/4$ for $n = b$ if

$$\frac{b^2(b^2 - 1)^2\alpha^2}{b^2 + 1} = \frac{1}{4} \tag{3.3.34}$$

and if b is not an integer we consider the nearest integer N. This relation is shown in Figure 3.20. For a thin shell, a/h could not be less than 10. The curve shows that the "critical" N would not be less than 4 and will be considerably higher for thinner shells. Then in Equation (3.3.34) we may neglect unity in comparison with $b^2$ and obtain, with $\alpha^2 = h^2/12a^2$,

$$b = 1.316\ (a/h)^{1/2} \tag{3.3.35}$$

This appears in Figure 3.20 as the broken-line curve.

The (circular) frequency of the $n^{th}$ flexural mode is $\omega_n$, given in Equation (3.3.31), in terms of the dimensionless time $\tau$. The frequency of the hoop mode, Equation (3.3.29), is unity. From Equation (3.3.31), $\omega_n = 1/2$, which means that this critical flexural mode has a frequency one-half that of the hoop mode. This is, of course, the relation required if the circumferential hoop stress is always to encourage flexure. On successive compressive swings of the hoop oscillation, a

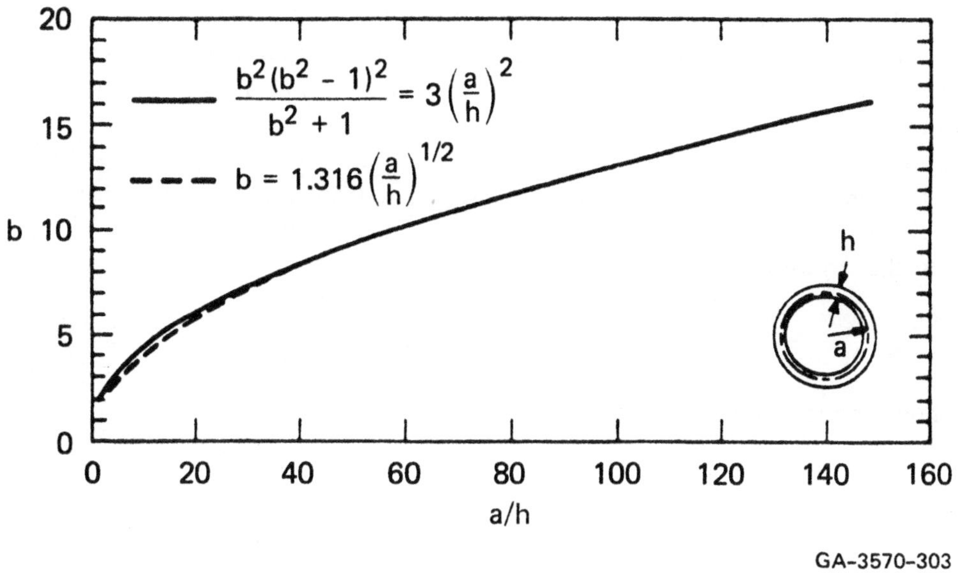


FIGURE 3.20 CRITICAL MODE *N* VERSUS RADIUS-TO-THICKNESS RATIO a/h

flexural "buckle" alternates between flexing inward and outward, and in either case the compression produces bending that increases the flexure. On the tensile swings, the hoop stress pulls on the buckle and increases its velocity as it passes through the undeformed position. This situation is similar to that in the column under an oscillating end load as discussed in Chapter 1. In the column and in the problem of a shell with applied oscillating pressure, the associated compression is maintained by the loading, which provides a source of energy for the flexural oscillations. In the present problem, the energy must come from the hoop mode. This type of instability is called autoparametric instability because the energy comes from the structure itself. To follow the energy transfer, the nonlinear terms in Equation (3.3.27) must be retained.

Before considering this energy transfer, it is useful to rewrite the flexural mode Equation (3.3.30), appropriate for initial growth, in terms of the parameter p introduced in Equation (3.3.33). With unity neglected in comparison with $n^2$ as already observed, Equation (3.3.30) then becomes

$$\ddot{a}_o + p^2\lambda^2(\lambda^2 - \sin\tau)\,a_n = p^2\lambda^2\,\alpha\sqrt{12}\,\bar{\delta}_n \sin\tau \qquad (3.3.36)$$

where

$$p = v_o/c\alpha \quad , \qquad \lambda = n/s \quad , \qquad \text{and} \qquad s^2 = p/\alpha \qquad (3.3.37)$$

This form shows clearly that the character of the solution is governed entirely by the parameter p, which we call the stability parameter. The only other parameter in Equation (3.3.36), other than the initial imperfection amplitude $\bar{\delta}_n$, is $\lambda^2$, and we have already seen that the critical value of n, and hence of $\lambda$, is known if $\alpha$ and p are known.

If other initial conditions are considered, a similar parameter appears. For example, if the shell is suddenly released from an initial hoop stress $\sigma_o$ uniform around the circumference, then

$$a_o(0) = \sigma_o/E_1 \quad , \qquad \dot{a}_o(0) = 0 \tag{3.3.38}$$

The hoop motion is then

$$a_o(\tau) = \frac{\sigma_o}{E_1} \cos \tau \tag{3.3.39}$$

and the flexural equation, from Equation (3.3.28) with the $a_n^3$ term neglected for early growth, becomes

$$\ddot{a}_n + p^2\lambda^2(\lambda^2 - \cos \tau)\, a_n = p^2\lambda^2\alpha\sqrt{12}\,\bar{\delta}_n \cos \tau \tag{3.3.40}$$

where now

$$p = \sigma_o/E_1\alpha \tag{3.3.41}$$

Observe that for either Equation (3.3.37) or (3.3.41), p is simply the peak hoop strain divided by the shell thickness parameter $\alpha = h/a\sqrt{12}$. As the peak hoop strain increases or the shell becomes thinner, the coefficient of $a_n$ in Equation (3.3.36) or (3.3.40) becomes more negative during the compressive phase of hoop motion and the instability is more severe. We will see in the following sections that, as p increases to values near 4, the motion becomes so unstable that the hoop energy is transferred to flexure during the first compressive phase of hoop motion, resulting in pulse buckling similar to that described in the preceding sections.

### 3.3.4 Small Initial Velocity--Autoparametric Vibrations[4]

Return now to the question of energy transfer for small values of p (values less than 1/2, as we will see). For this range, the second-order coupling term $a_n^2$ in the hoop mode Equation (3.3.27) must be retained but the third-order term $a_n^3$ in the flexural mode Equation (3.3.28) can be omitted without serious error. Figure 3.21 shows the results of numerical integration of coupled Equations (3.3.27) and

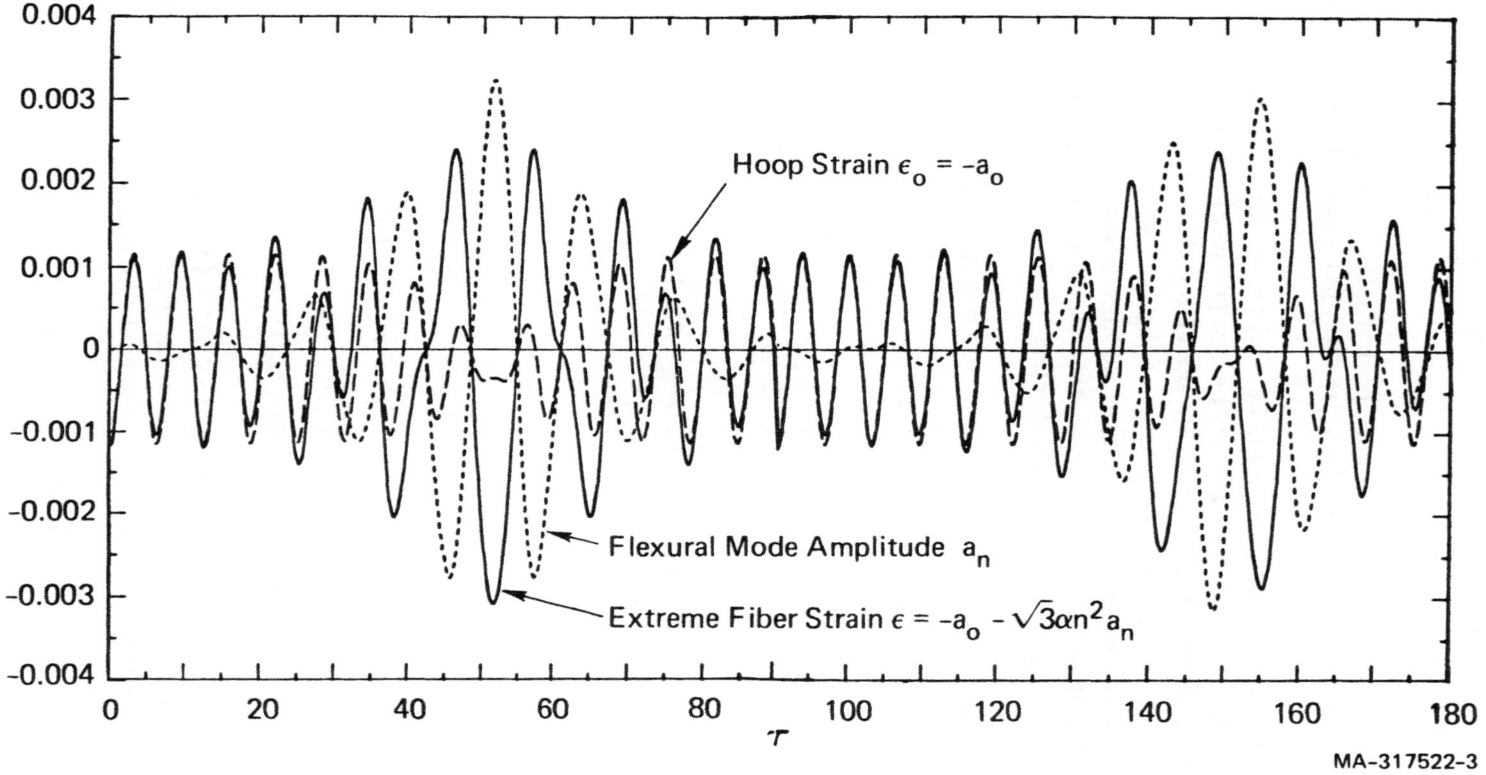


FIGURE 3.21 TRANSFER OF ENERGY BACK AND FORTH BETWEEN HOOP AND FLEXURAL MODES ($p = 0.4$, $\beta = 0.0$, $a/h = 100$, $n = 13$, $\bar{\delta}_n = 0.02$)

(3.3.28) with initial conditions in the form of Equation (3.3.39), with $p = 0.4$, $a/h = 100$, and $\bar{\delta}_n = 0.02$. The mode with most growth was found to be $n = 13$, which agrees with Equation (3.3.35).

The dominant feature of response is the transfer of energy back and forth between the hoop mode (dashed curve) and the flexural mode (dotted curve). At $\tau = 0$ and $\tau = 105$, all or nearly all the energy is in the hoop mode. At $\tau = 50$ and $\tau = 155$, nearly all the energy is in the flexural mode. Each transfer of energy from one form to the other takes place during 8 oscillations of the hoop mode. For smaller values of p, the number of cycles for energy transfer is larger. Also shown (solid curve) is the strain in an extreme fiber of the shell wall. When the flexural mode reaches its maximum, so does the extreme fiber strain. The peak amplitude of the extreme fiber strain is more than twice that of the initial hoop strain. This is a general result for energy transfer with no damping.

The peak amplitude of the extreme fiber strain is easily calculated by assuming complete energy transfer from the hoop mode to a flexural mode. For energies small enough that the $a_n^3$ term in Equation (3.3.28) can be neglected, the kinetic and potential energy expressions in Equations (3.3.22) and (3.3.23) become

$$T^* = 2\dot{a}_o^2 + \dot{a}_n^2 \tag{3.3.42}$$

$$U^* = 2a_o^2 + \left(n^4\alpha^2 - n^2 a_o\right) a_n^2 \tag{3.3.43}$$

where $T^* = 2T/\pi E_1 ha$, $U^* = 2U/\pi E_1 ha$, $n^2 >> 1$, and we assume that a single flexural mode dominates the response. For the initial velocity condition corresponding to Equation (3.3.29), the total energy in the shell is

$$T^*(0) = 2\dot{a}_o^2(0) = 2(v_o/c)^2 = 2p^2\alpha^2 \tag{3.3.44}$$

where $p = v_o/c\alpha$ from Equation (3.3.37). For the initial compression condition of Equation (3.3.38), the total energy in the shell is

$$U^*(0) = 2a_o^2(0) = 2(\sigma_o/E_1)^2 = 2p^2\alpha^2 \tag{3.3.45}$$

where now $p = \sigma_o/E_1\alpha$ from Equation (3.3.41). In either case, as already observed, p is simply the peak hoop strain (in unperturbed motion) divided by $\alpha$. With a general definition of p in terms of the total shell energy,

$$p^2 = [T^*(0) + U^*(0)]/2\alpha^2 \tag{3.3.46}$$

the following result applies to any hoop mode initial condition.

Define a stress amplification $A_\sigma$ as the ratio between the maximum total stress (at an inner or outer fiber of the shell) and the peak unperturbed hoop stress $\sigma_o$. Since the ring is linearly elastic, this amplification is given by the strain ratio. The flexural strain at an extreme fiber resulting from a flexural displacement $a_n$ is given by

$$\epsilon_b = \frac{h}{2}\,\frac{\partial^2 w}{\partial (a\theta)^2} = \frac{h}{2a}\cdot n^2 a_n = \sqrt{3}\,\alpha\, n^2 a_n \tag{3.3.47}$$

With complete energy transfer to the flexural mode, the peak amplitude $a_n$ is found from Equation (3.3.43) with $a_o = 0$. Also, at the peak we have $\dot{a}_n = 0$, and, for complete energy transfer, $\dot{a}_o = 0$. Then $T^* = 0$ and $U^*$ is equal to the total energy $2p^2\alpha^2$. Equation (3.3.43) is then

$$2p^2\alpha^2 = n^2\alpha^2 a_n^2 \tag{3.3.48}$$

With $a_n$ from Equation (3.3.48) substituted into Equation (3.3.47), the peak flexural strain is

$$\epsilon_b = \sqrt{6}\, p\alpha \tag{3.3.49}$$

The peak unperturbed hoop strain is $p\alpha$ so that the stress amplification is

$$A_\sigma = \frac{\sqrt{6}\, p\alpha}{p\alpha} = \sqrt{6} = 2.45 \tag{3.3.50}$$

independent of the mode number n.

This result (derived by Goodier and McIvor[4]) shows that a brittle shell (or ring) set into simple in and out breathing oscillations would eventually fail by brittle flexural fracture for initial hoop stresses exceeding the fracture stress divided by $\sqrt{6} = 2.45$. With damping, some energy is lost during the energy transfer oscillations and the allowable hoop stress is larger. Also, since real materials have some ductility, the bending strain for fracture is larger than for this elastic theory. Ductility also introduces a form of damping by extracting energy each time the flexural mode amplitude nears a peak. However, as p increases, damping mechanisms have less time to act because the number of oscillations for energy transfer decreases. To study potential failures of highly damped or ductile materials, we therefore study stability for larger values of p.

### 3.3.5 Intermediate Initial Velocity--Onset of Pulse Buckling[6]

If Equations (3.3.27) and (3.3.28) are integrated in coupled form without the $a_n^3$ term for values of p exceeding ½, the amplitudes of motion are found to increase without bound. This result can be deduced without numerical integration by using the energy Equations (3.3.42) and (3.3.43) written in the form

$$a_n^2 = \frac{T^* + U^* - 2\dot{a}_o^2 - \dot{a}_n^2}{n^2\alpha^2 - n^2 a_o} \qquad (3.3.51)$$

Thus, even with the total energy $T^* + U^*$ fixed, the maximum flexural amplitude is undefined when the denominator becomes zero or negative, i.e., when $a_o \geqslant n^2\alpha^2$, because the numerator is always positive if the hoop plus flexural kinetic energy is to remain less than the total energy.

The condition $a_o \geqslant n^2\alpha^2$ is expressed in terms of p by recalling from Equation (3.3.34) that the critical mode for most rapid growth is near $n^2\alpha = ½$ and that the peak value of $a_o$ is $p\alpha$. Then $a_o \geqslant n^2\alpha^2$ becomes

$$p \geqslant ½ \qquad (3.3.52)$$

This condition has a readily apparent physical interpretation. The peak hoop thrust corresponding to $p = ½$ is the static thrust that causes buckling, i.e., $\Omega_n = \mu_n$ in Equation (3.3.30). As in low-order static buckling theory, when the hoop compression reaches the critical buckling load, the amplitude of the buckling is undefined.

For $p > ½$, a higher-order theory must therefore be used to solve the coupled equations. Thus we use the fourth-order term in the potential energy Equation (3.3.23), so that Equation (3.3.43) becomes

$$U^* = 2a_o^2 + (n^4\alpha^2 - n^2 a_o)\, a_n^2 + \frac{3}{16} n^4 a_n^4 \qquad (3.3.53)$$

and the energy conservation equation $T^* + U^* =$ constant becomes

$$2\dot{a}_o^2 + \dot{a}_n^2 + 2a_o^2 + (n^4\alpha^2 - n^2 a_o)\, a_n^2 + \frac{3}{16} n^4 a_n^4 = 2p^2\alpha^2 \qquad (3.3.54)$$

Estimates for stress amplification with Equation (3.3.54) are more complicated than for the small p theory. In the following paragraphs several bounding estimates are made. Then we give examples of coupled motion found by numerical integration, to illustrate rapid growth for large p.

The simplest estimates are made in the same way as in the small p theory, by assuming complete energy transfer to a flexural mode. With $\dot{a}_o = a_o = \dot{a}_n = 0$, Equation (3.3.54) gives the following expression to determine $a_n$:

$$\frac{3}{16} n^4 a_n^4 + \alpha^2 n^4 a_n^2 - 2p^2\alpha^2 = 0 \tag{3.3.55}$$

The stress amplification is defined as before, which with Equation (3.3.47), becomes

$$A_\sigma = \frac{\epsilon_b}{p\alpha} = \frac{\sqrt{3}\,\alpha n^2 a_n}{p\alpha} = \frac{\sqrt{3}\,\lambda^2}{\alpha} \cdot a_n \tag{3.3.56}$$

It is apparent that in the high-order theory, with the $a_n^4$ term in Equation (3.3.55), the amplification is no longer independent of n. To solve for $A_\sigma$, assume for the moment that the buckling mode is the same as that for small p so that, from Equation (3.3.35), $n_{cr} = 1/\sqrt{2\alpha}$. Then Equations (3.3.55) and (3.3.56) yield

$$A_\sigma \min = \frac{\sqrt{2}}{p} \left[(1 + 6p^2)^{1/2} - 1\right]^{1/2} \tag{3.3.57}$$

A plot of stress amplification from Equation (3.3.57) is given as the dashed curve in Figure 3.22. For small values of p, the amplification approaches the value $\sqrt{6}$ from the low-order theory. As p increases, the amplification decreases. The flexural mode grows more rapidly [because the coefficient of $a_n$ in Equation (3.3.28) becomes more negative], but the maximum flexural amplitude is smaller in proportion to $p\alpha$ because energy is absorbed into the $a_n^4$ term.

At these large values of p, however, the mode number of the flexural mode with most rapid growth diverges from its value $n_{cr} = 1/\sqrt{2\alpha}$ for small p. (The loci of flexural modes in the Mathieu diagram extends farther into the unstable region, Figure 3.18, so many modes are unstable.) If instead of $n_{cr}$ being held fixed, it is adjusted at each value of p for maximum rate of flexural growth, then the stress amplification never falls below $A_\sigma = 1.62$, as shown by the upper curves in Figure 3.22. These critical values of n (or, more fundamentally, of dimensionless wave number $\lambda$) are found by numerical integration of Equations (3.3.27) and (3.3.28). The critical modes are found to be given reasonably well by the empirical formula

$$\lambda_{cr}^4 = \frac{1}{4p^2} + (0.6)^4 \tag{3.3.58}$$

This formula is constructed to give the proper asymptotic values $\lambda_{cr} \rightarrow 1/\sqrt{2p}$, corresponding to $n_{cr} = 1/\sqrt{2\alpha}$, as $p \rightarrow 0$ (weak instability of autoparametric vibra-

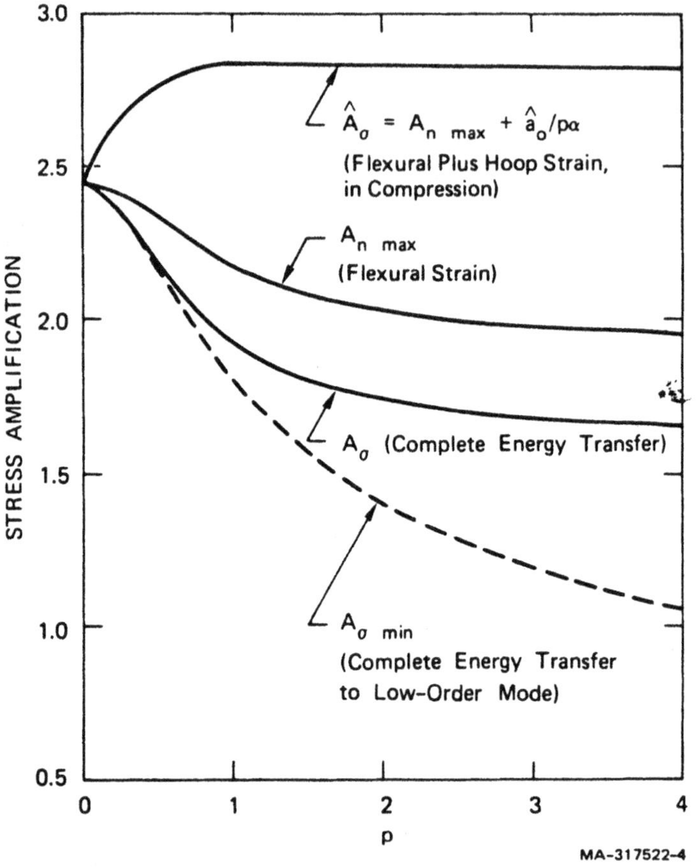


FIGURE 3.22 BOUNDING ESTIMATES FOR STRESS AMPLIFICATION FROM HIGHER-ORDER THEORY

tion buckling, as described in Section 3.3.4), and $\lambda_{cr} = 0.6$ as $p \rightarrow \infty$ (strong instability of pulse buckling, as described in the next section, Section 3.3.6). For $\lambda$ fixed at 0.6, Equations (3.3.55) and (3.3.56) give $A_\sigma = 1.62$, independent of p or $\alpha$. When $\lambda$ is varied according to Equation (3.3.58), it is convenient to write the solution for $A_\sigma$ from Equations (3.3.55) and (3.3.56) in the form

$$A_\sigma = 2\lambda \left[ (4\lambda^4 + 6)^{1/2} - 2\lambda^2 \right]^{1/2} \tag{3.3.59}$$

in which appropriate roots have been taken. Stress amplification from Equations (3.3.58) and (3.3.59) is given by the lowermost solid curve in Figure 3.22. It is clear that even with the proper critical mode at each value of p, the stress amplification falls well below the value $\sqrt{6}$ from the low-order theory.

However, these estimates of stress amplification are based on *complete* energy transfer, as in the small p theory. They do not account for the possibility that with finite p the higher-order theory can allow greater amplification when the hoop amplitude $a_o$ remains finite rather than vanishing altogether. This possibility arises because of the $n^4\alpha^2 - n^2 a_o$ coefficient in Equation (3.3.54), which is the coefficient that led to unbounded growth for $p \geqslant 1/2$ in the lower-order theory. For finite p, the maximum stress amplification is found by again setting $\dot{a}_o = \dot{a}_o = 0$ in Equation (3.3.54), but by then differentiating the equation and setting $da_n^2/da_o = 0$ to maximize $a_n^2$ with respect to $a_o$. This gives for $a_o$ at the maximum

$$\hat{a}_o = n^2 a_n^2/4 \tag{3.3.60}$$

This term is of second order but cannot be neglected for large p. The resulting maximum amplification is found by substituting this expression for $\hat{a}_o$ into Equation (3.3.54) with $\dot{a}_o = \dot{a}_n = 0$. With definition (3.3.56), the result is

$$A_{n\ max} = 2\sqrt{3}\ \lambda \left[(4\lambda^4 + 2)^{1/2} - 2\lambda^2\right]^{1/2} \tag{3.3.61}$$

The subscript n denotes that this expression includes only flexural stress, neglecting for the moment the stress from $\hat{a}_o$.

Amplifications obtained with this expression, again with $\lambda_{cr}$ from Equation (3.3.58), are given as the second curve from the top in Figure 3.22. For finite p, these amplifications are larger than in the curve from Equation (3.3.59) for complete energy transfer. Furthermore, with $a_o$ now nonzero at the maxima, the total bending plus hoop strain in an outer fiber is larger still. The maximum total strain amplification is

$$\begin{aligned} \hat{A}_\sigma &= A_{n\ max} + \hat{a}_o/p\alpha \\ &= A_{n\ max} + (A_{n\ max})^2/12\lambda^2 \end{aligned} \tag{3.3.62}$$

This expression is plotted as the top curve in Figure 3.22. Because $\hat{a}_o$ is always positive, see Equation (3.3.60), and the hoop strain is $\epsilon_o = -a_o$, this amplification refers always to compressive strain, as noted in the figure. Also, this maximization for strain assumes that in the actual coupled motion the phasing between the hoop and flexural modes will at some time align itself so that both $a_o$ and $a_n$ have a maximum at the same time. Numerical integration of the coupled Equations (3.3.27) and (3.3.28) shows that a very close alignment actually does take place when n is chosen properly [approximately according to Equation (3.3.58)]. This is demonstrated in the examples in Figures 3.21 and 3.23. Compressive stress amplifications from such integration lie surprisingly close to the curve from Equation (3.3.62), usually within a few percent.

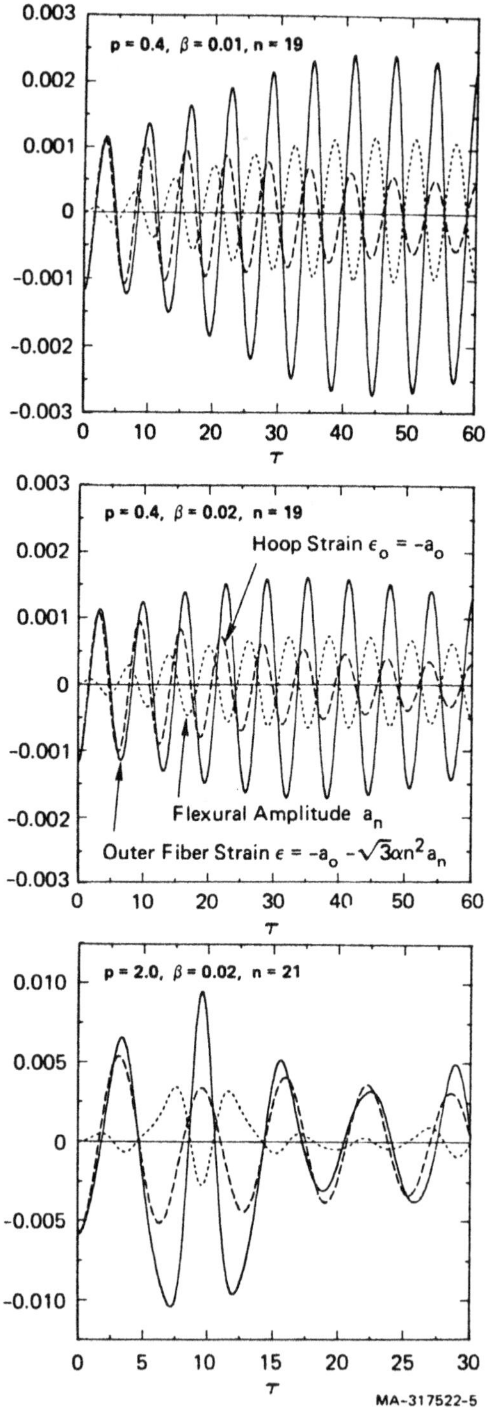


FIGURE 3.23 REPRESENTATIVE TIME HISTORIES OF FLEXURAL MODES AND EXTREME FIBER STRAINS FOR FINITE p AND $\beta$ ($a/h = 100$, $\bar{\delta}_n = 0.02$)

Figure 3.23 also shows the very rapid energy transfer that takes place for large values of p. In the example for p = 2, the extreme fiber strain reaches its largest maximum during the second compressive swing of the hoop mode. Subsequent maxima are greatly reduced because in these examples damping was introduced. This was done by adding a viscous term $2\beta\dot{a}_o$ to the hoop mode Equation (3.3.27) and a similar term $2\beta\omega_n\dot{a}_n$ to the flexural mode Equation (3.3.28). The parameter $\beta$ is the fraction of critical damping, taken to be the same in both equations. The upper two examples in Figure 3.23 show that for moderately small p (p = 0.4), increasing the damping from 1% to 2% greatly reduces the maximum extreme fiber strains because of the large number of oscillations required for the flexural motion to grow.

Figure 3.24 shows stress amplifications from many such numerical integrations for several values of damping $\beta$. Smooth curves were drawn through the

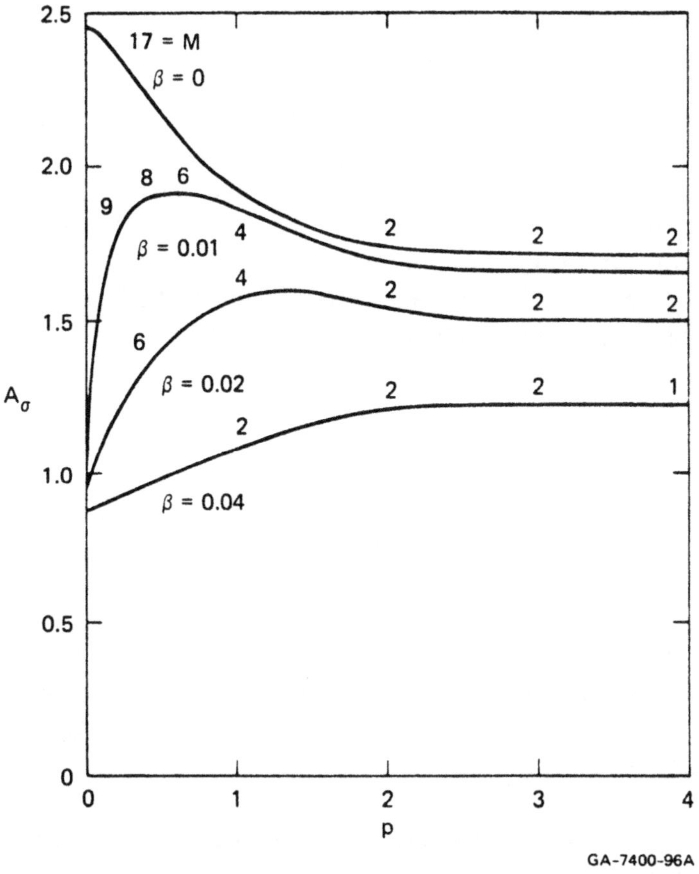


FIGURE 3.24 STRESS AMPLIFICATION (PEAK TENSION) IN A RING SUDDENLY RELEASED FROM SYMMETRIC HOOP COMPRESSION

points from each coupled motion calculation. In these calculations, the initial conditions were with the ring at rest under a uniform hoop compression, as in the examples in Figure 3.23. Shell parameters were also as shown in Figure 3.23; a/h = 100 and $\bar{\delta}_n$ = 0.02. Amplification in Figure 3.24 is for tensile peaks. As already mentioned, for $\beta$ = 0 the compressive amplifications from these calculations gave the same curve as predicted by the energy derivation, the top curve in Figure 3.22. The tensile amplification curve for $\beta$ = 0, in Figure 3.24, lies slightly above the complete energy transfer curve, labeled $A_\sigma$ in Figure 3.22. This is as one would expect, because the contribution to total strain from $\hat{a}_o$ (the difference between the upper two curves in Figure 3.22) is compressive and is much larger than the increase from $A_\sigma$ to $A_{n\ max}$. Tensile strain is therefore maximized by nearly complete energy transfer rather than by maximizing the flexural amplitude with finite $a_o$.

For each set of parameters in Figure 3.24, coupled motion was calculated for several values of n and the largest amplification was plotted. Values of n for largest amplification were always close to $n_{cr}$ given by Equation (3.3.58). The amplification points were always within a band of ± 0.1 in amplification about the smooth curves. Calculations with a/h = 200 gave very nearly the same amplifications as for a/h = 100, demonstrating that the primary influence of $\alpha = h/a\sqrt{12}$ is contained in p. Calculations with $\bar{\delta}_n$ = 0.005 and 0.05 showed an increase in amplification with increasing imperfections for finite $\beta$, but the increase was small enough that the curves for the nominal value $\bar{\delta}_n$ = 0.02 can be used for practical estimates of stress. Experiments described in Sections 2.2.2 and 2.2.10 suggest that equivalent single-mode imperfections are near this value.

Also given on the curves in Figure 3.24 are values of M, the number of compressive swings of the hoop mode at the time the maximum stress occurs, counting the initial compression as M = 1. With no damping ($\beta$ = 0), M can be quite large and is in fact indeterminant, since energy transfer in a beating fashion back and forth between the hoop and flexural mode can repeat itself indefinitely, Figure 3.21. For finite $\beta$, M is the number of cycles to the maximum of the first beat, since energy is being dissipated continuously and for subsequent beats the energy available for transfer is reduced. For $\beta$ = 0.01, M = 9 at p = 0.2, and M = 8 at p = 0.4 (see Figure 3.23, top) and so forth, as shown in Figure 3.24. For p > 2, energy transfer is quite rapid and is complete by M = 2 (see Figure 3.23, bottom). For large p and $\beta$, M = 1.

Figure 3.25 gives similar curves of stress amplification versus p, but for an initial symmetric inward radial velocity. The curve for $\beta$ = 0 is the same as for an initial hoop compression because without damping only the initial energy is important, not the initial phase of motion. Even with $\beta$ = 0.01 and 0.02 the curves are quite similar to those for initial compression; amplification drops markedly with increasing damping for small p, but $\beta$ has little effect for $p > 2$. For $\beta = 0.04$, the amplifications for initial compression and initial velocity differ more widely and the effect of damping is felt even for large p in both cases.

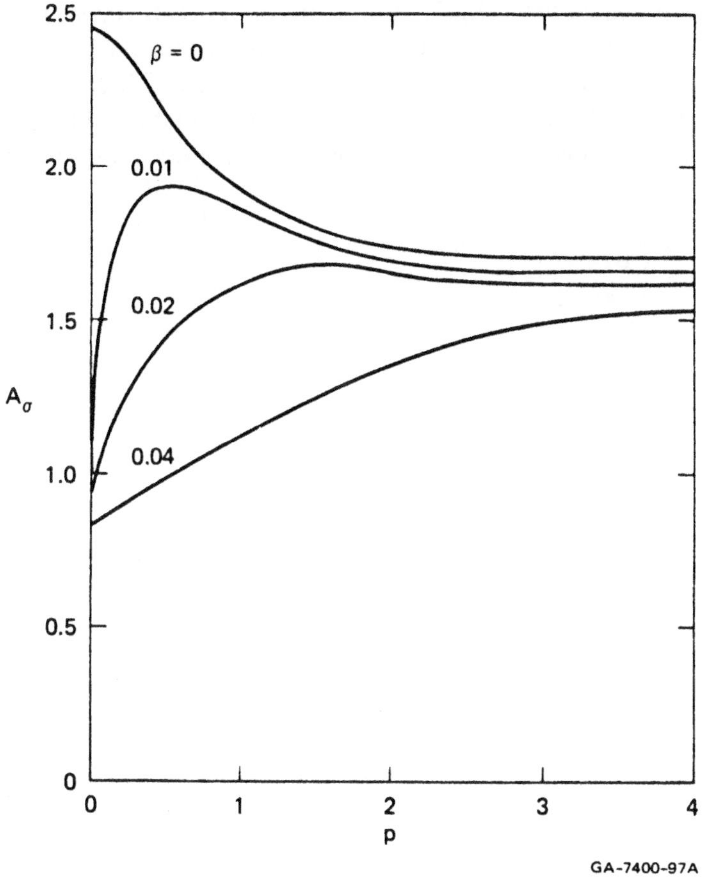


FIGURE 3.25 STRESS AMPLIFICATION (PEAK TENSION) IN A RING HAVING AN INITIAL SYMMETRIC INWARD RADIAL VELOCITY

For either initial condition, at large values of p the energy transfer to flexure takes place during the first compressive swing of the hoop mode. This is pulse buckling as described in the main body of this monograph. The transition from vibration buckling to pulse buckling is at about $p = 4$ for initial compression and at about $p = 2$ for initial radial velocity, but the distinction is not precise. For the initial radial velocity condition, the distinction carries over strongly to the physical result of the buckling because the initial kinetic energy can exceed the elastic hoop capacity of the shell. This excess energy results in fracture (brittle materials[6]) or permanent plastic wrinkles (ductile materials[3]) at wavelengths determined by the dynamic elastic instability.

### 3.3.6 High Initial Velocity--Pulse Buckling[3]

For large values of p, we must return to initial growth of the flexural modes, because the nature of the growth is distinctly different from that for the small values considered in Section 3.3.3. For large p, the parabolas given by Equation (3.3.32), which define the loci of flexural modes in the Mathieu diagram, extend into large regions of unstable motion. This is shown in Figure 3.26, in which the Mathieu diagram of Figure 3.18 is redrawn with axes that extend to $\Omega = 30$ and $\mu = 40$. The question now is not which mode or few modes fall into the region near

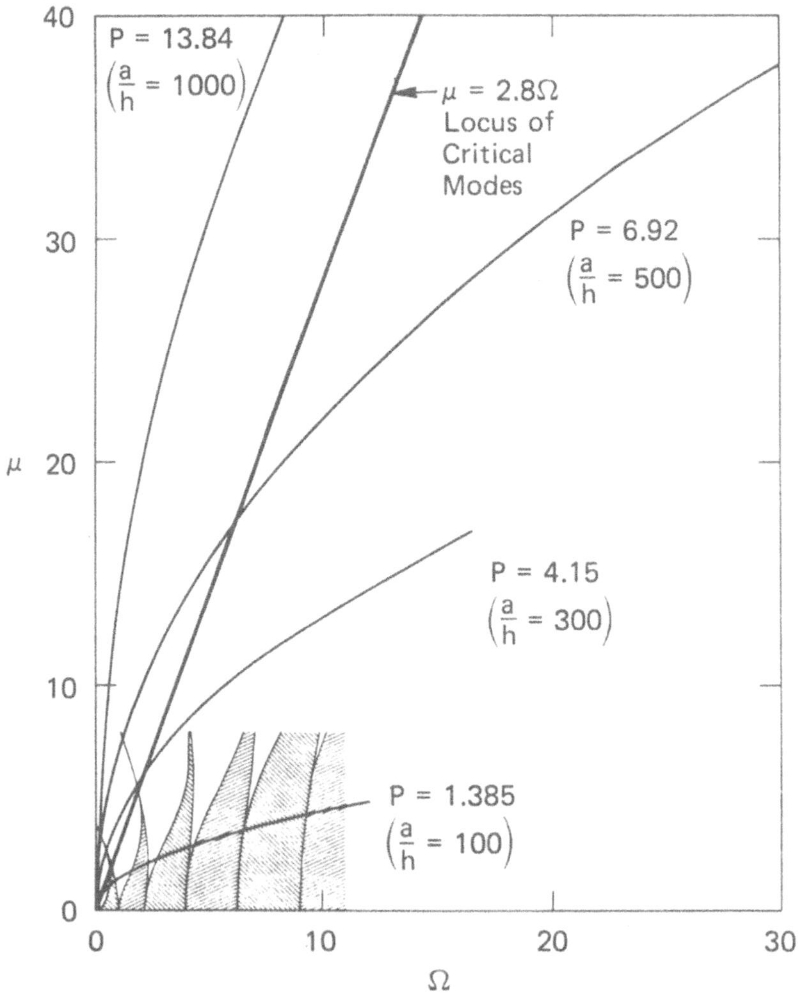


FIGURE 3.26 LOCI OF FLEXURAL MODES ON MATHIEU STABILITY CHART

Shaded areas are the stable regions.

$\Omega = 1/4$, but what is the relative magnitude of growth of the large number of modes that are unstable and is the growth rapid enough to cause wrinkling on the first swing of the hoop mode.

In Figure 3.26, several parabolas are drawn with various values of the stability parameter p. The combination of small $\alpha$ (thin cylinder) and large $v_o/c$ gives parabolas extending well into the unstable region. For unperturbed elastic hoop motion, $v_o/c = \sigma_o/E_1$ must be less than the yield strain of the material, which for engineering metals ranges from 0.002 to 0.005. To indicate the approximate minimum radius-to-thickness ratio at which elastic pulse buckling can occur, values of a/h associated with $v_o/c = 0.004$ are given in addition to the basic parameter p.

The range of modes that could conceivably grow rapidly enough to cause such buckling can be determined by considering the coefficient of $a_n$ in Equation (3.3.30). If $a_n$ is to grow to a significant value in one swing, then this coefficient must be negative at least part of the time. The largest value of n for which this can occur is at $\sin \tau = 1$. For $n^2 >> 1$, this maximum value is given by*

$$n_{max}^2 = \frac{v_o}{c\alpha^2} \equiv s^2 \tag{3.3.63}$$

It is convenient to use the dimensionless wave number $\lambda = n/s$. The range of interest is then $0 < \lambda < 1$, and $\lambda$ is treated as a continuous variable. These are the same parameters introduced in Equation (3.3.37). For the initial symmetric radial velocity condition, the flexural modes are governed by Equation (3.3.36), which is repeated here with $\bar{\delta}_n = 0$.

$$\ddot{a}_\lambda + p^2\lambda^2(\lambda^2 - \sin \tau)\, a_\lambda = 0 \tag{3.3.64}$$

For the present, we consider initial velocity perturbations

$$\dot{a}_n(0) = \gamma_n v_o/c \tag{3.3.65}$$

where $\gamma_n$ are Fourier coefficients that describe a small perturbation in uniformity of $v_o$ around the shell.

For pulse buckling, the solutions of Equation (3.3.64) have an essentially hyperbolic behavior. To emphasize this, and also to relate elastic buckling to the plastic flow buckling of Section 3.2, we consider first the equation obtained by replacing $\sin \tau$ with 1 in Equation (3.3.64); i.e., we assume that the hoop thrust is

---

*If $v_o/c$ is replaced by its equivalent, $\sigma_o/E_1$, this definition of $s^2$ is seen to be analogous to $s^2 = \sigma_y/E_h\alpha^2$, as given in Section 3.2 for plastic flow buckling.

constant at its maximum unperturbed value $\sigma_o = E_1 v_o/c$. For initial conditions (3.3.65), the solutions to this equation are

$$a_\lambda(\tau) = \frac{v_o \gamma_n}{c} \frac{\sinh r_n \tau}{r_n} \tag{3.3.66}$$

where

$$r_n = p\lambda (1 - \lambda^2)^{1/2} , \qquad 0 < \lambda < 1 \tag{3.3.67}$$

The maximum value of the amplification factor $(\sinh r_n \tau)/r_n$ occurs at the maximum value of $r_n$. The most amplified mode is therefore $\lambda = 1/\sqrt{2} = 0.71$. The mode number is the integer n nearest $s/\sqrt{2}$. This is essentially the solution given in Section 2.2, with s interpreted as noted earlier.

The solution to Equation (3.3.64) has basically the same character, but it must be found numerically. As an example, consider the thin shell in Figure 1.5, with $a/h = 480$ and $v_o = 800$ inch/s = 20 m/s. With $c = 200{,}000$ inch/s (5100 m/s) and $\nu = 0.35$ for the aluminum 6061-T6 material, then $p = 6.6$ and $s = 105$. Results of numerical integration of Equation (3.3.64) for these parameters and initial conditions Equation (3.3.65) are given in Figure 3.27. Many modes experience substantial growth. During the early motion ($\tau < 1.5$), growth is almost uniform from $n = 2$ to $n \approx 75$. Later, growth is very rapid and modes near $n = 63$ are much more amplified than others. The corresponding dimensionless wave number is $\lambda = 0.60$. This value is not much different from the value $\lambda = 0.71$ if the hoop thrust were constant at its peak value. The growth with time of the most amplified mode is shown in the lowermost plot of Figure 3.27. It reaches a maximum at $\tau = 3.1$ and then begins an oscillation. In the real shell, Figure 1.5, this oscillation never occurred because plastic hinges were formed during the time of rapid growth, from $\tau = 2$ to 3. The wavelength of observed wrinkles has mode numbers from $n = 50$ to 100, in the range of the theory.

To determine how the critical value of $\lambda$ varies with p, we integrated Equation (3.3.64) with initial conditions Equation (3.3.65) for a sequence of values of $\lambda$ at each of several values of p. The results are summarized in Table 3.5. Over the wide range of amplification, from $ca_{cr}/v_o\gamma_n = 5.3$ to 11,910, the dimensionless wave number $\lambda_{cr}$ ranges only from 0.630 to 0.585. In the midrange, where the onset of buckling is likely to occur, $\lambda_{cr} = 0.60$. The critical value of p for buckling is certainly in the range from 2 to 12 given in the table. For example, if we take $\gamma_n = 0.01$ (i.e., take an equivalent single critical mode perturbation of 1% of the initial velocity $v_o$), the range in amplitude of the critical mode is from 5% to 119 times the unperturbed hoop amplitude $v_o/c$. The locus of critical modes for $\lambda_{cr} = 0.6$ is the straight line $\mu = 2.8\ \Omega$ given in Figure 3.26. This follows from the relations $\mu = p^2\lambda^2$ and $\Omega = p^2\lambda^4$.

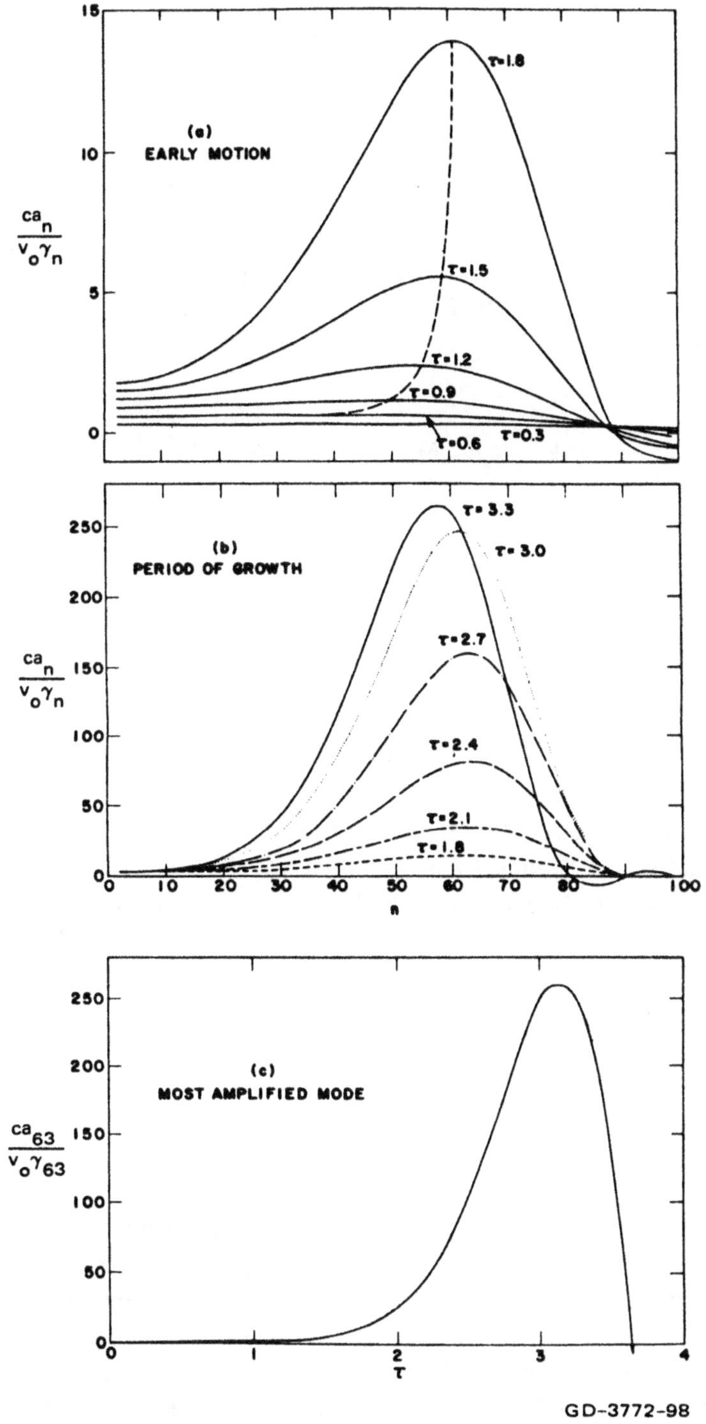


FIGURE 3.27 GROWTH OF FOURIER COEFFICIENTS FOR ELASTIC PULSE BUCKLING OF A THIN SHELL FROM RADIAL IMPULSE

($v_0/c$ = 0.004, a/h = 480)

Table 3.5

EFFECT OF p ON $\lambda_{cr}$ AND AMPLIFICATION OF PREFERRED MODE

| p | $\lambda_{cr}$ | $ca_{cr}/V_o\gamma_n$ (at $\tau$ = 2.7) |
|---|---|---|
| 2 | 0.630 | 5.3 |
| 4 | 0.615 | 20.7 |
| 6 | 0.605 | 96.6 |
| 8 | 0.595 | 474.0 |
| 10 | 0.585 | 2370.0 |
| 12 | 0.585 | 11910.0 |

Similar numerical integration was performed for initial imperfections in the shape of the shell, represented by $\bar{\delta}_n$ in Equation (3.3.36). It was again found that $\lambda_{cr}$ = 0.60 for p in the range from 3 to 12 and that growth is large and essentially exponential in this range. Peak amplification is close to empirical fit

$$G_{max} \equiv \left.\frac{a_n^*}{\bar{\delta}_n}\right|_{max} = 1.2\ e^{1.0p} \tag{3.3.68}$$

where $a_n^* = (a/h)a_n$ is the amplitude of the most amplified mode expressed as a multiple of the wall thickness, as is $\bar{\delta}_n$.

Equation (3.3.68) can be used to calculate critical conditions for incipient formation of plastic hinges at the elastic buckle wavelengths. First observe in the bottom graph in Figure 3.27 that the amplitude of the flexural mode is very small at $\tau = \pi/2 = 1.57$ compared with the amplitude at $\tau = 3.1$. (This graph is for growth of velocity imperfections, but it has the same form as for shape imperfections.) Thus, when the hoop mode reaches its maximum value at $\tau = \pi/2$, the flexural amplitude is small, and when the flexural mode reaches its maximum at $\tau = 3.1$, the hoop mode amplitude is small [the unperturbed hoop amplitude is zero at $\tau = \pi$, Equation (3.3.29)]. Incipient yield from hoop and flexural motion can therefore be described separately.

To determine the critical condition for incipient flexural yield, Equation (3.3.68) is used with Equation (3.3.47) for extreme fiber flexural strain:

$$\epsilon_b = \sqrt{3}\,\alpha n^2 a_n$$

$$= \sqrt{3}\,\alpha \cdot \frac{\lambda^2 p}{\alpha} \cdot a_n^* \frac{h}{a}$$

$$= \sqrt{3}\,\lambda^2 p\,(1.2\ \bar{\delta}_n\, e^p)\,\frac{h}{a} \qquad (3.3.69)$$

This expression is now set equal to the yield strain $\epsilon_y$, with the result

$$\frac{a}{h} \cdot \frac{\epsilon_y}{\bar{\delta}_n} = 0.748\ pe^p \qquad (3.3.70)$$

in which the critical wave number $\lambda = 0.60$ has been used. This expression is plotted in Figure 3.28. Also plotted are lines defining the limits of elastic hoop motion. These are found by setting the peak hoop strain, $\sigma_o/E_1 = v_o/c$, equal to the yield strain* $\epsilon_y$.

$$\frac{v_o}{c} = p\alpha = \frac{ph}{\sqrt{12}\,a} = \epsilon_y \qquad (3.3.71)$$

To plot this relation in Figure 3.28, (3.3.71) is rearranged into

$$\frac{a}{h} \cdot \frac{\epsilon_y}{\bar{\delta}_n} = \frac{p}{\sqrt{12}\ \bar{\delta}_n} \qquad (3.3.72)$$

Thus, in Figure 3.28, a line must be drawn for each assumed value of shape imperfection $\bar{\delta}_n$ in order to have the line appear on the same plot as the general curve for flexural yielding.

Figure 3.28 is used by entering with a value of the known shell quantity $a\epsilon_y/h\bar{\delta}_n$ and then examining for hoop yield or for flexural yield from dynamic buckling. Permitted values of p for elastic hoop motion lie below the lines for each specific value of $\bar{\delta}_n$. Values of p above these lines are not attainable for purely elastic response. Thus, for $\bar{\delta}_n = 0.01$, hoop yield will occur before dynamic buckling yield for all shells with $a\epsilon_y/h\bar{\delta}_n < 104$. For thinner shells (larger a/h), first yield will occur by dynamic buckling. The upper abscissa gives example values of

---

*The relationship between this yield strain and the yield strain in simple tension or compression depends on the yield function. For Tresca yield, their ratio is simply $1 - \nu^2$. For a ring, we replace $E_1$ with E and yield is by simple compression.

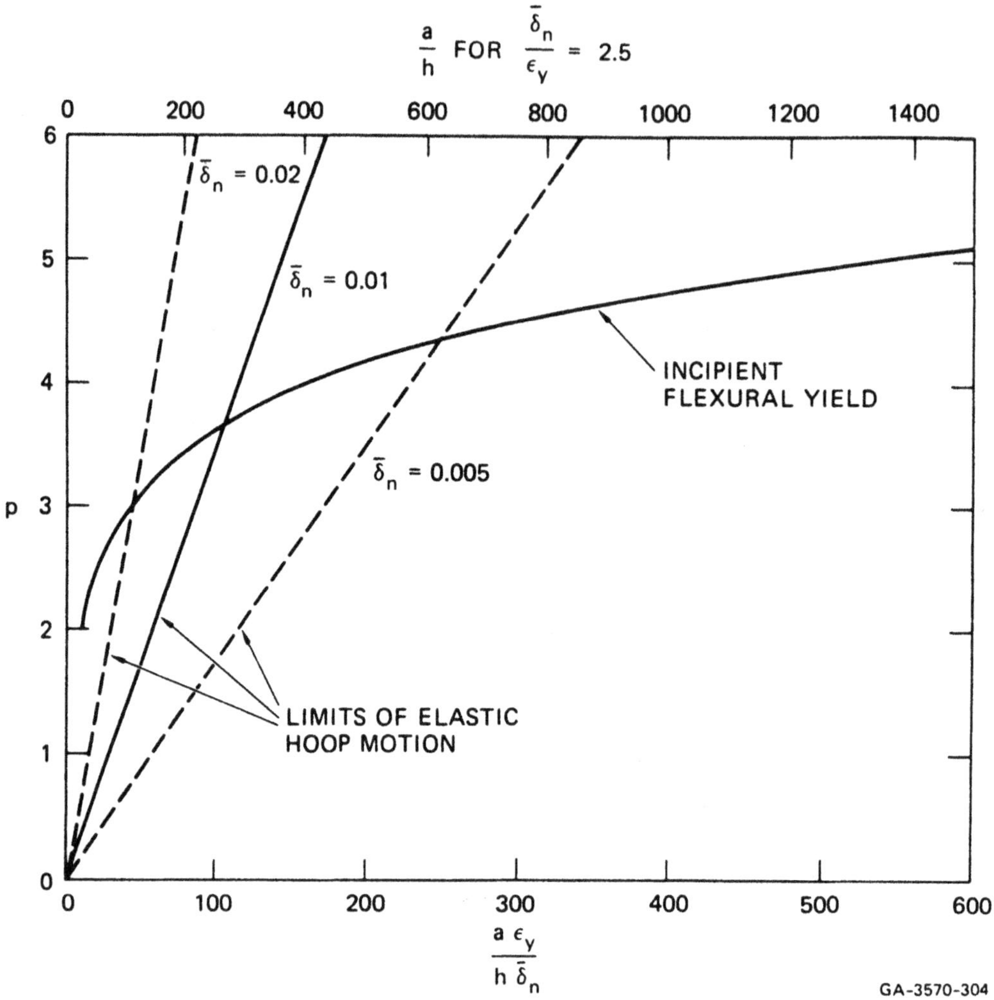


FIGURE 3.28 CRITICAL CONDITIONS FOR FLEXURAL YIELD FROM DYNAMIC ELASTIC PULSE BUCKLING IN THIN SHELLS OR RINGS

a/h for $\bar{\delta}_n/\epsilon_y = 2.5$, which for $\bar{\delta}_n = 0.01$ gives $\epsilon_y = 0.004$, a typical value for engineering metals. For these parameters, the thickest shell that will buckle to incipient flexural yield without first yielding in hoop compression has a/h = 260, a very thin shell.

Critical values of p for buckling with $\bar{\delta}_n = 0.01$ must be larger than $p_{cr} = 3.66$, the value at the intersection of the hoop and flexural yield curves. Observe also that critical values of p are never much larger than this value because the buckling curve increases exponentially with p, Equation (3.3.70). For the thinnest shells of practical interest, p at first flexural yield will not exceed about 5. The assumed value for $\bar{\delta}_n$ can also vary widely without causing large changes in the minimum value of p that can cause yield from buckling. For $\bar{\delta}_n = 0.02$ this value is $p = 3.0$, and for $\bar{\delta}_n = 0.005$, $p = 4.3$. From a practical viewpoint, the critical condition for buckling can therefore be taken as a critical value of p near the center of the range from $p = 3$ to $p = 5$, which covers essentially all cases of interest. For the simple formulas below, we take $p = 4$. From Equation (3.3.68), this is equivalent to taking the critical condition for buckling to be a critical value of amplification of initial imperfections; in this case, $G_{max} = 65$ at $p = 4$. At $p = 3$, $G_{max} = 24$; and at $p = 5$, $G_{max} = 178$.

From the definition $p = v_o/c\alpha$, the impulse $I = \rho h v_o$ that produces the critical condition for elastic buckling is

$$I_e = \rho hc\alpha p = \rho hc \frac{h}{\sqrt{12}\,a} \cdot 4$$

$$I_e = 1.15\ \rho ca \left( \frac{h}{a} \right)^2 \tag{3.3.73}$$

This expression is valid as long as the maximum hoop strain remains elastic. With $p = 4$, this condition requires

$$\frac{v_o}{c} = p\alpha = \frac{4h}{\sqrt{12}\,a} \leqslant \epsilon_y$$

or

$$\frac{a}{h} \geqslant \frac{4}{\sqrt{12}\,\epsilon_y} = \frac{1.15}{\epsilon_y} \tag{3.3.74}$$

For $\epsilon_y = 0.004$, $a/h \geqslant 288$.

## 3.4 CRITICAL RADIAL IMPULSES FOR ELASTIC AND PLASTIC-FLOW BUCKLING OF RINGS OR LONG CYLINDRICAL SHELLS

### 3.4.1 Approach

The ideas used to develop the simple formula in Equation (3.3.73) for critical impulses for elastic buckling of rings or long cylindrical shells can be extended to include plastic flow buckling, which occurs for thicker shells. At the end of Section 3.2, a lower bound on impulse for plastic flow buckling was given simply as the impulse needed to bring the ring or shell to hoop yield. For thick shells this is far too conservative because appreciable plastic flow in the hoop mode can take place before any serious buckle amplitudes are observed. In this section critical impulses are determined under the assumption that amplifications of initial shape imperfections by a factor of about 20 can be safely allowed without objectional deformation. Thus, for an equivalent shape imperfection in the critical buckling mode equal to 1% of the wall thickness, the buckle amplitudes would never exceed 20% of the wall thickness.

As in the plastic flow theory given in Section 3.2, it is assumed that no strain rate reversal occurs during the buckling motion, so that material points are always loading, and stresses and bending moments can be calculated with the tangent modulus in plastic compression. In calculating buckle thresholds, this is never true during the terminal motion because one must calculate right up to the point where the inward hoop motion stops, and beyond for very thin shells as just seen in Section 3.3. With finite flexure, points near the buckle crests on the tensile side of the flexure must therefore unload elastically before the hoop motion ceases. Since the elastic modulus is much larger than the tangent modulus, use of the tangent modulus theory is conservative because it allows more flexure than would actually be experienced. In Section 3.4.6, examples from a numerical analysis that includes strain rate reversal and elastic unloading show that results from the simple theory are nevertheless reasonably accurate. Experimental results for threshold plastic flow buckling are also given, in Section 3.4.4.

For intermediate length shells another resisting moment, called the directional moment, must be included in the theory. Also, in strain-rate sensitive materials viscoplastic moments can dominate over hardening (tangent modulus) moments. These effects are considered in Chapter 4. Because hardening moments dominate in many practical problems, the results in the following sections for the simplest plastic theory have considerable practical importance.

### 3.4.2 Strain Hardening in Engineering Metals

To study the complete spectrum of pulse buckling from thick to thin shells, it is necessary to use a more accurate description of strain hardening than the constant hardening modulus $E_h$ used in Section 3.2. Tests on several aircraft alloys show that an accurate description of strain hardening is given by

$$\frac{\sigma}{E_t} = \begin{cases} \dfrac{\sigma}{E} = \epsilon & 0 \leqslant \epsilon \leqslant \epsilon_y \\ \epsilon_y + K(\epsilon - \epsilon_y) & \epsilon \geqslant \epsilon_y \end{cases} \tag{3.4.1}$$

where $E_t$ is the tangent modulus, $\epsilon_y$ is the yield strain, and K is a parameter that describes the post-yield shape of the stress-strain curve. Integration of Equation (3.4.1) gives

$$\sigma = \begin{cases} E\epsilon & 0 \leqslant \epsilon \leqslant \epsilon_y \\ E\epsilon_y^{(K-1)/K}\left[\epsilon_y + K(\epsilon - \epsilon_y)\right]^{1/K} & \epsilon \geqslant \epsilon_y \end{cases} \qquad (3.4.2)$$

Examples of stress-strain curves from this formula are given in Figure 3.29. Large values of K describe curves with abrupt yield, for which the tangent modulus drops quickly after yield. Small values describe more rounded curves with substantial strain hardening. The curve for K = 30 was measured from tensile tests on aluminum 6061-T6 sheet and small diameter tube stock. The curve for K = 10 is representative of compression tests on magnesium alloys. The curves were deduced from the test data by graphical differentiation of stress-strain curves continuously recorded on a strip chart, which gave E and $E_t(\epsilon)$. Plots were then made of $\sigma/E_t$ versus $\epsilon$, from which K was determined. The yield strain $\epsilon_y$ was then determined so that Equation (3.4.2) matched the measured stress at about 1% strain. When the resulting parameters are substituted into Equation (3.4.2), the

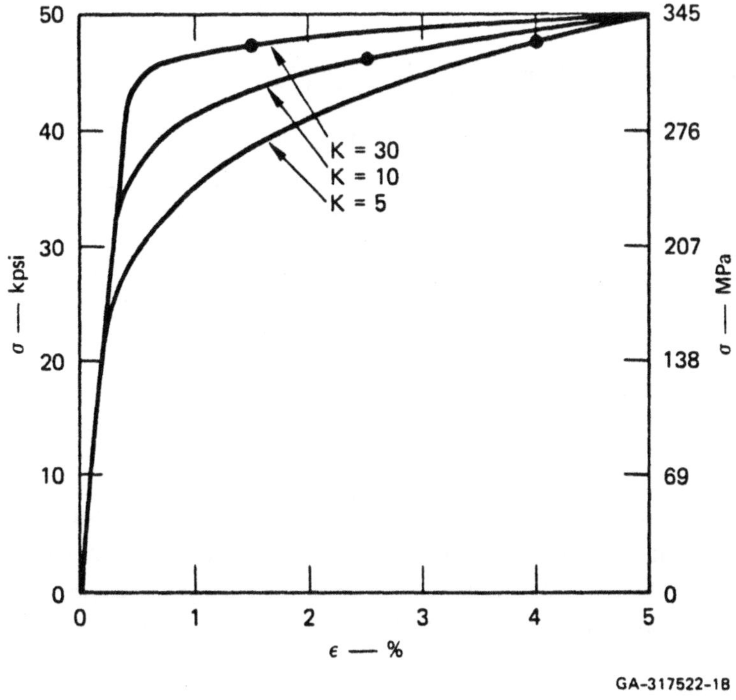


FIGURE 3.29 STRESS-STRAIN CURVES

entire stress-strain curve is matched to within the few percent experimental accuracy. Table 3.6 gives parameters found in this way for alloys of aluminum, magnesium, and titanium. The linear fit of Equation (3.4.1) to the differentiated stress-strain data was generally accurate to strains of at least 5%.

Table 3.6

ELASTIC AND PLASTIC PARAMETERS FOR A FEW AIRCRAFT ALLOYS*

| Alloy | $\rho$ kg/m$^3$ | E GPa | E $10^6$ psi | K | $\epsilon_y$ | $\hat{\epsilon}_y$† | $D = K^{-1/4}\epsilon_y^{1/2}$ |
|---|---|---|---|---|---|---|---|
| Aluminum 6061-T6 | 2730 | 69 | 10.0 | 30 | 0.0039 | 0.0043 | 0.0280 |
| Aluminum 6061-T6 (compression) | 2730 | 69 | 10.0 | 15 | 0.0038 | 0.0045 | 0.0341 |
| Magnesium AZ31B | 1820 | 41 | 6.0 | 14 | 0.0048 | 0.0059 | 0.0397 |
| Magnesium ZK60A | 1820 | 41 | 6.0 | 30 | 0.0065 | 0.0069 | 0.0355 |
| Titanium 6Al-4V | 4470 | 106 | 15.5 | 30 | 0.0070 | 0.0075 | 0.0370 |

* From tensile tests to $\epsilon$ = 5% strain, except as noted.

† Average stress to 3% strain, divided by elastic modulus E.

### 3.4.3 Equations of Motion

The equations of motion to be used in this section are those derived in Section 3.2; however, with the stress-strain curves of Equation (3.4.2), the coefficients are no longer constants.[7] Equation (3.2.10) is then written

$$u'''' + (1 + s^2)u'' + s^2 u + \frac{\ddot{u}}{\alpha^2\beta^2}$$

$$= - s^2(1 + u_i + u''_i) + \frac{aP}{\alpha^2\beta^2 Eh} \qquad (3.4.3)$$

where u = w/a, primes indicate differentiation with respect to $\theta$, and dots indicate differentiation with respect to $\tau$. In order to treat a continuously changing tangent modulus, which varies from the elastic modulus at small strains to much smaller values given by Equation (3.4.1) for plastic strains, the dimensionless time $\tau$ is that used in the elastic problem and additional dimensionless parameters $\beta^2$ and $s^2$ have been introduced as follows:*

$$\tau = \frac{ct}{a} \quad , \quad \beta^2 = \frac{E_t}{E} \quad , \quad s^2 = \frac{\sigma_\theta}{\alpha^2 E_t} \tag{3.4.4}$$

and $\alpha^2 = h^2/12a^2$ is the shell parameter as before. The parameter $s^2$ is the same as that in Equation (3.2.7) but with the fixed hoop stress S/h and hardening modulus $E_h$ replaced by their continuously changing counterparts. Thus, as the shell moves symmetrically inward in the hoop mode, both $\beta^2$ and $s^2$ change with time. The last term in Equation (3.4.3) is from an applied pressure $P(\theta, \tau)$, for later extension in Section 3.5 to finite duration pulses.

The initial shape imperfection $u_i(\theta)$ is expanded into a Fourier cosine series, as are the deformation $u(\theta, \tau)$ and applied pressure $P(\theta, \tau)$:

$$u_i(\theta) = \frac{h}{a} \sum_{n=2}^{\infty} \bar{\delta}_n \cos n\theta \tag{3.4.5}$$

$$u(\theta,\tau) = u_o(\tau) + \sum_{n=2}^{\infty} u_n(\tau) \cos n\theta \tag{3.4.6}$$

$$P(\theta,\tau) = P_o(\tau) \left[ 1 + \sum_{n=2}^{\infty} \gamma_n \cos n\theta \right] \tag{3.4.7}$$

There is no difficulty in adding sine terms, but in what follows response is analyzed for a sequence of modes as in the previous sections, so there is no need to include these terms explicitly. The pressure terms $\gamma_n P_o(\tau)$ are regarded as small nonuniformities whose effect will be compared with the effect of shape imperfections $\bar{\delta}_n$, expressed as in Section 3.3 as a fraction of the wall thickness h. Substitution of these expansions into Equation (3.4.3) gives the following equations for the series coefficients:

$$\ddot{u}_o + \frac{\sigma_\theta}{E} (1 + u_o) = \frac{aP_o(\tau)}{Eh} \tag{3.4.8}$$

$$\ddot{u}_n + \alpha^2\beta^2(n^2 - 1)(n^2 - s^2)u_n$$

$$= \alpha^2\beta^2 s^2(n^2 - 1)\bar{\delta}_n \frac{h}{a} + \frac{a\gamma_n P_o(\tau)}{Eh} \tag{3.4.9}$$

*In this section, $\beta$ no longer refers to viscous damping ratio.

In the more general case, with the load applied as a finite pressure pulse, the shell is taken initially at rest in the unstressed condition, giving initial conditions

$$u_n(0) = \dot{u}_n(0) = 0 \qquad n = 0, 2, 3, \ldots \tag{3.4.10}$$

For an ideal impulse,

$$I(\theta) = \lim_{\substack{P(\theta,0+)\to\infty \\ T\to 0}} \int_0^T P(\theta,\tau) \frac{a}{c} \, d\tau \tag{3.4.11}$$

these initial conditions are replaced by

$$u_n(0) = 0 \quad , \quad \dot{u}_n(0) = \frac{\gamma_n I_o}{\rho ch} \qquad n = 0, 2, 3, \ldots \tag{3.4.12}$$

and the pressure terms in Equations (3.4.8) and (3.4.9) are dropped.

In either case, the solutions to Equations (3.4.8) and (3.4.9) are found by numerical integration. The procedure begins by solving for the symmetric hoop motion $u_o(\tau)$ from Equation (3.4.8). Having neglected the small influence of flexure on the midsurface strain $\epsilon$ in the formulation of Equation (3.4.3), we take $u_o = w_o/a = \epsilon$. Then $\sigma_o(\epsilon)$ in Equation (3.4.8) is determined by using the material stress-strain relation, Equation (3.4.2). Similarly, $\epsilon(\tau)$ is used in Equations (3.4.1) and (3.4.2) to calculate the time-varying coefficients $\beta^2$ and $s^2$ in the flexural mode Equation (3.4.9). With these values at each time step, the flexural equations are also solved by numerical integration.

### 3.4.4 Plastic Flow Buckling[8]

Before performing this numerical integration, it is useful to obtain approximate solutions by extending the results in Section 3.2 found for constant s and $E_h$. Then $\sigma_\theta(\epsilon)$ is replaced by a constant average stress $\hat{\sigma}_y$, and the parameters $s^2 = \hat{\sigma}_y/\alpha^2 E_h$ and $\beta^2 = E_h/E$ are also constants. When these simplifications are introduced into Equation (3.4.9), its solution with initial conditions from Equation (3.4.12) is simply

$$u_n = \frac{hs^2\bar{\delta}_n}{a(s^2 - n^2)} [\cosh p_n\tau - 1] + \frac{\gamma_n I_o}{\rho chp_n} \sinh p_n\tau \tag{3.4.13}$$

where $n < s$ and

$$p_n = \alpha\beta \left[(n^2 - 1)(s^2 - n^2)\right]^{1/2} \tag{3.4.14}$$

The first term in Equation (3.4.13) shows the exponential growth of shape imperfections. For $n^2 >> 1$, the term is of exactly the same form as the amplification function in Equation (2.2.50) for the bar under axial impact. All the conclusions for growth of random imperfections in the bar can therefore be carried over to the ring or long shell. A band of high-order modes is highly amplified. The preferred mode number, near the maximum amplification, is the integer closest to $n = s/\sqrt{2}$. With the reasonable assumption of random imperfections, buckling thresholds can be estimated by studying motion in this mode, with an equivalent single imperfection coefficient $\bar{\delta}_n$ that represents the integrated effect of modes with mode numbers near the critical mode number. The magnitude of the amplification is governed by the duration of the inward motion. This will be discussed in more detail later.

The second term in Equation (3.4.13) is from buckling precipitated by imperfections in the nearly uniform inward velocity. A plot of this amplification, in Figure 3.11, has the same general features as for shape imperfections. It has a maximum at the integer closest to $n = s/\sqrt{2}$. The relative importance of shape and velocity imperfections can therefore be determined by calculating response for $n = s/\sqrt{2}$. For this mode, $n^2 >> 1$, $p_n$ is maximum and is given by

$$p_{n\ max} = \frac{\alpha\beta s^2}{2} = \frac{\hat{\sigma}_y}{2\alpha\sqrt{EE_h}} \equiv Q \tag{3.4.15}$$

The ratio of the maximum displacements caused by shape and velocity imperfections is then

$$\frac{u_{shape}}{u_{vel}} = \frac{\bar{\delta}_n}{\gamma_n} \cdot \frac{h\hat{\sigma}_y}{a\alpha v_o(\rho E_h)^{1/2}} \cdot \frac{\cosh Q\tau - 1}{\sinh Q\tau}$$

$$\approx \frac{\bar{\delta}_n}{\gamma_n} \frac{2\sqrt{3}\ \hat{\sigma}_y}{v_o(\rho E_h)^{1/2}} \tag{3.4.16}$$

where $v_o = I_o/\rho h$ is the initial inward velocity of the shell wall.

Equation (3.4.16) can be interpreted directly in terms of the idealized compressive stress-strain curve $\sigma = \sigma_y + E_h\epsilon$ used in the analysis. To do this, the initial kinetic energy in the shell wall is equated to the plastic work of membrane straining to a final strain $\epsilon_f$ at average stress $\hat{\sigma}_y$:

$$\frac{\rho v_o^2}{2} = \hat{\sigma}_y\epsilon_f = \hat{\sigma}_y \cdot \frac{\sigma_h}{E_h} \tag{3.4.17}$$

in which $\sigma_h$ is the increment in stress due to strain hardening. This equation is now used to eliminate $v_o$ in Equation (3.4.16), with the result

$$\frac{u_{shape}}{u_{vel}} = \left(\frac{6\hat{\sigma}_y}{\sigma_h}\right)^{1/2} \cdot \frac{\bar{\delta}_n}{\gamma_n} \tag{3.4.18}$$

The strain hardening increment $\sigma_h$ is a small fraction of the average yield stress $\hat{\sigma}_y$ for most metals of interest (see Figure 3.29 for K = 10 and 30). Equation (3.4.18) shows that in these materials shape imperfections are likely to dominate over velocity imperfections if shape imperfections as a function of wall thickness ($\bar{\delta}_n$) are comparable to fractional perturbations in the nominally uniform load ($\gamma_n$). For example, from formulas to be given later in this section, a shell with a/h = 24 made of aluminum 6061-T6, for which K = 15, buckles at a plastic strain increment of $\epsilon_p$ = 2.0%. With $E_h$ = 400,000 psi (2.8 GPa) from Figure 3.29, this gives $\sigma_h$ = 8000 psi (55 MPa). With an average stress $\hat{\sigma}_y$ = 40,000 psi (276 MPa) this gives $(6\hat{\sigma}_y/\sigma_h)^{1/2} = 5.5$. Thus, velocity imperfections would have to be about five times larger than shape imperfections to produce the same deformation.

The simple solution in Equation (3.4.13) for constant parameters also suggests a procedure for finding an approximate formula for critical impulse at the threshold of allowable buckling. The criterion for buckling is taken as a critical amplification of initial shape imperfections. Experimental results will be given to demonstrate that, as for the thin bar buckling in Section 2.2, shape imperfections are proportional to wall thickness; thus this criterion corresponds to buckle amplitudes equal to some fraction of the wall thickness, a reasonable criterion.

Amplification in Equation (3.4.13) depends mainly on the argument $Q\tau$ of the hyperbolic term, since imperfections are typically very small, so concern is with large amplifications in which hyperbolic growth dominates. It is then reasonable to assume that a critical amplification criterion corresponds closely to $Q\tau$ reaching a critical value, that is,

$$Q\tau_f = B \tag{3.4.19}$$

where B is a constant to be determined and $\tau_f$ is the dimensionless duration of the inward membrane plastic flow in the hoop mode. In real time this duration is approximately

$$t_f = \frac{I_o a}{\hat{\sigma}_y h} \tag{3.4.20}$$

which is the time to bring a unit axial length of wall element, of area $had\theta$, to rest from velocity $v_o = I_o/\rho h$ at constant deceleration $\hat{\sigma}_y hd\theta/\rho had\theta = \hat{\sigma}_y/\rho a$. With the definition $\tau = ct/a$, Equations (3.4.15), (3.4.19), and (3.4.20) are combined to give the following expression for critical impulse:

$$I_{cr} = 2\alpha h(\rho\hat{\sigma}_y)^{1/2}\left(\frac{E_h}{\hat{\sigma}_y}\right)^{1/2}\cdot B \tag{3.4.21}$$

This formula can be used directly for materials with a constant hardening modulus $E_h$. In this case, the amplification of shape imperfections, from Equation (3.4.13) with $\gamma_n = 0$ and $n^2 = s^2/2$, is

$$G_{max} = \left.\frac{u_n^*}{\bar{\delta}_n}\right|_{max} = 2(\cosh p_n\tau - 1)$$

$$= 2(\cosh B - 1) \approx e^B \tag{3.4.22}$$

where $u_n^* = w_n/h = (a/h)u_n$. It will be shown in the following pages that a reasonable value for $\bar{\delta}_n$ is about 0.01, just as for the thin strips in Section 2.2. Then a reasonable critical value for $G_{max}$ is 20, which then gives deformations of 0.2 h, which are perceptible but generally not unacceptable. For $G_{max} = 20$, $B = 3.0$. Other choices for $G_{max}$ do not change B appreciably; for $G_{max} = 50$, $B = 3.9$. With $B = 3$, Equation (3.4.21) becomes

$$I_{cr} = 6\alpha h(\rho E_h)^{1/2}$$

$$= \sqrt{3}\ a(\rho E_h)^{1/2}\left(\frac{h}{a}\right)^2$$

$$= \sqrt{3}\ \rho ca\left(\frac{E_h}{E}\right)^{1/2}\left(\frac{h}{a}\right)^2 \tag{3.4.23}$$

This formula is similar to Equation (3.3.73) for elastic buckling but applies only for shells with a/h small enough that buckling occurs only during substantial plastic flow. The two formulas differ simply by the factor $(1.15/\sqrt{3})(E/E_h)^{1/2}$. Typically, $E/E_h \approx 100$ so this factor is about 6.60. Thus, for a given a/h, Equation (3.4.23) gives much smaller impulses than Equation (3.3.73). However, Equation (3.4.23) must, of course, give larger impulses than Equation (3.3.73) because it is to be applied to thicker shells. With this simple constant $E_h$ theory, there is obviously a range of a/h for which neither the elastic nor the simple plastic theory applies. For $E/E_h = 100$, this range is at least from $a/h = 288$, the lower limit of Equation (3.3.73), to $a/h = 288/\sqrt{6.6} = 112$, the value for which Equation (3.4.23) gives the same critical impulse as from the elastic theory for the thinner shell, a contradiction in physical expectation.

This range in a/h for which neither theory applies is a range of considerable practical importance. This is why the more accurate description of material behavior was introduced in Equations (3.4.1) and (3.4.2), along with equations of motion with variable parameters. Even with these added complications, it is possible to derive a simple critical impulse formula for plastic flow buckling that covers the entire range of interest and is surprisingly accurate.

To derive a critical impulse formula for varying $\sigma$ and $E_t$, Equations (3.4.8) and (3.4.9) for motion are first integrated numerically for several examples. The results show that most of the growth of $u_n$ occurs near the end of the hoop motion because $\sigma$ increases and $E_t$ decreases with increasing $\epsilon$ and hence with increasing $\tau$. An indication of why this has a strong effect on growth is apparent in Equation (3.4.15), which shows how $\hat{\sigma}_y$ and $E_h$ enter directly into the hyperbolic argument $Q\tau$. To obtain a simple formula for critical impulse, $\sigma$ and $E_t$ are therefore taken to be constant at their final values, and Equation (3.4.21) is used from the constant parameter solution.

These final values are determined by substituting the final hoop strain $\epsilon_f$ into the material stress-strain relation in Equation (3.4.1). The final hoop strain is found by setting the kinetic energy imparted by $I_o$ equal to the strain energy to bring the shell wall to rest, with the small energy diverted into flexure neglected.

$$\frac{1}{2}\rho h \left(\frac{I_o}{\rho h}\right)^2 = h \int_0^{\epsilon_f} \sigma(\epsilon)\, d\epsilon \tag{3.4.24}$$

For this integration, the material is taken to be elastic, perfectly plastic, with the result

$$I_o^2 = 2\rho h^2 \hat{\sigma}_y \left(\epsilon_f - \frac{\hat{\epsilon}_y}{2}\right) \tag{3.4.25}$$

The hat is used in $\hat{\epsilon}_y$ to distinguish this yield strain, adjusted to fit the elastic, perfectly plastic model, from the yield strain $\epsilon_y$ as defined for the actual stress-strain relation in Equations (3.4.1) and (3.4.2). Inspection of Figure 3.29 shows that $\hat{\epsilon}_y > \epsilon_y$ and that $\hat{\epsilon}_y \rightarrow \epsilon_y$ as $K \rightarrow \infty$. For $\epsilon_f >> \hat{\epsilon}_y/2$, the term $\epsilon_f - \hat{\epsilon}_y/2$ in Equation (3.4.25) can be replaced by $\epsilon_f - \epsilon_y$ to simplify the expression for $\sigma/E_t$

from Equation (3.4.1), with the result

$$\frac{\sigma}{E_t} = \frac{KI_o^2}{2\rho h^2 \hat{\sigma}_y} \tag{3.4.26}$$

in which it has also been assumed that K >> 1 so that the $\epsilon_y$ term in Equation (3.4.1) can be neglected. Table 3.6 shows that this is a reasonable assumption. With $\sigma \approx \hat{\sigma}_y$ as already assumed in calculating $\epsilon_f$, Equation (3.4.26) substituted into Equation (3.4.21) gives the desired simple formula for critical impulse:

$$I_o = B^{1/2} a \left(\frac{2}{3K}\right)^{1/4} (\rho \hat{\sigma}_y)^{1/2} \left(\frac{h}{a}\right)^{3/2} \tag{3.4.27}$$

The corresponding final strain, from Equations (3.4.25) and (3.4.27), is

$$\epsilon_f = \frac{1}{2}\left[B\left(\frac{2}{3K}\right)^{1/2} \frac{h}{a} + \hat{\epsilon}_y\right] \tag{3.4.28}$$

Numerical integration of the equations of motion Equations (3.4.8) and (3.4.9) demonstrates that the simple formula in Equation (3.4.27) is accurate to within 10% for K ranging from 10 to 30 and a/h ranging from 20 to 200. This degree of accuracy is remarkable, considering the approximations that led to the formula. It is certainly within the accuracy of the assumptions that underlie our entire approach to the analysis of plastic flow buckling.

As in the constant $E_h$ formula in Equation (3.4.23), the choice for B in Equation (3.4.27) corresponds to a choice for the allowable amplification of imperfections. In the following paragraphs we give experimental results[9] from which the choice of $G_{max}$ = 20 was made. Experiments were run on a sequence of 6-inch-diameter (152 mm) aluminum 6061-T6 shells with a/h = 24. Impulses nearly uniform around the circumference were applied by sheet explosive placed over a ¼-inch-thick (6.3 mm) layer of foam neoprene rubber as described in Section 3.2, Figure 3.12. Buckled profiles from five experiments at increasing values of impulse are given in Figure 3.30. The upper records show that buckling is well developed at impulses greater than 5900 taps (1 tap = 1 dyne-sec/cm$^2$ = 0.1 Pa-s). Very small ripples can also be seen for the shells tested at 3900 and 2800 taps (390 and 280 Pa-s). Additional profiles of these two shells at a higher magnification are given by the lower two traces. These show very distinct buckling over the entire circumference at 3900 taps and over portions of the shell at 2800 taps. The amplitude of buckles in these two shells is so small that in visual inspection, without the aid of the recording device to magnify deformations, no buckling could be observed.

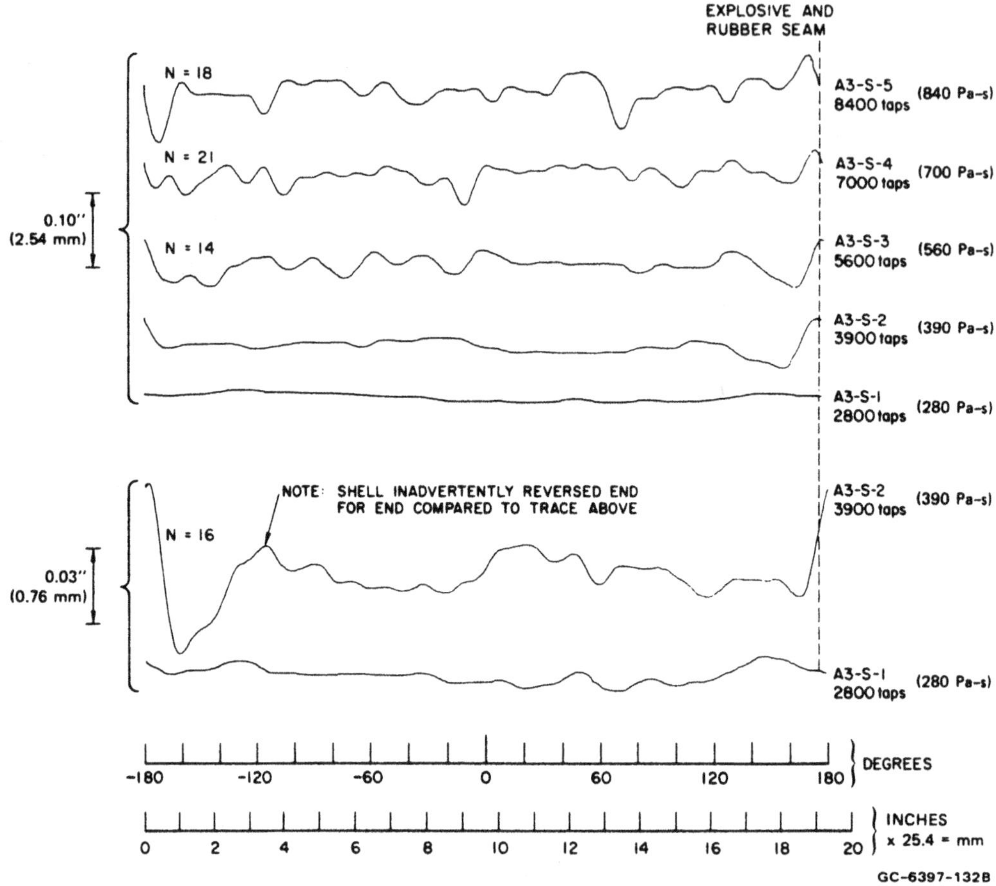


FIGURE 3.30 BUCKLE PROFILES AT SEVERAL MAGNITUDES OF SYMMETRIC RADIAL IMPULSE

Aluminum 6061-T6 cylinders, a/h = 24, h = 1/8 inches = 3.2 mm.

Peak deformation (half the maximum peak-to-peak deformation away from the explosive seam) in each shell is given in a semilog plot against impulse in Figure 3.31. The points falls near a straight line, indicating an exponential increase in amplitude with impulse, just as in the theory.

The theoretical curves of amplification versus impulse were calculated by numerical integration of Equations (3.4.8) and (3.4.9) for the three stress-strain curves in Figure 3.29. These show that the experimental exponent (slope of the curve) is matched by the K = 5 theoretical curve. The K = 10 curve, which corresponds to a stress-strain curve near that for aluminum 6061-T6 in compression, is at a slope a factor of 1.6 larger than the slope of the experimental line. This difference can be attributed mainly to effects of strain rate reversal, neglected in the analytical theory, as will be shown in Section 3.4.7.

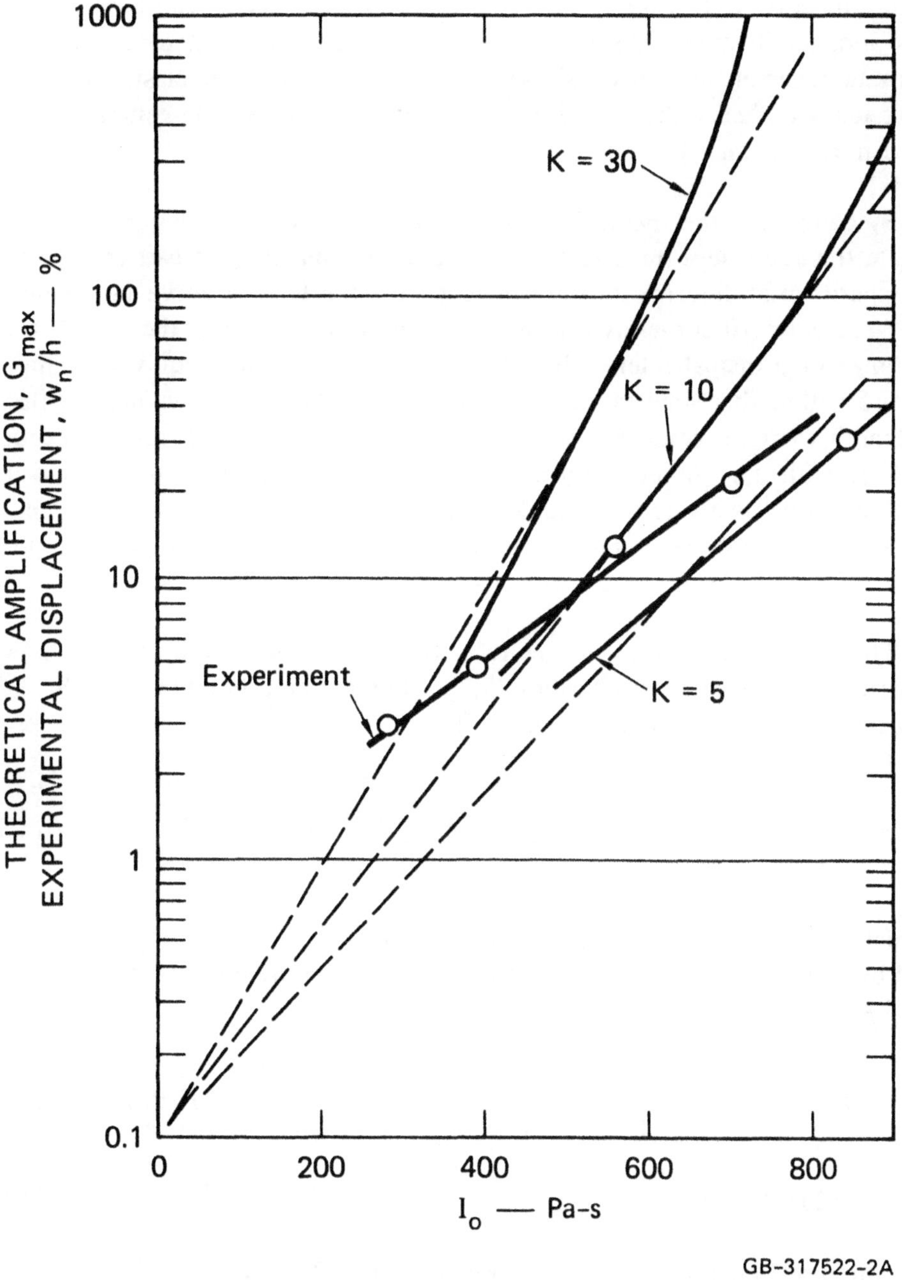


FIGURE 3.31 THEORETICAL AMPLIFICATION AND EXPERIMENTAL PEAK DISPLACEMENT VERSUS IMPLUSE FOR SHELLS IN FIGURE 3.30

The theoretical mode number of the most amplified mode is n = 28 for K = 10 and n = 27 for K = 5. These are larger than the total number of waves on the shell profiles in Figure 3.30 ($n \approx 18$), but are close to the wave numbers of many of the individual waves. Again, the difference can be attributed to the effects of strain-rate reversal, as will be shown. The high modulus of elastic unloading in flexure tends to freeze the buckle motion, suppressing growth in general, and freezing the motion sooner for high mode numbers.

By comparing the theoretical and experimental curves in Figure 3.31, we can estimate the equivalent imperfections as we did for buckling of bars in Figure 2.21. The experimental deformations are matched for the K = 5 curve by taking $\bar{\delta}_n$ = 1.7%. The K = 10 curve gives a match within a factor of 2 in the $5\% < w_n/h < 50\%$ range of principal interest by taking $\bar{\delta}_n$ = 0.6%. These equivalent imperfections are within the same range as was found for elastic bar buckling, at the long wavelength extreme compared with wall thickness. We are therefore encouraged that, even with this limited experimental data, the assumption of $\bar{\delta}_n$ = 1% will give critical buckling loads to reasonable accuracy. Figure 3.31 shows that, if $\bar{\delta}_n$ is changed above and below this value by a factor of 2, critical impulse to produce $w_n/a$ = 20% with the K = 10 curve changes only by ± 14%.

The numerical results in Figure 3.31 can be used to determine how the parameter B varies with amplification $G_{max}$. The amplification curve for K = 10 is first approximated by the exponential function (the straight line in the semilog plot):

$$G_{max} = 0.10\ e^{0.0086\ I_o} \tag{3.4.29}$$

in which $I_o$ is in Pa-s. The relationship between $I_o$ and $B^{1/2}$ for this example is found by substituting the cylinder parameters for this numerical example into Equation (3.4.27):

$$I_o = B^{1/2}\ (0.0762\ \text{m}) \left( \frac{2}{3 \times 10} \right)^{1/4} \left(2730\ \frac{\text{kg}}{\text{m}^3} \cdot 300 \times 10^6\ \text{Pa}\right)^{1/2} \left( \frac{1}{24} \right)^{3/2}$$

$$= 298\ B^{1/2}\ \text{Pa-s} \tag{3.4.30}$$

Substitution of Equation (3.4.30) into Equation (3.4.29) gives

$$G_{max} = 0.10\ e^{2.56\ B^{1/2}}$$

or

$$B^{1/2} = 0.39\ \ln(10\ G_{max}) \tag{3.4.31}$$

Observe also that if $K = 5$ or $K = 30$ is used in Equation (3.4.27) in place of $K = 10$, Equation (3.4.31) remains unchanged if the coefficient of $I_o$ in the exponent of Equation (3.4.29) is changed from 0.0086 to $0.0086/\sqrt{2}$ or $0.0086\sqrt{3}$, respectively. These three coefficients substituted into Equation (3.4.29) give a fan of three lines in Figure 3.31 passing through the point $I_o = 0$, $G_{max} = 0.10$, as shown. These lines are reasonably good approximations to the amplification curves for $K = 5$, 10, and 30 in the neighborhood of $G_{max} = 20$. This shows that the variation of $I_o$ with K as given by Equation (3.4.27) is essentially correct near the terminal motion, for which the equation was derived.

Equation (3.4.31) shows the expected result that the value of $B^{1/2}$ to be used in the critical impulse formula depends on the amplification one specifies for the buckling criterion. For $G_{max} = 20$, $B^{1/2} = 2.07$. Changing to $G_{max} = 10$ or $G_{max} = 50$ changes $B^{1/2}$ to 1.80 or 2.42, respectively. These values are only −13% and +17% below and above the value for $G = 20$. Reasonable estimates for threshold buckling to produce observable wrinkles (peak deformation of about 20% of the wall thickness for $\bar{\delta}_n = 1\%$ as discussed above) can therefore be made by using $B^{1/2} = 2$. Equations (3.4.27) and (3.4.28) then become

$$I_o = 2a\left(\frac{2}{3K}\right)^{1/4}(\rho\hat{\sigma}_y)^{1/2}\left(\frac{h}{a}\right)^{3/2} \tag{3.4.32}$$

$$\epsilon_f = \left(\frac{8}{3K}\right)^{1/2}\frac{h}{a} + \frac{\hat{\epsilon}_y}{2} \tag{3.4.33}$$

### 3.4.5 Summary of Formulas for Critical Impulse

The formulas just given for plastic flow buckling are applicable only to shells thick enough that buckling takes place during plastic flow. At the other extreme, the formula for threshold bending yield from elastic buckling in Equation (3.3.73) is applicable only for shells thin enough that there is no plastic flow during the inward hoop motion, as specified by Equation (3.3.74). For an interim range of wall thicknesses (radius-to-thickness ratios), neither set of assumptions leading to these critical impulse formulas is applicable. For practical use, however, this range is covered by simply adopting the conservative procedure of using whichever formula gives the smaller impulse. With this procedure, the transition radius-to-thickness ratio is the value for which the plastic flow and elastic buckling formulas give the same impulse. (This will be demonstrated graphically in Figure 3.32.) To calculate this value, it is useful to divide Equation (3.4.32) by $\rho ca$, which gives

$$\frac{I_o}{\rho ca} = 2\left(\frac{2}{3K}\right)^{1/4}\left(\frac{\rho\hat{\sigma}_y}{\rho^2 c^2}\right)^{1/2}\left(\frac{h}{a}\right)^{3/2}$$

$$= 1.807\ K^{-1/4}\ \hat{\epsilon}_y^{1/2}\ (h/a)^{3/2} \qquad (3.4.34)$$

in which $\hat{\epsilon}_y = \hat{\sigma}_y/\rho c^2 = \hat{\sigma}_y/E_1$. When this expression is equated to the elastic formula $I_o/\rho ca = 1.15\ (h/a)^2$, the resulting expression gives the transition radius-to-thickness ratio,

$$\frac{a}{h} = 0.405\ K^{1/2}\ \hat{\epsilon}_y^{-1} = 0.405/D^2 \qquad (3.4.35)$$

where $D = K^{-1/4}\ \hat{\epsilon}_y^{1/2}$ is a material parameter. Values are given in Table 3.6 for a few aircraft alloys. For these materials, Equation (3.4.35) gives values of a/h somewhat larger than the value $1.15/\epsilon_y$ given by Equation (3.3.74) for incipient hoop yield. This makes this procedure for calculating critical impulse definitely conservative for $1.15/\epsilon_y < a/h < 0.405/D^2$ because the plastic flow buckling curve lies below the elastic buckling curve for $a/h < 0.405/D^2$. This will be made clear by a graphical example. Thus, Equations (3.3.73) and (3.4.32) can be summarized by

$$\frac{I_o}{\rho ca} = \begin{cases} 1.807\ D(h/a)^{3/2}, & a/h \leqslant 0.405/D^2 \\ 1.15\ (h/a)^2\ , & a/h \geqslant 0.405/D^2 \end{cases} \qquad (3.4.36)$$

Figure 3.32 gives a plot of threshold impulse from these formulas for 6-inch-diameter (152 mm) shells of 6061-T6 aluminum. Observe first that the intersection of the elastic and plastic flow buckling branches forms a bend at point A that is concave downward. The dashed lines extended each way from the intersection demonstrate the earlier statement that Equations (3.4.36) are constructed to give impulse from whichever formula gives the smaller value. Observe also that if the bend at A occurs to the right of the intersection point B between the hoop yield and elastic buckling lines, then the plastic buckling line lies below the elastic buckling line between points A and B, as stated in the preceding paragraph.

The important result demonstrated by Figure 3.32 is that, for thick shells, the critical impulse for buckling from this analysis, which allows a finite amount of buckling motion, is well above the impulse for incipient hoop yield, the conservative criterion suggested at the end of Section 3.2. For the experimental shells with a/h = 24, the theoretical and experimental threshold impulse is a factor of 4.2 above the incipient yield impulse.

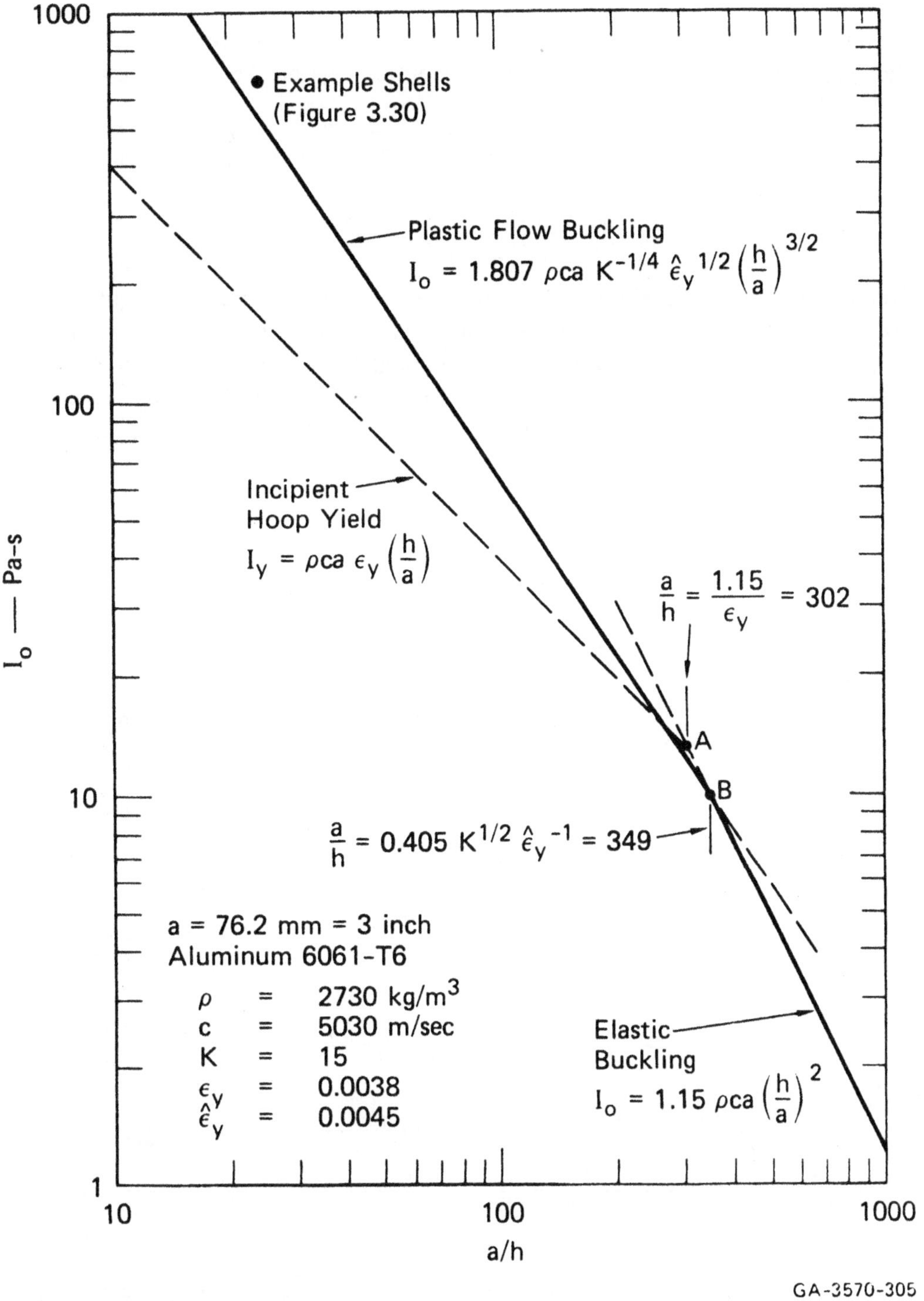


FIGURE 3.32 CRITICAL IMPULSE FOR THRESHOLD BUCKLING FROM SYMMETRIC RADIAL IMPULSE

### 3.4.6 Buckling With a Cosine Impulse Distribution

In many practical problems impulses are imparted asymmetrically around the shell circumference. One asymmetric distribution that has been studied is

$$I = \begin{cases} I_c \cos \theta, & -\frac{\pi}{2} \leqslant \theta \leqslant \frac{\pi}{2} \\ 0 & \frac{\pi}{2} < \theta < \frac{3\pi}{2} \end{cases} \qquad (3.4.37)$$

Profile recordings of shells buckled from this loading distribution are given in Figure 3.33. The shells are of the same material and dimensions as those in Figure 3.30 for symmetric loads. Buckling is confined to the area near $-60 < \theta < 60$ degrees and is of the same form as in Figure 3.30. Based on this observation, buckling thresholds for this loading are calculated by the same basic procedure just outlined for symmetric loads.

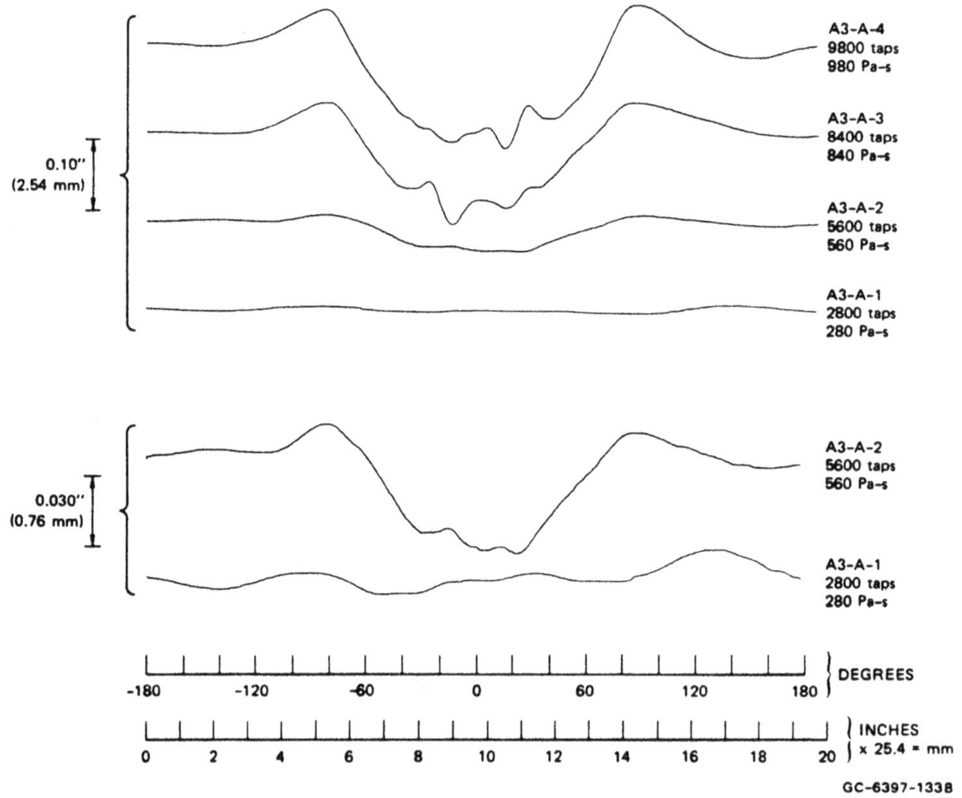


FIGURE 3.33 BUCKLE PROFILES AT SEVERAL MAGNITUDES OF COSINE IMPULSE

Same type shells as in Figure 3.30.

The first step is to calculate the membrane solution for $\sigma_\theta$ and $\epsilon$ as functions of time under the peak load by using a finite element code. These quantities are then entered into Equation (3.4.9) under the assumption that the wavelengths are short enough that $\sigma_\theta$ and $\epsilon$ do not vary significantly over a few buckling wavelengths. This condition was borne out by results of the example that follows. Calculations were also made for symmetric impulse so that the ratio of critical cosine and symmetric impulses could be plotted aginst a/h in Figure 3.34. For relatively thick shells ($20 < a/h < 200$), in which buckling takes place during plastic flow, the peak cosine impulse that causes buckling is only about 20% greater than the symmetric impulse. This is consistent with the experimental results in Figures 3.30 and 3.33. The uniform impulse at a peak buckle amplitude of $w_n/h = 0.2$ is 700 Pa-s. The cosine impulse for this same amplitude is 840 Pa-s, an increase of only 20%. A similar comparison can be made at 390 and 560 Pa-s, a load increment of about 50%. These comparisons are limited, but the result is clear: the critical asymmetric load is only slightly larger than the critical symmetric load for plastic flow buckling.

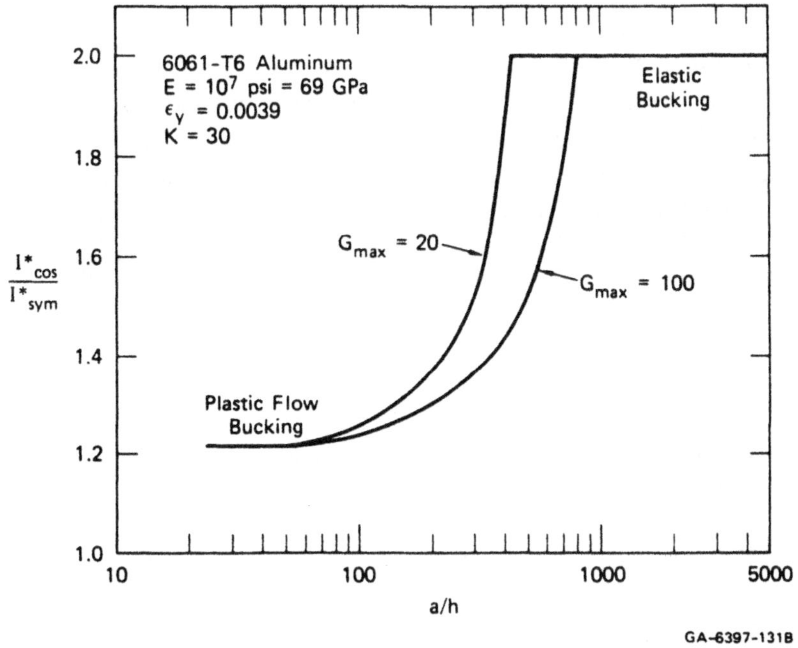


FIGURE 3.34 RATIO OF CRITICAL PEAK COSINE AND SYMMETRIC IMPULSE VERSUS RADIUS-TO-THICKNESS RATIO

As the shells become thinner, however, the ratio between the cosine and symmetric critical loads in Figure 3.34 increases abruptly with a/h until at a/h > 700 the buckling is entirely elastic for either distribution and the ratio is 2. The elastic ratio can be calculated directly by using the membrane solution given by Payton[10] for the cosine load defined by Equation (3.4.37). From Payton's results, the first inward compressive strain under the peak load, which causes buckling, is closely approximated by

$$\epsilon_{\cos}(\tau) = \frac{\sqrt{2}}{2}\frac{I_{\cos}}{\rho c h}\sin\sqrt{2}\,\tau \quad , \qquad 0 \leqslant \sqrt{2}\,\tau \leqslant \pi \qquad (3.4.38)$$

The strain for a shell under symmetric impulse is, from Equation (3.3.29),

$$\epsilon_{sym}(\tau) = \frac{I_{sym}}{\rho c h}\sin\tau \qquad (3.4.39)$$

According to our procedure for examining buckling only under the peak load, the equations of motion for the flexural modes are identical for either case and are given by Equation (3.3.20) with the appropriate forcing terms from Equation (3.4.38) or Equation (3.4.39). The membrane and flexural responses are therefore physically similar and a dimensional analysis can be performed in which the neighborhood of the cosine-loaded shell near the peak impulse is taken to be equivalent to a symmetrically loaded shell of the same thickness but having a different radius and applied impulse. The equivalent radius $a_e$ and impulse $I_e$ are determined so that the duration and peak strain of the inward motion in the equivalent shell are the same as in the cosine-loaded shell.

The hoop strain in the equivalent shell is, from Equation (3.4.39),

$$\epsilon_e = \frac{I_e}{\rho c h}\sin\tau_e \quad , \qquad \tau_e = \frac{ct}{a_e} \qquad (3.3.40)$$

If the period in this shell is to be the same as in Equation (3.4.38) for the cosine load, then $\sqrt{2}\tau = \tau_e$, or

$$\frac{\sqrt{2}\;ct}{a} = \frac{ct}{a_e} \qquad (3.4.41)$$

which gives $a_e = a/\sqrt{2}$. Similarly, if the peak strain in the equivalent shell is to be the same as in Equation (3.4.38), then

$$\frac{\sqrt{2}}{2}\frac{I_{\cos}}{\rho c h} = \frac{I_e}{\rho c h} \qquad (3.4.42)$$

which gives $I_e = I_{\cos}/\sqrt{2}$.

Since the equivalent shell is under symmetric impulse, we can use Equation (3.3.73) to find its critical buckling impulse $I_e^*$. (The asterisk is used in this discussion to denote critical values for threshold buckling.) This yields

$$I_e^* = 1.15\ \rho c a_e \left(\frac{h}{a_e}\right)^2 = 1.15\ \rho c a \left(\frac{h}{a}\right)^2 \cdot \sqrt{2} = I_{sym}^* \cdot \sqrt{2} \qquad (3.4.43)$$

where $I_{sym}^*$ is the critical impulse for the original shell under symmetric impulse. Also, we have shown that $I_e = I_{cos}/\sqrt{2}$ in general, so that in particular $I_e^* = I_{cos}^*/\sqrt{2}$. Substituting this into Equation (3.4.43) yields the desired relation between critical cosine impulse and symmetric impulse in the original ring:

$$I_{cos}^* = 2I_{sym}^* \qquad (3.4.44)$$

This simple analogy can also be used to calculate the radius-to-thickness ratios over which $I_{cos}^*/I_{sym}^*$ increases abruptly to 2.0. This increase is caused by the change from elastic to plastic-flow buckling and is therefore related to the yield strain $\epsilon_y$. For symmetric impulse, we have already shown that buckling becomes entirely elastic at $a/h = 1.15/\epsilon_y$, which gives $a/h = 288$ for $\epsilon_y = 0.004$. The asymmetric impulse that produces first membrane yield, from Equation (3.4.38), is $I_{cos}^* = \sqrt{2}\ \rho c h \epsilon_y$. Equating this to the critical cosine impulse for buckling gives $\sqrt{2}\ \rho c h \epsilon_y = 2.30\ \rho c a (h/a)^2$ or $a/h = 1.63/\epsilon_y$. For $\epsilon_y = 0.004$, $a/h = 406$. Thus, for $288 < a/h < 406$, the shell under cosine impulse buckles partly during elastic motion and partly during plastic flow, while the shell under symmetric impulse buckles entirely elastically. Because the tangent modulus becomes much smaller during plastic flow, the growth during the plastic motion is large compared with elastic growth and hence reduces the ratio of critical cosine and symmetric impulse. When both the cosine and symmetrically loaded rings buckle plastically, the ratio is still lower, but this must be explained by the peculiarities of the plastic motion.

Because this transition between elastic and plastic buckling under symmetric and cosine loading takes place over a narrow range of a/h, the curve in Figure 3.34 is very steep as $I_{cos}^*/I_{sym}^*$ approaches 2. Also, the transition depends on the choice for $G_{max}$ in the buckling criterion, i.e., on the amplitude of imperfections in the shell and the amplitude of acceptable buckle deformations. This is shown by the second curve in Figure 3.34, calculated for $G_{max} = 100$. The relationship between buckling from symmetric and asymmetric loads is therefore quite tenuous for shells in the range $300 < a/h < 700$.

### 3.4.7 Effects of Strain Rate Reversal

All the preceding critical load theory for plastic-flow buckling neglects the effects of inevitable strain rate reversal as the hoop motion ceases and bending tensile strains overtake the hoop compressive strain. To explore these effects, we analyzed[11] the shells of Figure 3.30 with the numerical code SABOR/DRASTIC 6. In the numerical analysis, responses in modes n = 20 and n = 30 were calculated for a symmetric impulse I = 840 Pa-s with initial velocity imperfections $\gamma_n = 0.01$, that is, with initial velocities in the $n^{th}$ harmonic equal to 1% of the initial symmetric radial velocity, as defined by Equation (3.4.12). The mode n = 20 corresponds to the average wavelength of buckles with appreciable amplitude in the shells of Figure 3.30. The mode n = 30 is near the most amplified mode found by using Equations (3.4.8) and (3.4.9) from the simple theory without strain rate reversal, again with initial velocity imperfections $\gamma_n = 0.01$ in Equation (3.4.12).

Results of these calculations are given in Figure 3.35, in which buckle amplitude is plotted against time for each of the four calculations. At early times, before strain-rate reversal or significant transfer of hoop energy to buckling, the results from the simple theory and from SABOR/DRASTIC 6 are reasonably close. As time increases, results from the code and from the simple theory diverge. As expected, the simple theory overestimates the buckle amplitude, both because the energy transferred to buckling is neglected and because strain-rate reversal tends to increase the bending stiffness and therefore resistance to buckling. The decrease in time $t_f$ at the end of the buckle motion [defined by $\dot{u}_o(t_f) = 0$] in the code result as compared with the simple theory is a direct result of the neglected extraction of buckle energy from the hoop motion.

The more amplified mode number from the code result is n = 20 rather than n = 30, in closer agreement with the experiment. The preference for lower mode numbers in the more complete theory in the code is attributed to the sharper curvature and hence greater bending in the higher mode number. With strain-rate reversal included, this curvature leads to unloading elastically and reloading in tension, thus quickly retarding the motion, as indicated by the leveling out of the curves from the code calculations. This leveling out also occurs for the lower mode number n = 20, but at a later time and to a lesser extent.

The simple theory and code results from n = 20 do not differ greatly until the very final motion, and then differ only by a factor of about 2 in buckle deformation. The large error, by almost an order of magnitude, occurs when the effect of the wrong mode number selection n = 30 is superimposed on the wrong terminal motion. Both effects, of course, are caused by the error in terminal motion calculation.

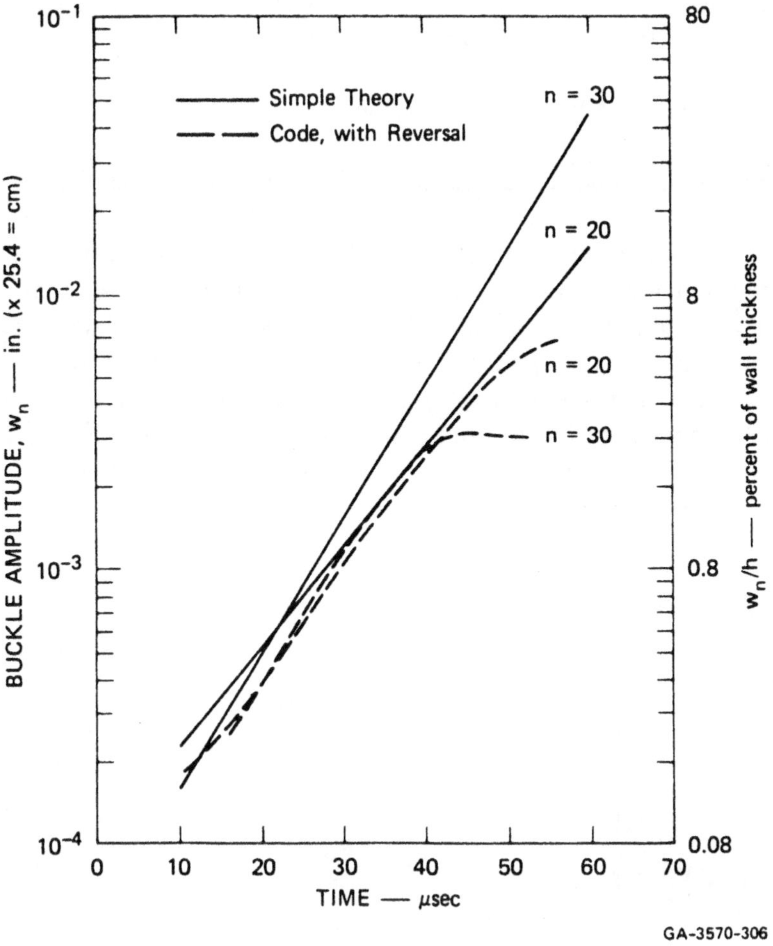


FIGURE 3.35 COMPARISON OF BUCK'.E GROWTH WITH AND WITHOUT STRAIN RATE REVERSAL

Nevertheless, because the buckle amplitudes increase exponentially with the time (initially, at least) and impulse, and because in the simple theory equivalent imperfections are tied to experiment, these errors in amplitude have little effect on the simple theory threshold buckling predictions. Thus the simple theory critical impulse formulas give useful estimates for practical applications. The results in Figure 3.35 show that, for plastic-flow buckling, actual imperfections are probably several times larger than the values used in the simple theory.

## 3.5 DYNAMIC PULSE BUCKLING OF CYLINDRICAL SHELLS FROM TRANSIENT RADIAL PRESSURE[8]

### 3.5.1 Approach and Equations of Motion

Examples of cylindrical shells buckled in the more general case of transient radial pressures of varying duration are shown in Figure 3.36. The first shell was subjected to an impulsive pressure as in the problems just treated and is buckled into a high-order wave pattern with n = 45 waves around the circumference. The next shell was subjected to a load of lower pressure but longer duration and is buckled into a lower-order pattern with buckles corresponding to n = 13. In the last shell, the pressure was reduced still further, and the duration was increased to a value much longer than the response time of the shell. This shell is buckled with wavelengths corresponding to n = 7, close to the n = 6 pattern for static radial pressure uniform around the shell.

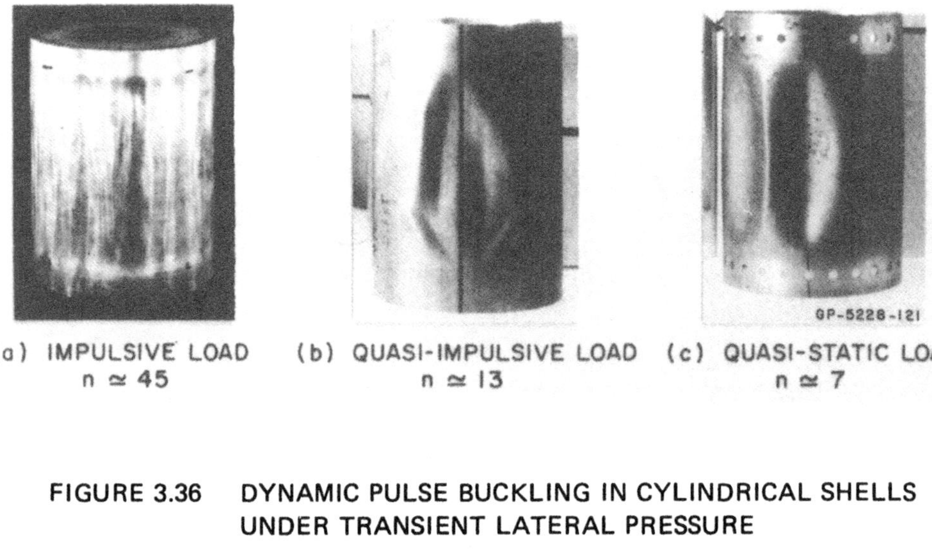


(a) IMPULSIVE LOAD $n \simeq 45$ (b) QUASI-IMPULSIVE LOAD $n \simeq 13$ (c) QUASI-STATIC LOAD $n \simeq 7$

FIGURE 3.36 DYNAMIC PULSE BUCKLING IN CYLINDRICAL SHELLS UNDER TRANSIENT LATERAL PRESSURE

Aluminum 6061-T6, a/h = 100, L/2a = 1.

A general theory to cover this entire range of buckling would be very difficult because end effects must be included in addition to the plastic flow already considered for short duration loads. Buckling under long duration loads depends directly on the length of the shell. Near each extreme in duration, however, simplifying assumptions can be made to remove one complication or the other. For loads of short duration, plastic effects must be treated, but the buckle wavelengths are short enough that the end conditions can be neglected. We therefore simply allow finite pulse durations in the equations already used to treat ideal impulse. At very long durations, the added complexity of having to include the end conditions is offset by the buckling's being elastic; allowing plastic flow is out of the question for static loads. Critical buckling loads for intermediate duration are estimated by extrapolation of the results from each extreme.

The two simplified theoretical models are therefore called the tangent modulus model and the elastic model. The equations of motion for the tangent modulus model are those already given by Equations (3.4.8) and (3.4.9), with initial conditions from Equation (3.4.10). For the elastic model we use shallow-shell Donnell equations for the finite length shell shown in Figure 3.37.

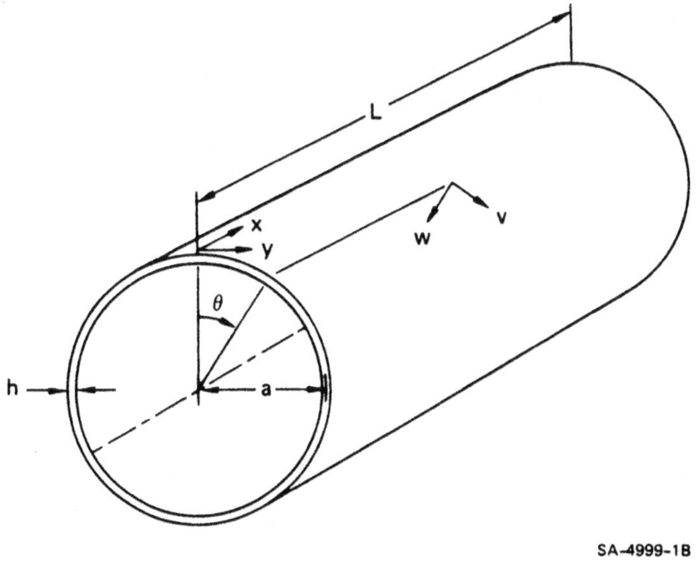


FIGURE 3.37 COORDINATES AND SHELL NOMENCLATURE

The Donnell equations are derived in Section 3.5.2. Then static buckling solutions with these equations are obtained in Section 3.5.3. Dynamic solutions for both the Donnell elastic model and the tangent modulus model are obtained in Section 3.5.4 for loads of decreasing duration and increasing peak pressure. Results are presented as plots of peak pressure versus impulse at critical combinations to produce buckling over the entire range of load durations. In Section 3.5.5 simple formulas are given for these curves. Experimental results and comparisons with theory are given in Section 3.5.6.

### 3.5.2 Donnell Equations for Elastic Buckling

The Donnell equations are simple and compact equations because they take advantage of simplifying approximations that can be made for shell bending in high-order modes, with $n \geqslant 4$. In these modes, each section of shell between nodes behaves as a shallow shell, that is, a shell with a span much larger than its height. For $n = 4$, the arc length between nodes is $360/8 = 45$ degrees, with a span of $L = 2a \sin 22.5°$ and a height $H = a(1 - \cos 22.5°)$. The height-to-span ratio is 0.10. This arc length is shown in Figure 3.38. For a shallow shell, the equilibrium equation is simply that of a flat plate with the addition of a term because of the shell curvature. For the hoop mode, $n = 0$, these equations reduce to the same simple form as for more complete thin shell equations. Thus, the equations are valid for $n = 0$ and for $n \geqslant 4$, which are the modes of concern in the analyses that follow.

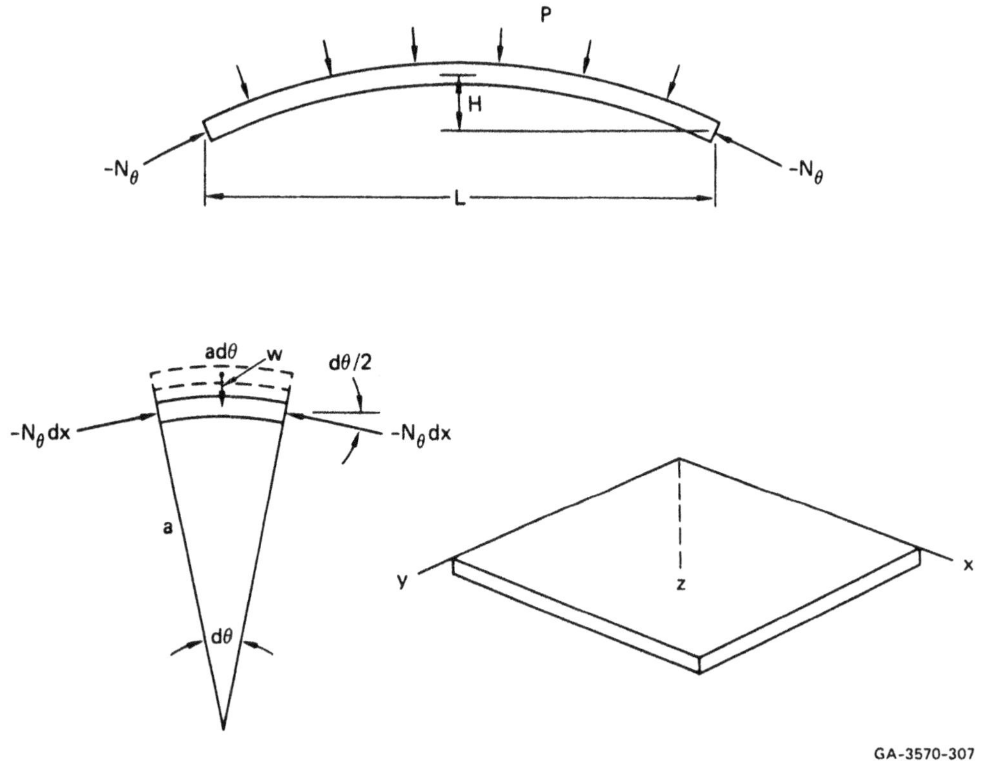


FIGURE 3.38 ELEMENTS OF SHALLOW SHELL THEORY

The added force normal to the shell wall because of curvature is found from equilibrium of the shell element $ad\theta$ in Figure 3.38:

$$F_z = 2N_\theta dx \frac{d\theta}{2} = N_\theta dx \frac{dy}{a} \qquad (3.5.1)$$

in which we have used $dy = ad\theta$. The stress resultant $N_\theta$ is taken positive in tension to conform with conventional plate theory. The element also shows the circumferential strain associated with radial displacement,

$$\epsilon_\theta = \frac{a d\theta - (a + w) d\theta}{a d\theta} = -\frac{w}{a} \tag{3.5.2}$$

The remainder of the Donnell equations come from plate theory. Thus, we first derive the equations for lateral motion for a plate with coordinates as shown in the lower right sketch of Figure 3.38. The shell terms of Equations (3.5.1) and (3.5.2) will be added later.

Moment equilibrium of the plate element in Figure 3.39 about an axis parallel to the x axis gives

$$\left(M_y + \frac{\partial M_y}{\partial y} dy\right) dx - M_y dx + \left(M_{xy} + \frac{\partial M_{xy}}{\partial x} dx\right) dy - M_{xy} dy - Q_y dx dy = 0$$

which simplifies to

$$\frac{\partial M_y}{\partial y} + \frac{\partial M_{xy}}{\partial x} = Q_y \tag{3.5.3}$$

Similarly, moment equilibrium about an axis parallel to the y axis requires that

$$\frac{\partial M_x}{\partial x} + \frac{\partial M_{yx}}{\partial y} = Q_x \tag{3.5.4}$$

and force equilibrium normal to the element requires that

$$\frac{\partial Q_x}{\partial x} + \frac{\partial Q_y}{\partial y} = -P - \rho h \frac{\partial^2 w}{\partial t^2} \tag{3.5.5}$$

With $Q_x$ and $Q_y$ from Equations (3.5.3) and (3.5.4) and $M_{xy} = M_{yx}$, the force equilibrium equation (3.5.5) becomes

$$\frac{\partial^2 M_x}{\partial x^2} + 2\frac{\partial^2 M_{xy}}{\partial x \partial y} + \frac{\partial^2 M_y}{\partial y^2} = -P + \rho h \frac{\partial^2 w}{\partial t^2} \tag{3.5.6}$$

To find the relationships between these moments and the displacement w, we use Hooke's law and the kinematics of deformation. To a good approximation $\sigma_z = 0$, so we use Hooke's law for plane stress.

$$\epsilon_x = \frac{1}{E}(\sigma_x - \nu\sigma_y) \quad , \qquad \epsilon_y = \frac{1}{E}(\sigma_y - \nu\sigma_x)$$

$$\gamma_{xy} = \frac{1}{G}\tau_{xy} = \frac{2(1 + \nu)}{E}\tau_{xy} \tag{3.5.7}$$

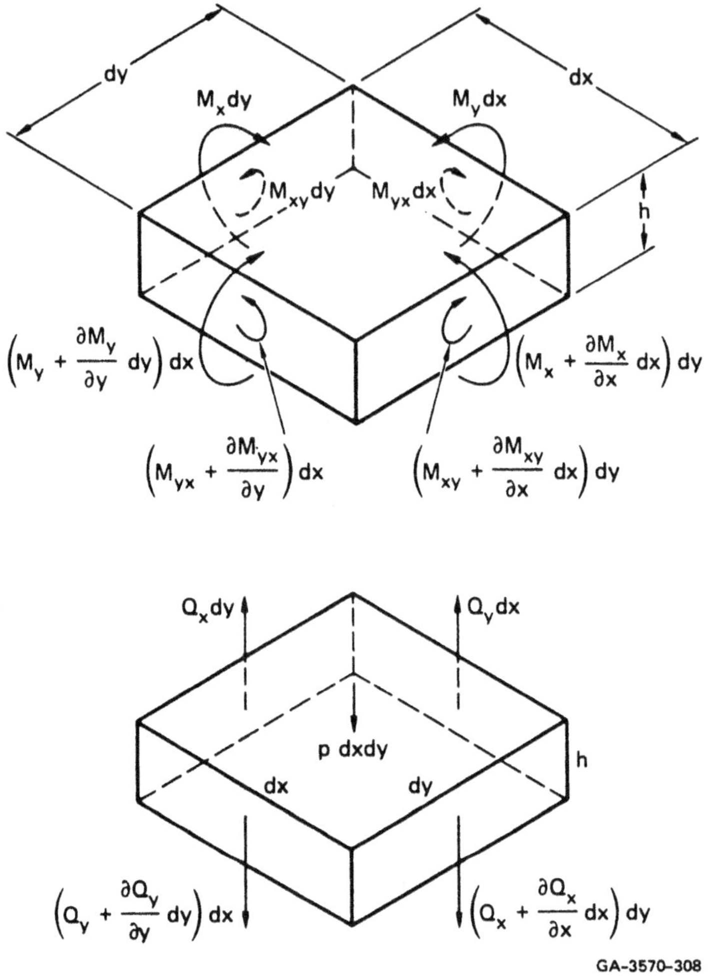


FIGURE 3.39 MOMENT AND FORCE EQUILIBRIUM OF A PLATE ELEMENT

Consider now the bending of an element acted on only by a bending moment $M_x$ as shown in Figure 3.40a. For this form of displacement over lengths large with respect to the plate thickness h, the strain $\epsilon_y$ in the y direction is zero. With $\epsilon_y = 0$ in Equation (3.5.7) we have

$$\epsilon_y = 0 \quad , \qquad \sigma_y = \nu \sigma_x \tag{3.5.8}$$

and therefore

$$\sigma_x = E\epsilon_x + \nu^2 \sigma_x = \frac{E}{1 - \nu^2} \epsilon_x \tag{3.5.9}$$

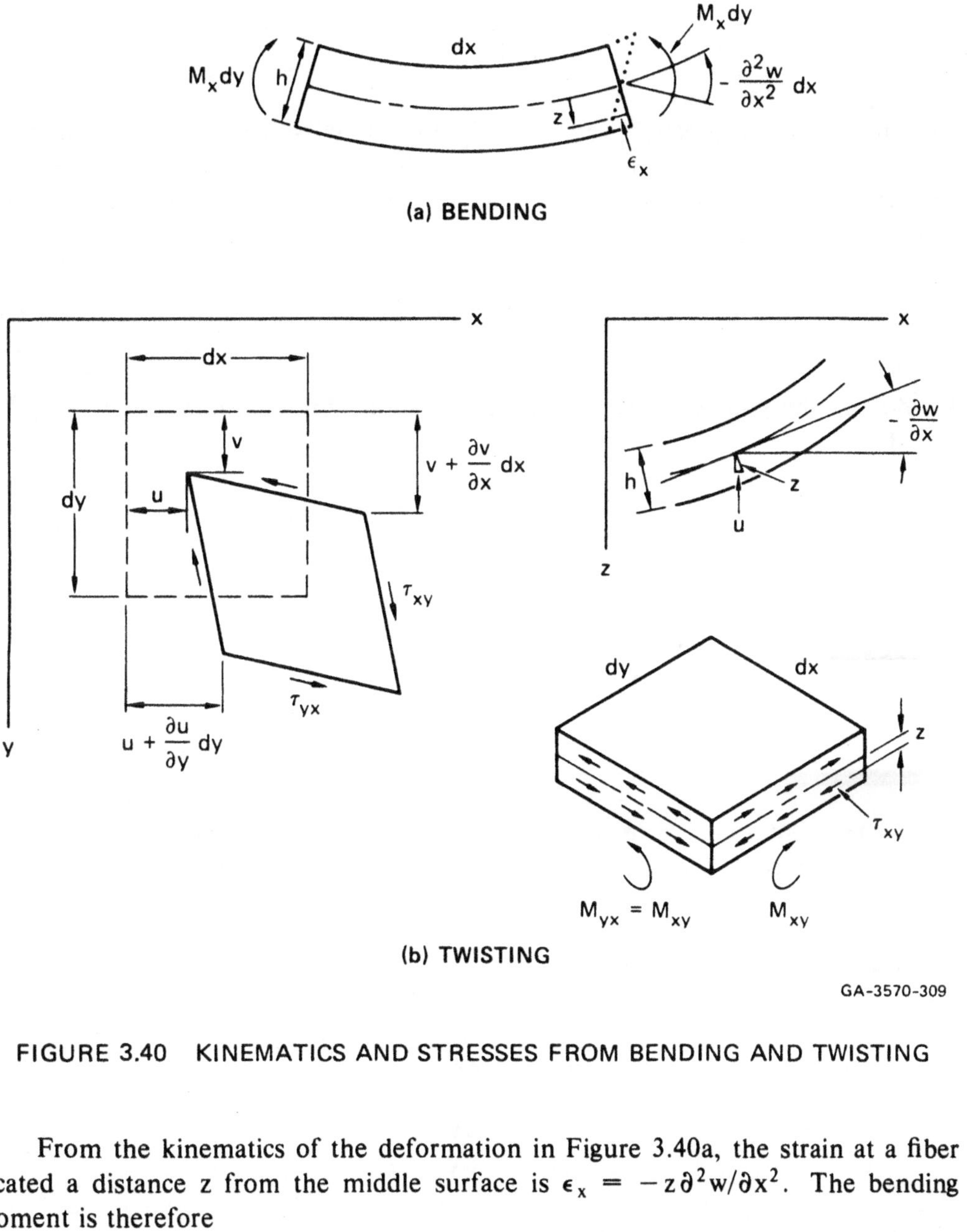


FIGURE 3.40 KINEMATICS AND STRESSES FROM BENDING AND TWISTING

From the kinematics of the deformation in Figure 3.40a, the strain at a fiber located a distance z from the middle surface is $\epsilon_x = -z\partial^2 w/\partial x^2$. The bending moment is therefore

$$M_x\, dy = -\int_{-h/2}^{h/2} z \cdot \frac{E}{1-\nu^2}\, z\, \frac{\partial^2 w}{\partial x^2}\, dz\, dy$$

$$M_x = \frac{-E}{1-\nu^2}\,\frac{\partial^2 w}{\partial x^2}\,\frac{h^3}{12} = -D\,\frac{\partial^2 w}{\partial x^2} \qquad (3.5.10)$$

where

$$D = \frac{Eh^3}{12(1 - \nu^2)} \tag{3.5.11}$$

The stress in the y direction, caused by the Poisson effect and the zero strain condition, also gives rise to a bending moment. With $\sigma_y = \nu\sigma_x$ from Equation (3.5.8), this moment is of the same form as in Equation (3.5.10), so we can write immediately

$$M_y = -\nu D \frac{\partial^2 w}{\partial x^2} \tag{3.5.12}$$

Bending in the y direction is similar. The total bending-curvature relationships are therefore

$$M_x = -D \left( \frac{\partial^2 w}{\partial x^2} + \nu \frac{\partial^2 w}{\partial y^2} \right) \tag{3.5.13}$$

$$M_y = -D \left( \frac{\partial^2 w}{\partial y^2} + \nu \frac{\partial^2 w}{\partial x^2} \right) \tag{3.5.14}$$

The relationship between the twisting moment and shear deformation is shown in Figure 3.40b. The originally rectangular element with sides dx and dy represents a thin sheet of material at a distance z from the middle surface of the plate. In-plane displacements are u and v in the x and y directions, respectively. The shear strain of this element is

$$\begin{aligned} \gamma_{xy} &= \frac{\partial u}{\partial y} dy \cdot \frac{1}{dy} + \frac{\partial v}{\partial x} dx \cdot \frac{1}{dx} \\ &= \frac{\partial u}{\partial y} + \frac{\partial v}{\partial x} \end{aligned} \tag{3.5.15}$$

The corresponding shear stress is

$$\tau_{xy} = G \left( \frac{\partial u}{\partial y} + \frac{\partial v}{\partial x} \right) \tag{3.5.16}$$

The section view through the plate along a plane parallel to the x axis shows that a rotation of the middle surface through an angle $-\partial w/\partial x$ produces a displacement at z equal to

$$u = -z\frac{\partial w}{\partial x} \tag{3.5.17}$$

Similarly, the v displacement is

$$v = -z\frac{\partial w}{\partial y} \tag{3.5.18}$$

With these expressions for u and v, the shear stress in Equation (3.5.16) is given by

$$\tau_{xy} = -2Gz\frac{\partial^2 w}{\partial x \partial y} \tag{3.5.19}$$

Integration of this shear stress through the wall thickness gives the twisting moment.

$$M_{xy} = \int_{-h/2}^{h/2} z\tau_{xy}dz = -\frac{Gh^3}{6}\frac{\partial^2 w}{\partial x\partial y} = -D(1-\nu)\frac{\partial^2 w}{\partial x\partial y} \tag{3.5.20}$$

Equations (3.5.13), (3.5.14), and (3.5.15) for bending and twisting moments are now substituted into equilibrium Equation (3.5.6) to obtain an expression involving only the lateral displacement w and load P.

$$-D\frac{\partial^2}{\partial x^2}\left(\frac{\partial^2 w}{\partial x^2} + \nu\frac{\partial^2 w}{\partial y^2}\right) - 2D(1-\nu)\frac{\partial^4 w}{\partial x^2\partial y^2} - D\frac{\partial^2}{\partial y^2}\left(\frac{\partial^2 w}{\partial y^2} + \nu\frac{\partial^2 w}{\partial x^2}\right)$$

$$= -P + \rho h\frac{\partial^2 w}{\partial t^2}$$

After simplification, this becomes

$$D\frac{\partial^4 w}{\partial x^4} + 2D\frac{\partial^4 w}{\partial x^2\partial y^2} + D\frac{\partial^4 w}{\partial y^4} + \rho h\frac{\partial^2 w}{\partial t^2} = P$$

or

$$D\nabla^4 w + \rho h\frac{\partial^2 w}{\partial t^2} = P \tag{3.5.21}$$

where

$$\nabla^4 = \nabla^2\nabla^2 = \left(\frac{\partial^2}{\partial x^2} + \frac{\partial^2}{\partial y^2}\right)^2 \tag{3.5.22}$$

This is the dynamic equilibrium equation for lateral motion of a thin, flat plate with no force resultants in the plane of the plate.

To treat buckling, we must also include thrust in the plane of the plate, just as we included thrust along the axis of a bar and circumferential hoop forces in a ring or long shell. We must also include in-plane forces that arise because of middle surface stretching and compression as the shell bends in the presence of its initial curvature [of radius a, see Equation (3.5.2)]. The general formulation for in-plane forces is first introduced for the flat plate, and then the shell terms of Equations (3.5.1) and (3.5.2) are added under the assumption that the region of the shell over which the equations are applied is shallow, as discussed earlier. In the general formulation, in-plane forces are taken positive in tension, according to the convention of plate theory. When in-plane compression is applied by external loads, these are introduced as negative quantities so that buckling thrusts are positive quantities, consistent with the convention throughout this text.

Equilibrium of in-plane stress resultants in the x direction in the lower sketch in Figure 3.41 requires that

$$\frac{\partial N_x}{\partial x} + \frac{\partial N_{xy}}{\partial y} = 0 \qquad (3.5.23)$$

Equilibrium in the y direction requires

$$\frac{\partial N_y}{\partial y} + \frac{\partial N_{yx}}{\partial x} = 0 \qquad (3.5.24)$$

From the upper sketch, the $N_x$-related forces introduce a net force in the z direction because of the small slope of the plate element and the small change in slope from one side of the element to the other. The net force in the z direction is

$$\left(N_x + \frac{\partial N_x}{\partial x} dx\right) dy \left(\frac{\partial w}{\partial x} + \frac{\partial^2 w}{\partial x^2}\right) dx - N_x dy \frac{\partial w}{\partial x}$$

With terms higher than second order omitted because they will vanish in the limiting process, this simplifies to

$$N_x \frac{\partial^2 w}{\partial x^2} dx\,dy + \frac{\partial N_x}{\partial x}\frac{\partial w}{\partial x} dx\,dy \qquad (3.5.25)$$

A similar argument for $N_y$-related forces leads to a net force in the z direction equal to

$$N_y \frac{\partial^2 w}{\partial y^2} dx\,dy + \frac{\partial N_y}{\partial y}\frac{\partial w}{\partial y} dx\,dy \qquad (3.5.26)$$

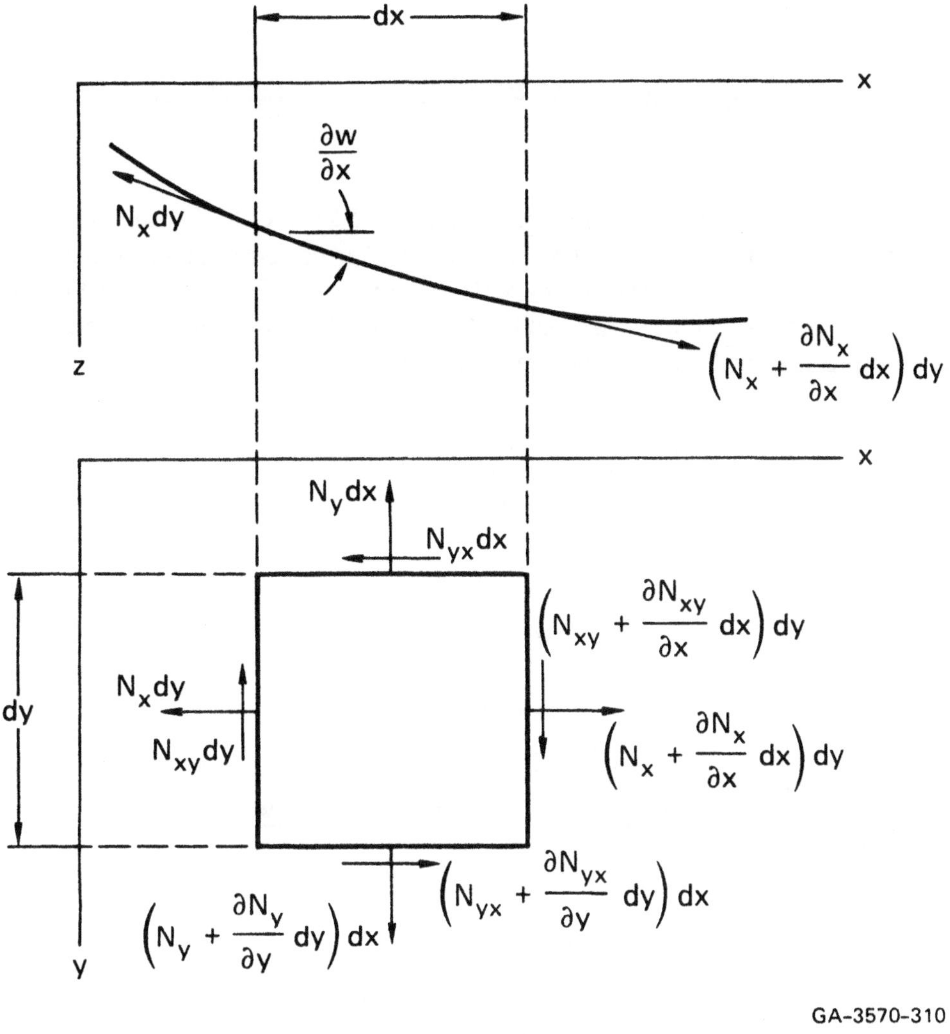


FIGURE 3.41 EQUILIBRIUM OF IN-PLANE STRESS RESULTANTS

The shear force $N_{xy}dy$ and its incremented counterpart acting on the opposite side of the element also give a net force in the z direction because they are acting at angles $\partial w/\partial y$ and $\partial w/\partial y + (\partial^2 w/\partial x \partial y)dx$, respectively. After simplification, this net force is given by

$$N_{xy} \frac{\partial^2 w}{\partial x \partial y} dx\, dy + \frac{\partial N_{xy}}{\partial x} \frac{\partial w}{\partial y} dx\, dy$$

The force $N_{yx}dx$ and its counterpart result in another force in the z direction given by an analogous expression. The total force in the z direction from shear (noting that $N_{yx} = N_{xy}$) is

$$2N_{xy}\frac{\partial^2 w}{\partial x\partial y}dx\,dy + \frac{\partial N_{xy}}{\partial x}\frac{\partial w}{\partial y}dx\,dy + \frac{\partial N_{xy}}{\partial y}\frac{\partial w}{\partial x}dx\,dy \qquad (3.5.27)$$

The sum of the forces from expressions (3.5.25), (3.5.26), and (3.5.27) is found to involve only the curvature and twisting terms, because the force system acting at a constant slope vanishes by virtue of the in-plane equilibrium Equations (3.5.23) and (3.5.24). This sum is added to the load P dxdy acting on the element. Equilibrium Equation (3.5.6) is then replaced by

$$\frac{\partial^2 M_x}{\partial x^2} + 2\frac{\partial^2 M_{xy}}{\partial x\partial y} + \frac{\partial^2 M_y}{\partial y^2} =$$

$$-\left(P + N_x\frac{\partial^2 w}{\partial x^2} + 2N_{xy}\frac{\partial^2 w}{\partial x\partial y} + N_y\frac{\partial^2 w}{\partial y^2}\right) + \rho h\frac{\partial^2 w}{\partial t^2} \qquad (3.5.28)$$

Similarly, Equation (3.5.21) is replaced by

$$D\nabla^4 w + \rho h\frac{\partial^2 w}{\partial t^2} = P + N_x\frac{\partial^2 w}{\partial x^2} + N_{xy}\frac{\partial^2 w}{\partial x\partial y} + N_y\frac{\partial^2 w}{\partial y^2} \qquad (3.5.29)$$

This is the flat plate equation including in-plane forces. It is modified into the shallow cylindrical shell equation by adding the curvature term $N_\theta/a$ from Equation (3.5.1) and changing to the x,$\theta$ shallow-shell coordinate notation shown in Figure 3.37, with $\partial y = a\partial\theta$. The result is immediately written

$$D\nabla^4 w - N_x\frac{\partial^2 w}{\partial x^2} - 2N_{x\theta}\frac{\partial^2 w}{a\partial\theta\partial x} - N_\theta\frac{\partial^2 w}{a^2\partial\theta^2} - \frac{N_\theta}{a}$$

$$+ \rho h\frac{\partial^2 w}{\partial t^2} = P \qquad (3.5.30)$$

It is now convenient to separate in-plane forces that result directly from applied loads and in-plane forces that arise from stretching as the shell deforms in the flexural modes. Thus, we define

$$
\begin{aligned}
N_x &= -\bar{N}_x + N_x^* \\
N_\theta &= -\bar{N}_\theta + N_\theta^* \\
N_{x\theta} &= -\bar{N}_{x\theta} + N_{x\theta}^*
\end{aligned}
\qquad (3.5.31)
$$

where the barred quantities are from applied loads and the starred quantities are from flexural stretching. As mentioned previously, the applied load terms are defined such that compression is positive.

The stretching terms are found by applying the compatibility condition to the middle surface strains. These strains are

$$
\epsilon_x = \frac{\partial u}{\partial x} \quad , \quad \epsilon_\theta = \frac{\partial v}{a\partial\theta} - \frac{w}{a} \quad , \quad \gamma_{x\theta} = \frac{\partial u}{a\partial\theta} + \frac{\partial v}{\partial x} \qquad (3.5.32)
$$

The terms from in-plane middle surface displacements u and v are analogous to those for a general sheet element within a plate or shallow-shell wall, as given in Figure 3.40. The additional circumferential strain $-w/a$, from Equation (3.5.2), occurs because of the initial shell curvature as shown in Figure 3.38. For a plate, this term is absent and the three strains in Equations (3.5.32) are defined by only the two displacements u and v. Hence the strains must be related in a special way so that all three of Equations (3.5.32) are satisfied. This relationship is the familiar compatibility condition of plane elasticity. Similarly, when the shell term $w/a$ is added (w is specified by the flexure) the strains must again be related, but in another way. Thus, the compatibility condition becomes an expression for determining the middle surface strains caused by shell flexure.

The form of Equations (3.5.32) suggests immediately that the compatibility condition can be expressed by twice differentiating each of the equations, the first with respect to $\theta$, the second with respect to x, and the third once with respect to x and once with respect to $\theta$ (in each $\theta$ differentiation we multiply by $1/a$ to maintain $\partial y = a\partial\theta$). The first of these differentiated equations is added to the second, and then the third is used to obtain finally

$$
\frac{\partial^2 \epsilon_x}{a^2 \partial\theta^2} + \frac{\partial^2 \epsilon_y}{\partial x^2} = \frac{\partial^2 \gamma_{xy}}{a\partial x\partial\theta} - \frac{1}{a}\,\frac{\partial^2 w}{\partial x^2} \qquad (3.5.33)
$$

Since our objective is to find the in-plane stress resultants, this compatibility relation is now expressed in terms of these resultants. With Hooke's law, Equation

(3.5.33) becomes

$$\frac{\partial^2}{a^2\partial\theta^2}(\sigma_x - \nu\sigma_\theta) + \frac{\partial^2}{\partial x^2}(\sigma_\theta - \nu\sigma_x)$$

$$= 2(1 + \nu)\frac{\partial^2\tau_{xy}}{a\partial x\partial\theta} - \frac{E}{a}\frac{\partial^2 w}{\partial x^2} \tag{3.5.34}$$

The stress resultants from flexure are $N_x^* = h\sigma_x$, $N_\theta^* = h\sigma_\theta$, and $N_{x\theta}^* = h\tau_{xy}$. It is convenient to eliminate the shear stress resultant by using the in-plane equilibrium Equations (3.5.23) and (3.5.24). The first is differentiated with respect to x, the second with respect to y, and then the two are added. The resulting equation (with $\partial y = a\partial\theta$) is substituted into Equation (3.5.34) to obtain finally

$$\nabla^2(N_x^* + N_y^*) = -\frac{Eh}{a}\frac{\partial^2 w}{\partial x^2} \tag{3.5.35}$$

where

$$\nabla^2 = \frac{\partial^2}{\partial x^2} + \frac{\partial^2}{a^2\partial\theta^2}$$

Compatibility Equation (3.5.35) and the equilibrium Equations (3.5.23) and (3.5.24) are sufficient to determine $N_x^*$, $N_\theta^*$, and $N_{x\theta}^*$. The equilibrium equations are satisfied identically when the stress resultants are expressed in terms of a stress function F:

$$N_x^* = \frac{\partial^2 F}{a^2\partial\theta^2}, \quad N_\theta^* = \frac{\partial^2 F}{\partial x^2}, \quad N_{x\theta}^* = \frac{\partial^2 F}{a\partial\theta dx} \tag{3.5.36}$$

When these expressions are substituted into Equation (3.5.35), we find that to satisfy compatibility F must be a solution of

$$\nabla^4 F = -\frac{Eh}{a}\frac{\partial^2 w}{\partial x^2} \tag{3.5.37}$$

The complete Donnell equations consist of this equation plus shallow shell Equation (3.5.30) with in-plane stress resultants from Equations (3.5.31) and (3.5.36). We now use the separation of these resultants into two parts, $-\bar{N}$ and $N^*$. From Equations (3.5.36) and (3.5.37), we see that the stress resultants $N^*$ from flexure are linear in w. Thus, when they are introduced into Equation (3.5.30), all produce terms of second-order in w except for the term $N_\theta^*/a$. The second-order terms are small and are omitted. Then Equations (3.5.31) substituted into Equations (3.5.30) give the second Donnell equation

$$D\nabla^4 w + \bar{N}_x \frac{\partial^2 w}{\partial x^2} + 2\bar{N}_{x\theta} \frac{\partial^2 w}{a\partial\theta\partial x} + \bar{N}_\theta \frac{\partial^2 w}{a^2\partial\theta^2} + \frac{\bar{N}_\theta}{a}$$

$$- \frac{1}{a}\frac{\partial^2 F}{\partial x^2} + \rho h \frac{\partial^2 w}{\partial t^2} = P \qquad (3.5.38)$$

The direct-loading term $\bar{N}_\theta$ is the circumferential force from motion in the hoop mode. Thus, from Equation (3.5.2) for n = 0 we have

$$\bar{N}_\theta = h\bar{\sigma}_\theta = \frac{Eh}{(1 - \nu^2)}\frac{w_o}{a} \qquad (3.5.39)$$

expressed as positive in compression according to our convention.

To consider buckling from an initially imperfect shape described by $w_i(x,\theta)$, Equation (3.5.38) becomes

$$D\nabla^4 w + \bar{N}_x \frac{\partial^2}{\partial x^2}(w + w_i) + 2\bar{N}_{x\theta}\frac{\partial^2}{a\partial\theta\partial x}(w + w_i)$$

$$+ \bar{N}_\theta \frac{\partial^2}{a^2\partial\theta^2}(w + w_i) + \frac{\bar{N}_\theta}{a} - \frac{1}{a}\frac{\partial^2 F}{\partial x^2} + \rho h \frac{\partial^2 w}{\partial t^2} = P \qquad (3.5.40)$$

The shape $w_i$ does not enter into the term $D\nabla^4 w$ because this term represents lateral force arising from flexural stresses, and by definition the shell is unstressed in the initially imperfect shape $w_i$. Similarly, the introduction of $w_i$ does not affect Equation (3.5.37) for stress function F.

### 3.5.3 Fourier Series Solution--Static Buckling

We proceed now to obtain both static and dynamic solutions to the Donnell equations subject to boundary conditions of simple supports at the ends of the cylindrical shell. With simple supports, the displacement w and axial bending moment $M_x$ vanish at x = 0 and x = L. Thus,

$$w = \partial^2 w/\partial x^2 = 0 \quad \text{at} \quad x = 0, L \qquad (3.5.41)$$

The curvature condition follows from Equation (3.5.13) and from the observation that, at the boundary, w is zero at every point around the circumference, so $\partial^2 w/a^2\partial\theta^2 = \partial^2 w/\partial y^2 = 0$.

Conditions (3.5.41) are satisfied by the Fourier series

$$w(x,\theta,t) = w_o(t) + \sum_{m=1}^{\infty}\sum_{n=1}^{\infty} w_{mn}(t)\,\sin\frac{m\pi x}{L}\,\sin n\theta \tag{3.5.42}$$

$$w_i(x,\theta) = \sum_{m=1}^{\infty}\sum_{n=1}^{\infty} \delta_{mn}\,\sin\frac{m\pi x}{L}\,\sin n\theta \tag{3.5.43}$$

$$P(x,\theta,t) = P_o(t) + \sum_{m=1}^{\infty}\sum_{n=1}^{\infty} P_{mn}(t)\,\sin\frac{m\pi x}{L}\,\sin n\theta \tag{3.5.44}$$

$$F(x,\theta,t) = \sum_{m=1}^{\infty}\sum_{n=1}^{\infty} F_{mn}\,\sin\frac{m\pi x}{L}\,\sin n\theta \tag{3.5.45}$$

The form chosen for P is appropriate to satisfy Equation (3.5.40). Similarly, the form for F is appropriate to satisfy compatibility Equation (3.5.37) and hence also Equation (3.5.40). Each Fourier term in Equation (3.5.42) describes a deformed shape consisting of m half-waves in the axial direction and n full waves around the circumference of the shell. Addition of $\sin(m\pi x/L)\cos n\theta$ terms, which also satisfy the boundary conditions and differential equations, would cause no increase in difficulty.

Substitution of Equation (3.5.42) into Equations (3.5.41) confirms that the boundary conditions are satisfied term by term except for the hoop mode term $w_o(t)$. For this term we neglect the effect of the end supports and assume that the hoop motion is uniform along the length of the shell. This is equivalent to neglecting axial flexure and calculating hoop motion with simple membrane theory. This is a reasonable approximation and is commonly made in shell buckling theory. Flexural stresses in the hoop mode induced by the end supports die off exponentially with distance from the ends, the decrease being more rapid for thinner shells. This approximation gives a hoop thrust $\overline{N}_\theta$ uniform along the length of the shell, which is an acceptable approximation because buckling deformation near the ends of the shell is small and an error in $\overline{N}_\theta$ in this region is unimportant.

The Fourier expansions are now substituted into differential Equations (3.5.37) and (3.5.40) to obtain equations of motion for the modal amplitudes $w_o$ and $w_{mn}$. After substitution, all terms multiplying each shape function $\sin(m\pi x/L)\sin n\theta$ are gathered together and set to zero, because the shape functions are independent (orthogonal). That is, since no shape function can be expressed in terms of the others, the complete equation is satisfied only if the coefficient of each function is identically zero. Similarly, the group of terms multiplying unity (the hoop mode shape function) is also set to zero.

For the hoop mode, the result is

$$\frac{Eh}{(1-\nu^2)}\frac{w_o}{a} + \rho h \frac{\partial^2 w_o}{\partial t^2} = P_o(t)$$

or

$$\frac{\partial^2 w_o}{\partial t^2} + \frac{E}{(1-\nu^2)\rho a^2} w_o = \frac{P_o(t)}{\rho h} \tag{3.5.46}$$

This equation is simplified by introducing dimensionless quantities

$$\tau = \frac{ct}{a} \quad , \quad u = \frac{w}{a} \quad , \quad \hat{P} = \frac{(1-\nu^2)a}{Eh} P \tag{3.5.47}$$

where $c^2 = E/(1-\nu^2)\rho$. Then Equation (3.5.46) becomes

$$\ddot{u}_o + u_o = \hat{P}_o \tag{3.5.48}$$

where dots indicate differentiation with respect to $\tau$. These dimensionless quantities will be used in formulating the final dynamic equations of motion, but for the present we retain the original dimensional variables.

For the m,n terms, the coefficients $F_{mn}$ are first determined in terms of $w_{mn}$ from compatibility Equation (3.5.37), which yields

$$\left(-\frac{m^2\pi^2}{L^2} - \frac{n^2}{a^2}\right)^2 F_{mn} = -\frac{Eh}{a}\left(-\frac{m^2\pi^2}{L^2}\right) w_{mn}$$

It is convenient to introduce a circumferential wave number parameter $\beta = nL/\pi a$. Then this equation becomes

$$F_{mn} = \frac{Eh}{a}\frac{L^2}{\pi^2} m^2(m^2+\beta^2)^{-2} w_{mn} \tag{3.5.49}$$

Equations (3.5.42), (3.5.43), (3.5.44), and (3.5.49) are now substituted into Equation (3.5.40) to obtain

$$D\frac{\pi^4}{L^4}(m^2+\beta^2)^2 w_{mn} + \frac{Eh}{(1-\nu^2)}\frac{w_o}{a}\left(-\frac{n^2}{a^2}\right)(w_{mn}+\delta_{mn})$$

$$-\frac{1}{a}\left(-\frac{m^2\pi^2}{L^2}\right)\frac{Eh}{a}\frac{L^2}{\pi^2} m^2(m^2+\beta^2)^{-2} w_{mn} + \rho h \frac{\partial^2 w_{mn}}{\partial t^2} = P_{mn} \tag{3.5.50}$$

These are the dynamic equations to be used for loads of long and intermediate duration. In the extreme of loads with very long duration and no sudden changes, they yield the static buckling solution. In this extreme, derivatives with respect to time can be neglected. For static buckling, the hoop mode Equation (3.5.46) is then simply

$$P_o = \frac{Eh}{(1 - \nu^2)} \frac{w_o}{a} \tag{3.5.51}$$

This is now substituted into Equation (3.5.50). After dividing through by $D\pi^4/L^4$, the coefficient of $w_{mn}$ in the resulting expression is

$$(m^2 + \beta^2)^2 + \frac{Eh}{Da^2} \frac{L^4}{\pi^4} m^4(m^2 + \beta^2)^{-2} - \frac{P_o a}{D} \frac{L^2}{\pi^2} \beta^2 \tag{3.5.52}$$

The other terms involve the shape and load perturbations $\delta_{mn}$ and $P_{mn}$. No matter how small the perturbations are, if the above coefficient of $w_{mn}$ vanishes, the modal amplitudes $w_{mn}$ become unbounded. The values of static loading pressures $P_o$ at which expression (3.5.52) vanishes are eigenvalues for static buckling. The lowest value is the static buckling load.

These eigenvalues can be expressed in terms of the two combinations of shell parameters that appear as coefficients in expression (3.5.52). The first group is a dimensionless shell parameter

$$\frac{Eh}{Da^2} \frac{L^4}{\pi^4} = \frac{12(1 - \nu^2)}{h^2 a^2} \frac{L^4}{\pi^4} = \frac{12}{\pi^4} Z^2 \tag{3.5.53}$$

where

$$Z = (1 - \nu^2)^{1/2} L^2/ah \tag{3.5.54}$$

The second group is a dimensionless pressure parameter

$$k_\theta = \frac{P_o a L^2}{D\pi^2} = \frac{\sigma_\theta h L^2}{D\pi^2} \tag{3.5.55}$$

The subscript $\theta$ indicates that the loading is lateral pressure and hence produces middle surface stress $\sigma_\theta$, in the $\theta$ direction. In terms of these parameters, the equation for critical pressure, found by setting expression (3.5.52) equal to zero, is

$$k_\theta = \frac{(m^2 + \beta^2)^2}{\beta^2} + \frac{12Z^2}{\pi^4} \frac{m^4}{\beta^2(1 + \beta^2)^2} \tag{3.5.56}$$

By inspection, $k_\theta$ is smallest when m is smallest. Hence, buckling modes of concern all have m = 1 half-wave in the axial direction and

$$k_\theta = \frac{(1 + \beta^2)^2}{\beta^2} + \frac{12Z^2}{\pi^4} \frac{1}{\beta^2(1 + \beta^2)} \tag{3.5.57}$$

The static buckling load is given by the minimum value of $k_\theta$ for a given shell parameter Z. Thus, the general static solution is found by minimizing Equation (3.5.57) with respect to $\beta$ and then expressing the minimum value of $k_\theta$ as a function of Z. For a given shell, of course, $\beta$ can take on only discrete values corresponding to integer values of n.

Figure 3.42 gives a graphical representation of Equation (3.5.57) for several realistic values of Z. The upper abscissa is $\beta$, with corresponding values of Z given on each curve. The lower abscissa gives example values of n for shells with L/a = 2, corresponding to experimental results to be given in Section 3.5.6. With L/a = 2, $n = \pi\beta/2$ and the values of Z are to be associated with values of radius-to-thickness ratio ranging from 25 to 500. The allowable values of $k_\theta$ for L/a = 2 are indicated by the solid points on the curves.

Two observations are apparent from Figure 3.42. First, the minimum load parameter $k_\theta$ does not occur at the lowest value of n. Even for the thickest shell, with a/h = 25, the minimum occurs at n = 4. As the shells become thinner, n increases. For a/h = 200, $n_{min} = 7$. The increase in n as the shells become thinner occurs because part of the strain energy of deformation is in bending and part is in stretching. To minimize the total strain energy, the thinner shells deform in higher modes because in these shells bending requires less energy than stretching. As the shells become longer, n is reduced because $n = a\beta/\pi L$ and the values of $\beta$ at the minima do not shift rapidly with Z (they vary as $Z^{1/4}$, as will be shown).

The second observation is that even though only discrete values of $\beta$ are allowed for a given shell, the minima are flat enough that the continuous curves give good approximations to the values of $k_\theta$ at the lowermost discrete points. Only for very long, thick shells, which buckle with n = 2 or 3, is this a poor approximation. These shells are not considered here because the Donnell equations are inaccurate for n > 4.

It is therefore useful to obtain an explicit expression for $k_\theta$ at the minima of the continuous curves. With $\beta$ considered as a continuous variable, Equation (3.5.57) can be differentiated to find the minima. Setting $dk_\theta/d\beta^2 = 0$ yields

$$(1 + \beta^2)^4 = \frac{12Z^2}{\pi^4} \left( \frac{1 + 3\beta^2}{\beta^2 - 1} \right) \tag{3.5.58}$$

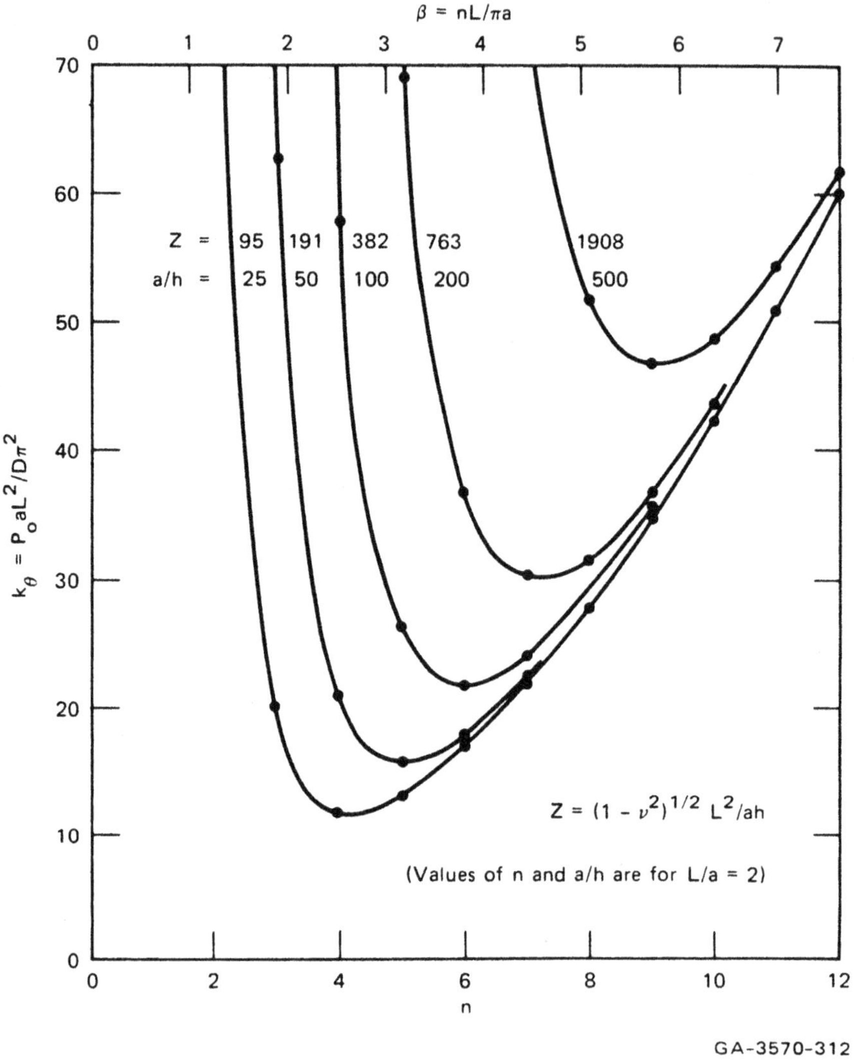


FIGURE 3.42 STATIC RADIAL PRESSURE FACTOR VERSUS $\beta$ WITH Z AS A PARAMETER

To obtain a simple explicit solution to Equation (3.5.58), observe in Figure 3.42 that for Z $\gtrsim$ 100, $\beta^2 >> 1$ at the minima of the curves. With this approximation, Equation (3.5.58) yields $\beta^2 = (6Z)^{1/2}/\pi$. This expression for $\beta^2$ substituted into Equation (3.5.57) gives for $k_\theta$ at the minima

$$k_\theta = \frac{4\sqrt{6}}{3\pi} Z^{1/2} = 1.040 \ Z^{1/2} \tag{3.5.59}$$

With the definitions of $k_\theta$ and Z in Equations (3.5.54) and (3.5.55), Equation (3.5.59) gives the static buckling pressure:

$$P_o = \frac{D\pi^2}{aL^2} \cdot \frac{4\sqrt{6}}{3\pi} \cdot \frac{(1 - \nu^2)^{1/4} L}{a^{1/2} h^{1/2}}$$

$$= \frac{\pi\sqrt{6}}{9(1 - \nu^2)^{3/4}} \cdot E\frac{a}{L}\left(\frac{h}{a}\right)^{5/2} = 0.92\ E\frac{a}{L}\left(\frac{h}{a}\right)^{5/2} \qquad (3.5.60)$$

The numerical value for the coefficient in the last of Equations (3.5.60) is for $\nu$ = 0.3. This formula is accurate for Z in the range

$$100 < Z < \left(\frac{a}{h}\right)^2 \qquad (3.5.61)$$

The lower limit assures that $\beta^2 >> 1$, as already discussed. The upper limit assures that the shells are not so long that they buckle into $n \leqslant 3$, beyond the range of the theory. To derive the upper limit, we use $\beta^2 = (6Z)^{1/2}/\pi$ with $\beta$ expressed in terms of n. Then

$$\left(\frac{nL}{\pi a}\right)^4 = \frac{6Z}{\pi^2}$$

or

$$Z = \frac{n^4}{6\pi^2}\left(\frac{L}{a}\right)^4 = (1 - \nu^2)^{1/2}\left(\frac{L}{a}\right)^2\left(\frac{a}{h}\right)$$

in which the final expression is the definition of Z. With n = 3 this gives an upper limit on allowable L/a for the present theory:

$$\frac{L^2}{a^2} < \frac{6\pi^2(1 - \nu^2)^{1/2}}{3^4}\left(\frac{a}{h}\right) = 0.70\left(\frac{a}{h}\right)$$

To express this inequality in terms of a/h and Z, we multiply it by $(1 - \nu^2)^{1/2}(a/h)$ so that the left side becomes Z and the right side becomes $0.66(a/h)^2$. In Equation (3.5.61) the coefficient 0.66 is rounded to unity because the error in $P_o$ is still small within this range.

### 3.5.4 Critical Pressure-Impulse Curves for Dynamic Buckling

We return now to the dynamic equations of motion and consider dynamic buckling from finite duration pressure pulses. Two representative pulses are the triangular and exponential pulses in Figure 3.43. The triangular pulse is defined by

$$P(t) = \begin{cases} P_m\,(1 - t/T) & 0 < t \leqslant T \\ 0 & t < 0 \;, \quad t \geqslant T \end{cases} \tag{3.5.62}$$

The exponential pulse is defined by

$$P(t) = \begin{cases} 0 & t \leqslant 0 \\ P_m\, e^{-t/T} & t > 0 \end{cases} \tag{3.5.63}$$

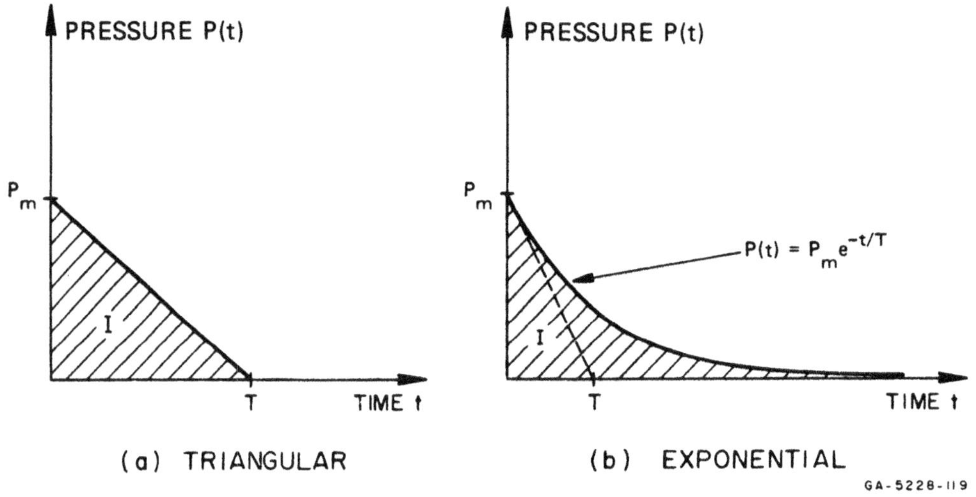


FIGURE 3.43 PULSE SHAPES

For either pulse, with the shape specified, any particular pulse can be characterized by two parameters that specify the pressure and time scale factors. Convenient parameters are the maximum pressure $P_m$ and the impulse intensity I per unit area, given by the area under the pressure-time history as indicated in Figure 3.43. For the triangular pulse, $I = P_m T/2$. For the exponential pulse $I = P_m T$.

These are convenient parameters because, at the extremes of very short and very long durations, one or the other retains physical meaning and is the dominant parameter. For very short durations, only the impulse is important and the buckling problem reverts to that for impulsive loads as discussed in earlier sections. For very long durations, only the maximum pressure is important.

Long duration buckling is not precisely equivalent to the static buckling of Section 3.5.3 because even for long durations we are considering pulses that have a sudden jump in pressure at $t = 0+$. However, we will see that for ductile materials the oscillations produced by the jump do not significantly change the critical pressure for permanent buckling. Therefore, as $T \rightarrow \infty$, the critical pressure approaches the static buckling pressure given by Equation (3.5.60).

For loads of intermediate duration, both $P_m$ and I are important; furthermore, buckling depends on pulse shape. However, the dependence on shape will be shown to be small as long as the shape is reasonably simple and does not differ widely from the shapes in Figure 3.43. For example, the two pulses shown have the same values $P_m$, I and the buckle motion for each will be found to be very similar.

For the dynamic problem it is convenient to use the dimensionless variables introduced in Equation (3.5.47). Then we use Equation (3.5.48) for the hoop mode, and Equation (3.5.50) for the flexural modes is similarly simplified. Also, from the result of the static solution we consider only modes with $m = 1$ half-wave along the shell length. The equations of motion for the dynamic problem are then

$$\ddot{u}_o + u_o = \hat{P}_o \tag{3.5.64}$$

$$\ddot{u}_n + \left[\alpha^2(n^2 + \pi^2a^2/L^2)^2 + (1 - \nu^2)(\pi a/L)^4(n^2 + \pi^2a^2/L^2)^{-2} - n^2u_o\right] u_n = \hat{P}_n + n^2u_o\delta_n \tag{3.5.65}$$

where $\alpha^2 = h^2/12a^2$ and the subscript 1n has been replaced by simply n. When the expression in the bracket multiplying $u_n$ is set to zero, the result is Equation (3.5.58) for static buckling.

The form of Equations (3.5.64) and (3.5.65) is similar to that for Equations (3.4.8) and (3.4.9) in the tangent modulus theory. Each set consists of a single equation for the hoop mode and a group of mutually exclusive equations for the flexural modes that contain the hoop mode amplitude as a parameter. Each set is integrated numerically for motion starting from rest, with initial conditions

$$u_n(0) = \dot{u}_n(0) = 0, \quad n = 0,\ 2,\ 3\ .\ .\ . \tag{3.5.66}$$

In the examples, imperfections are taken only in initial shape, described by $\delta_n$, because they are expected to dominate (for the tangent modulus theory, as shown in Section 3.4) or give results similar to those for imperfections in load (described by $\hat{P}_n$).

Results for the longer duration loads, for which Equations (3.5.64) and (3.5.65) apply, have the same basic feature as for impulsive loads as found with Equations (3.4.8) and (3.4.9). A group of modes becomes highly amplified as the load increases, and buckling can be characterized by motion in the most amplified mode. An essential difference is that the number of highly amplified modes for long duration loads is much smaller than for impulsive loads. This is shown in Figure 3.44 for an example shell of 6061-T6 aluminum with a/h = 100 and L/a = 2.

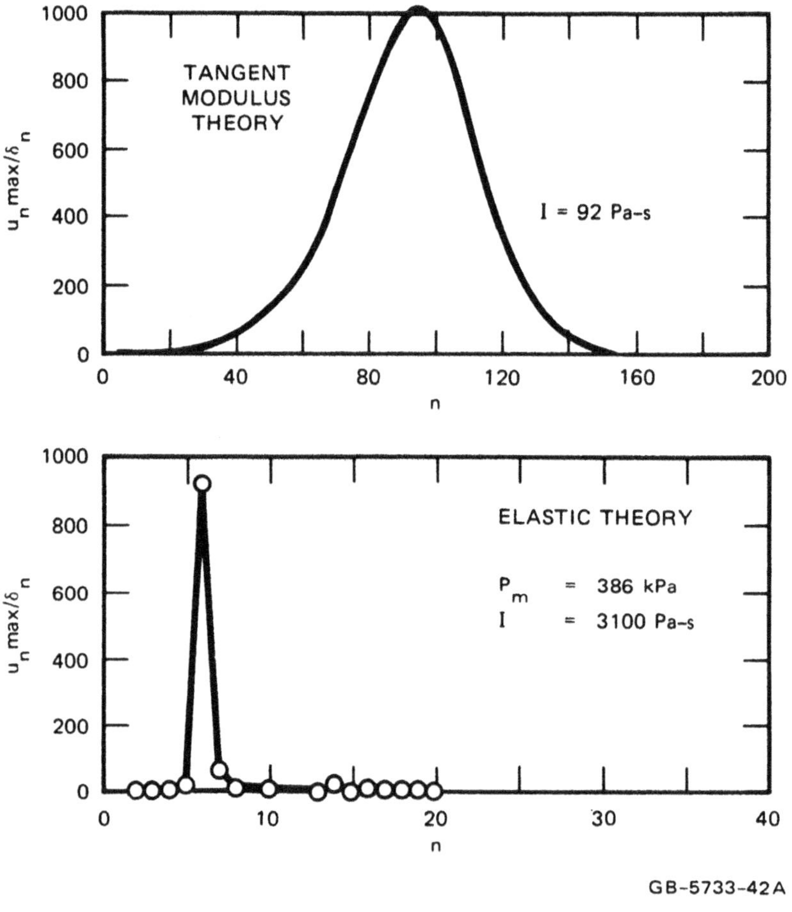


FIGURE 3.44 AMPLIFICATION FUNCTIONS FOR THE TANGENT MODULUS AND ELASTIC THEORIES (6061-T6 aluminum shell, a/h = 100, L/a = 2. Examples are for very high amplification, so impulse and mode numbers in tangent modulus theory are somewhat larger than those in Section 3.4.)

The amplification function for the tangent modulus theory has substantial values for mode numbers ranging from about 60 to 120. However, the amplification function for long duration loads is essentially zero except for n = 6 (which is the static buckling mode in this example). As the pulse duration is decreased and the peak pressure is increased, this mode number increases but only to about n = 8 within the elastic range of the long pulse theory. For thinner shells, several modes are highly amplified; however, the number of amplified modes is always very small compared with that for the tangent modulus theory because the parameter $\sqrt{\sigma_\theta / E_t}$, which determines the mode number, is about 10 times larger in the plastic range.

Another feature of buckling under long duration loads arises because of the step in pressure at t = 0+. This step causes oscillations in the hoop mode with a peak amplitude about twice the pseudo-static response to $P_o$. With no damping, these oscillations cause autoparametric vibration buckling similar to that described in Section 3.3.4 for impulsive loads in the elastic range. However, this vibration buckling motion cannot cause permanent plastic buckles because the motion is damped out by a small amount of plastic deformation during each oscillation of the flexural mode. In calculating critical loads for buckling, we therefore consider only modes that grow essentially monotonically (and exponentially) with time.

Figures 3.45 through 3.50 give critical curves of peak pressure and impulse intensity for pulse loads that cause buckling. Along each curve the amplification in the buckle mode with most growth is a constant. Figure 3.45 gives the complete critical load curve corresponding to the example amplification curves in Figure 3.44. The numbers along the curves are the mode numbers N of the most amplified modes. Along the branch for each theory, N increases as the peak pressure $P_m$ increases. The curves are dashed near the cusp-like intersection between the two branches because the assumptions for the theories break down in this region. On the elastic branch, the hoop stress exceeds the elastic limit when $P_m$ exceeds about half the static yield pressure. This is indicated by the tic marks on the curves. On the tangent modulus branch, as $P_m$ is reduced, the amount of plastic flow becomes so small that strain reversal occurs with very little buckle growth. Actual buckle motion near this cusp is very complex and requires a numerical theory. No rigorous calculations have been undertaken for this region.

Because the two approximate theories are either entirely elastic or entirely plastic, there is an order of magnitude change in mode numbers from one branch to the other, as already mentioned. This order of magnitude change is also observed experimentally, as shown by the many examples throughout Chapter 3, and in particular by the examples in Figure 3.36. The nature of the transition from one form of buckling to the other has not been observed experimentally because of the special facilities required and the expense of running experiments in the transition region. Limited experiments described in Section 3.5.6 show that buckling in the cusp region is biased toward the low-order buckling of the elastic theory, as one might expect because early strain rate reversal results in nearly elastic response.

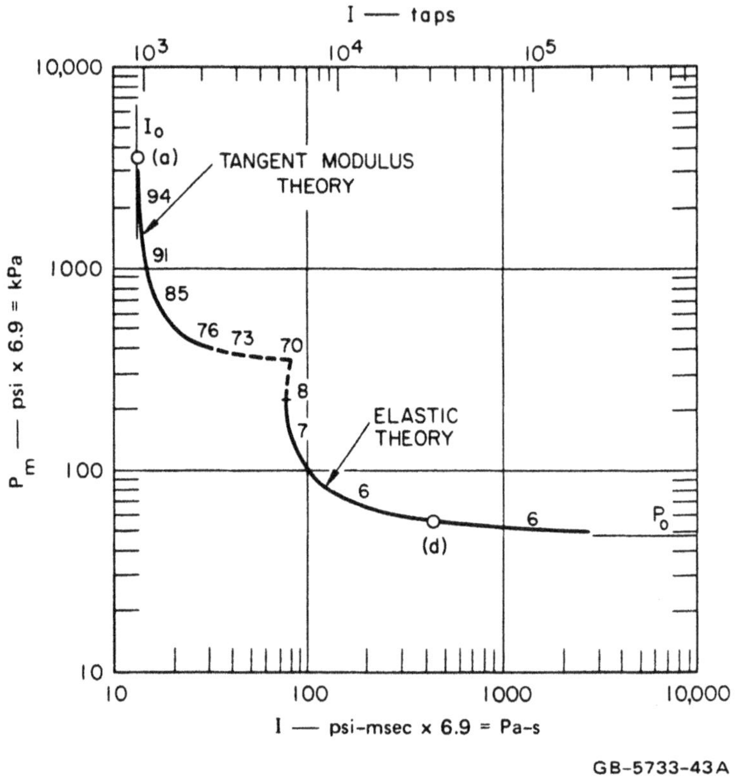


FIGURE 3.45 CRITICAL CURVE FOR BUCKLING OF SHELLS IN FIGURE 3.36 (exponential pulses, 6061-T6 aluminum, a = 3 inches = 76 mm, a/h = 100, L/a = 2)

Figure 3.46 shows that the value chosen for critical amplification does not strongly influence the critical load curves. The largest differences caused by changing the amplification by two orders of magnitude occur in the cusp region, where the theory must be extrapolated anyway because of the limitations just discussed. In the examples throughout this section, the nominal amplification for buckling is 1000 because the calculations were made several years before the experimental results on equivalent imperfections given in the preceding sections were available. More appropriate amplifications for threshold buckling are in a range near 20, but the conclusions are unchanged and critical loads in the examples are only slightly higher than in Section 3.4. Final formulas, given later, are based on an amplification of 20.

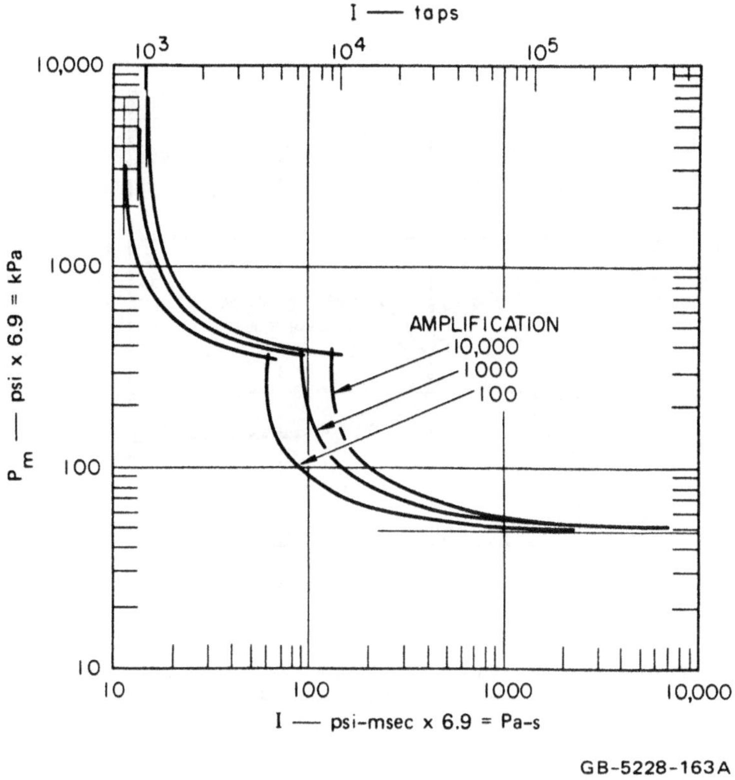


FIGURE 3.46 CRITICAL CURVES FOR BUCKLING AMPLIFICATIONS OF 100, 1000, AND 10,000 (exponential pulses, same shell as Figure 3.45)

Figure 3.47 demonstrates that changes in pulse shape have a minor effect on critical curves. (This is a general result for many types of response to pulse loads.)

Figures 3.48 and 3.49 show the effect of shell thickness. In Figure 3.48, a/h is varied from 24 to 250 and each critical curve is normalized to the critical ideal impulse $I_0$ and the critical step pressure $P_0$ for that value of a/h. In this form the great change in shape of the curves is made apparent. For the thickest shell, the curve remains close to the $P_0, I_0$ asymptotes for any pulse duration. (In this log-log plot, curves of constant pulse duration are straight lines at 45° slope.) As a/h is increased, the curves move away from the asymptotes over a much wider range of pulse durations. Also, the elastic branch becomes more dominant for the thinner shells. In Figure 3.49 the same curves are given in dimensional units. In this plot the broad ranges of peak pressures and impulses are apparent. Four orders of magnitude in both pressure and impulse must be spanned in order to display the curves (and to perform experiments).

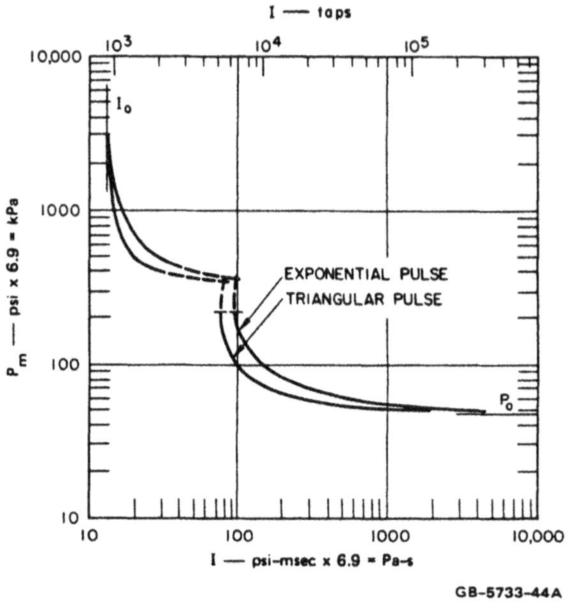


FIGURE 3.47 EFFECT OF PULSE SHAPE ON CRITICAL CURVES FOR BUCKLING (same shell as Figure 3.45)

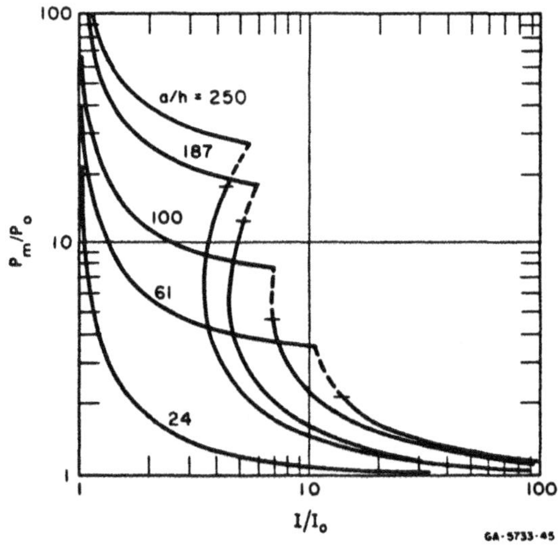


FIGURE 3.48 EFFECT OF a/h ON CRITICAL CURVES FOR BUCKLING (exponential pulses, 6061-T6 aluminum, L/a = 2, curves normalized to $P_o$ and $I_o$ for each shell)

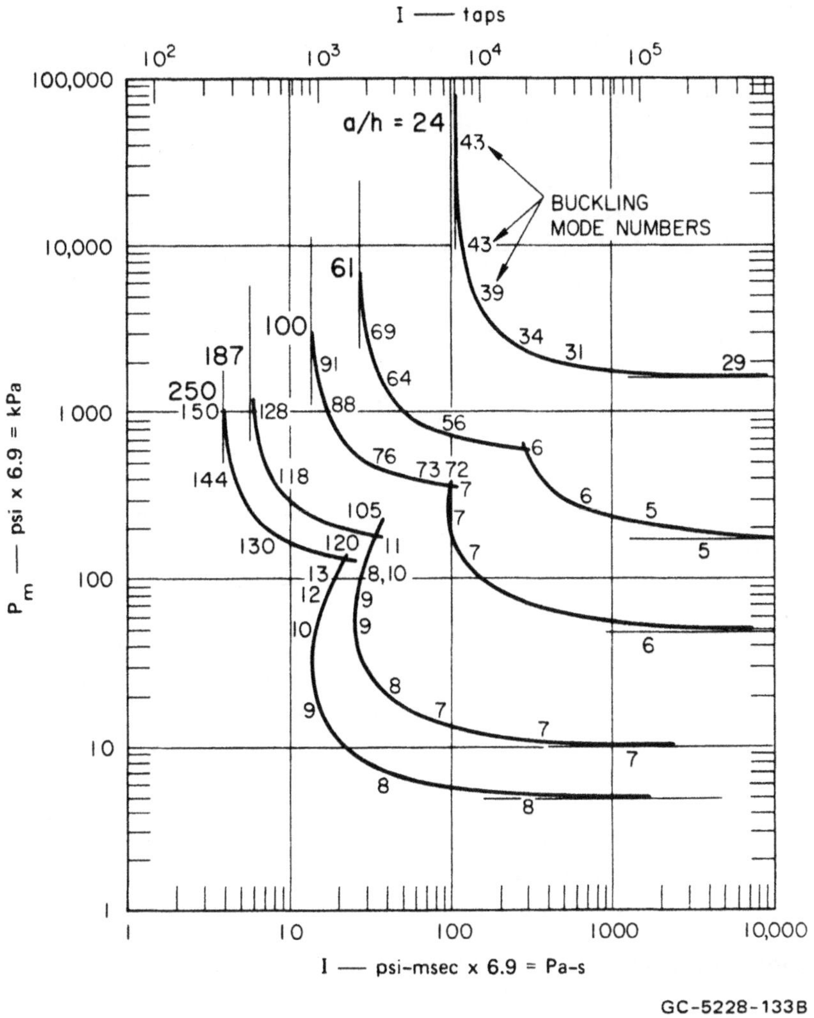


FIGURE 3.49 EFFECT OF a/h ON CRITICAL CURVES FOR BUCKLING (same as Figure 3.48, but without normalization to $P_o$ and $I_o$)

The final example, in Figure 3.50, shows the effect of shell length. Only the elastic branch is affected by shell length. The branch moves to lower pressures as L/a is increased from 1 to 5. The result is to broaden the P,I curve in a manner similar to the broadening with increasing a/h.

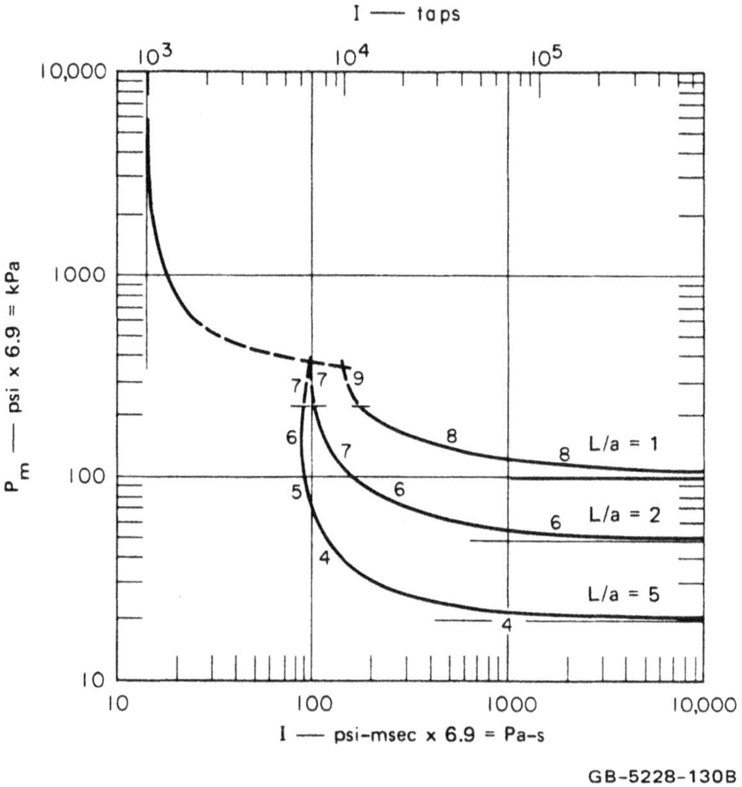


FIGURE 3.50 EFFECT OF L/a ON CRITICAL CURVES FOR BUCKLING (exponential pulses, 6061-T6 aluminum, a = 3 inches = 76 mm, a/h = 100)

### 3.5.5 Simple Formulas for Critical Curves

The general form of the critical buckling curves in the preceding examples is given in Figure 3.51 and can be described by a few simple formulas. The curves consist of two branches, one from each theoretical model, each of which can be approximated to an accuracy of about 20% by simple rectangular hyperbolas of the form

$$\left(\frac{P_m}{P_A} - 1\right)\left(\frac{I_m}{I_A} - 1\right) = 1 \tag{3.5.67}$$

where $P_A$ and $I_A$ are the asymptotic values of each hyperbola.

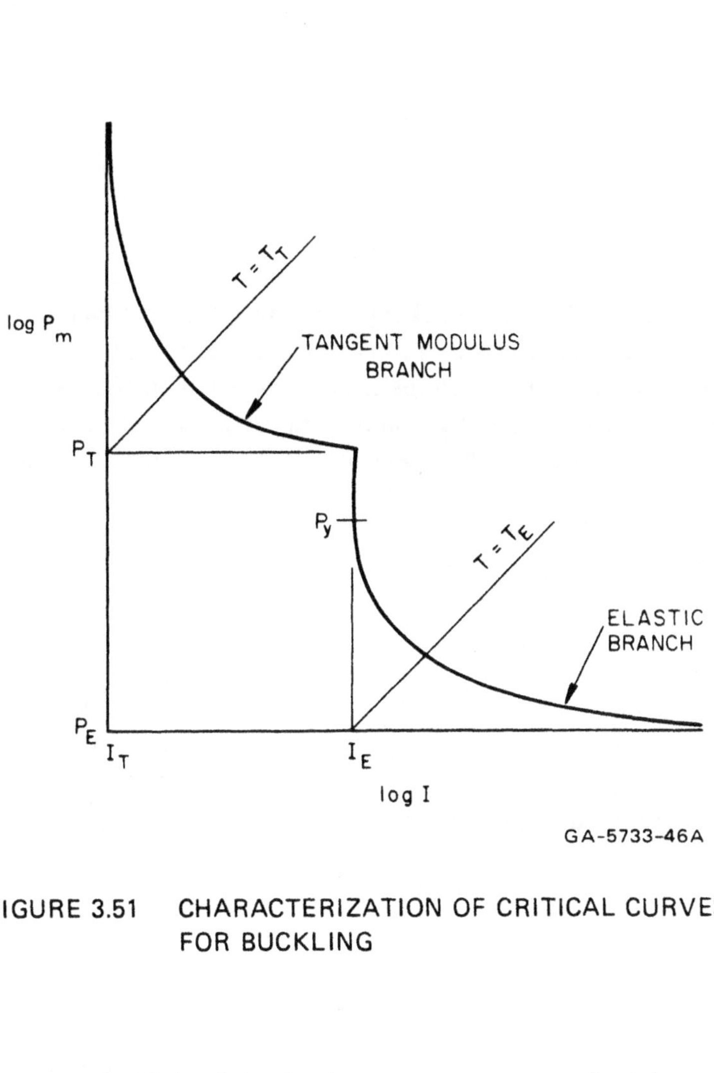


FIGURE 3.51 CHARACTERIZATION OF CRITICAL CURVES FOR BUCKLING

For the tangent modulus branch, these asymptotes are given by

$$P_T = 0.75\ \sigma_y\ h/a \tag{3.5.68}$$

$$I_T = \begin{cases} 1.807\ \rho ca\ \tilde{D}\,(h/a)^{3/2} & a/h \leqslant 0.405/\tilde{D}^2 \\ 1.15\ \rho ca\,(h/a)^2 & a/h \geqslant 0.405/\tilde{D}^2 \end{cases} \tag{3.5.69}$$

where $\tilde{D} = K^{-1/4}\hat{\epsilon}_y^{1/2}$ is a material property as defined in Section 3.4.5 (the tilde is added here to distinguish $\tilde{D}$ from the shell stiffness parameter D). The formula for $P_T$ is an empirical result of the numerical integration and gives a pressure asymptote equal to three-fourths of the yield pressure. The formula for $I_T$ is derived in Section 3.4.

For the elastic branch, the asymptotes are given by

$$P_E = 0.92\ E\left(\frac{a}{L}\right)\left(\frac{h}{a}\right)^{5/2} \tag{3.5.70}$$

$$I_E = 3\ \rho c a\left(\frac{h}{a}\right)^{2} \tag{3.5.71}$$

The formula for $P_E$ is the static buckling pressure given by Equation (3.5.60). The formula for $I_E$ is the result of the numerical integrations performed with the Donnell equations and gives amplifications near 20 rather than 1000 as in the examples in Figures 3.45 to 3.50. It has the same form as the elastic formula for $I_T$, which applies for very thin shells. The coefficient is 3 rather than 1.15 because, for $I_E$, formula (3.5.71) takes into account the finite length of the shell. Thus, formula (3.5.71) should include some dependence on shell length, but the dependence is small as can be seen in Figure 3.50 for $1 \leqslant L/a \leqslant 5$.

Lines at 45° in the log-log plot of Figure 3.51 are lines of constant pulse duration. For example, the exponential pulse has characteristic time $T = I/P$, so the lines of constant duration are $I = TP$. The two lines drawn through the intersections of the asymptotes of each branch define characteristic times $T_T$ and $T_E$ given by $T_T = I_T/P_T$ and $T_E = I_E/P_E$. From Equations (3.5.68) and (3.5.69) the characteristic time for the tangent modulus branch is

$$T_T = 2.4\ \frac{a\tilde{D}}{c\hat{\epsilon}_y}\left(\frac{h}{a}\right)^{1/2} = 2.4\ \frac{a}{c}\ K^{-1/4}\hat{\epsilon}_y^{-1/2}\left(\frac{h}{a}\right)^{1/2} \tag{3.5.72}$$

However, in the numerical examples it was found that variations in K moved the horizontal pressure asymptote slightly from the value given by Equation (3.5.68) in such a way as to compensate for the small variation in $T_T$ with K given by Equation (3.5.72). Thus, a better expression for $T_T$, with K in the range $10 \leqslant K \leqslant 60$ typical of engineering metals, is simply

$$T_T = \frac{a}{c}\left(\frac{h}{a}\right)^{1/2}\hat{\epsilon}_y^{-1/2} \tag{3.5.73}$$

Similarly, from Equations (3.5.70) and (3.5.71), the characteristic time for the elastic branch is

$$T_E = 3\ \frac{L}{c}\left(\frac{a}{h}\right)^{1/2} \tag{3.5.74}$$

From Figure 3.51 we see that, if the time constant T of the applied pulse is much shorter than $T_T$, the load appears impulsive to the shell, and if T is much larger than $T_E$, the load appears quasi-static. Loads with durations near or between $T_T$ and $T_E$ are quasi-impulsive, and both pressure and impulse are important to the response. As shells become moderately longer and thinner, $T_T$ and $T_E$ become more widely separated (see Figures 3.48 and 3.50) and the range of quasi-impulsive loads increases. Conversely, for short, thick shells, the tangent modulus and elastic curves move closer together and only a small range is quasi-impulsive. At the other extreme, if the shells are very thin, all buckling is elastic and only the elastic branch exists.

As an example, for 6061-T6 aluminum alloy, c = 5000 m/s and $\hat{\epsilon}_y^{1/2}$ = $(0.0045)^{1/2}$ = 0.067. Then, for a/h = 100, $T_T$ = 0.30 ms per meter of radius. Thus, even for a shell 2 meters in diameter, the pulse duration must be a fraction of a millisecond in order to be impulsive. At the long duration extreme, $T_E$ = 6.0 ms per meter of length. If the shell is 2 meters long and 2 meters in diameter, a load with $T \approx 5T_E = 5 \times 12 = 60$ ms appears quasi-static. The total pulse duration with sensible pressure for an exponential pulse is about 5T. The corresponding total sensible pulse duration is therefore 300 ms. Thus, pulses lasting only a fraction of a second are essentially static even for a shell large enough to stand in.

The range of quasi-impulsive loads for this shell extends somewhat beyond the range $0.30 < T < 6$ ms. From the shape of the P,I curve in Figure 3.47 for a/h = 100, this range, for which both peak pressure and impulse are important to shell response, extends over more than two orders of magnitude, from about 0.1 to 20 ms. Because of this broad range of quasi-impulsive response, many experiments are required to characterize the complete dynamic buckling behavior of shells.

### 3.5.6 Experimental Results and Comparison with Theory

A limited number of experiments have been performed to demonstrate that the essentials of the theory are correct. Experiments at the impulsive extreme are given in earlier sections of the monograph. Here we concentrate on mid- and long-duration loads. Dynamic loading was applied by explosives, and static loading was applied hydraulically. The explosive techniques are illustrated in Figure 3.52 for both asymmetric and symmetric loads. While the theory in the preceding sections is limited to symmetric loads, in application many loads are applied by blast waves that impinge laterally from one side of the shell.

The techniques shown cover load durations ranging from long (quasi-static), provided by shock tubes as shown in the upper photographs, to intermediate (quasi-impulsive), provided by free-standing spherical and imploding cylindrical

ASYMMETRIC SYMMETRIC

(a) SIDE-ON IN SHOCK TUBE (b) NOSE-ON IN CAPPED TUBE

QUASI-STATIC

(c) HE SPHERE (d) CYLINDRICAL CHARGE

QUASI-IMPULSIVE

(e) SIDE-ON CONTACT CHARGE (f) UNIFORM CONTACT CHARGE

IMPULSIVE

FIGURE 3.52 TECHNIQUES FOR ASYMMETRIC AND SYMMETRIC PULSE LOADING charges as shown in the central photographs, to very short (impulsive), provided by sheet explosive in direct contact with a rubber shock attenuator placed on the shell, as shown in the lower photographs. Results of the direct contact experiments are those given in Figures 3.30 and 3.33.

In all the experiments described here, the shells were made of 6061-T6 aluminum alloy and had an outside diameter of 6 inches (152 mm). The shells were held in place by clamping them to heavy end plugs with external straps, which can be seen in the upper photographs in Figure 3.52. The plugs were 1.5 inches (38 mm) thick, so for an unsupported length between plugs of 6 inches (152 mm) the shells were cut 9 inches (229 mm) long.

For asymmetric blast loading, the shells were held with a central shaft running through the end plugs, with the axis of the shell perpendicular to the direction of motion of the impinging blast. The resulting diffracted load on the shell had an approximately $\cos^2\theta$ distribution of pressure over the exposed side of the shell, with a peak pressure equal to the reflected pressure of the blast. The pressure on the leeward side of the shell was approximately equal to the incident pressure of the blast wave. For the high shock strengths in the tests here, this pressure was small (as small as one-seventh) compared with the reflected pressure. In the asymmetric contact charges, the explosive was arranged to provide a $\cos\theta$ distribution of peak pressure over one side and no loading on the opposite side.

For symmetric loading with the shock tube, the shell and central shaft were attached to an ogive nose tip (Figure 3.52b) and the assembly was pointed down the axis of the shock tube. The rear end of the assembly was attached to a 3-inch-thick (76 mm) steel plate that was capped over the shock tube. In this way the maximum loading on the model was the reflected pressure traveling back down the shock tube. Again, the incident pressure coming up the tube was always much smaller than the reflected pressure. Also, the duration of the incident pressure loading was very small compared with the total pulse duration. (The time between incident and reflected shock arrival was very short because the model was very short compared with the 6-m length of the shock tube.) The shock that was reflected back up the tube from the explosive driver end was negligibly small, because of shock decay through the length of the tube. A small opening around the reflecting cap at the top of the tube allowed the spent gases to escape.

The explosive charges for both asymmetric and symmetric loading in the shock tube were flat pancakes of primacord, which gave pressure pulses closely approximating the exponential shape used in the buckling theory. The cylindrical charges for imploding symmetric loading were also made of primacord, but strung out to form a shell coaxially surrounding the shell model. Pulse shape was again nearly exponential, as was also the shape for asymmetric loading from spherical charges. Detailed descriptions of these techniques and further shell response experiments are given in reference 9.

Figure 3.53 gives photographs of six shells buckled from asymmetric and symmetric long-duration loads obtained with the shock tube. All the shells have L/a = 2, with a/h = 100, 61, and 24. The asymmetrically loaded shells tend to be buckled in the same mode number as under symmetric loading but are buckled only over the loaded side. The number of buckle lobes in the symmetrically loaded shells agrees well with the theoretical mode numbers for long duration pulses, given in Figure 3.49.

Three views are given of each shell, arranged as in an engineering drawing. The black line on the asymmetrically loaded shells faced the oncoming blast and experienced the maximum pressure, which is the pressure recorded for the test. The numbers given with each shell are code numbers describing the shell, the peak pressure and impulse, and a damage code and peak deformation as a percentage of the radius. For example, the code 6(100)1-9 specifies that the shell is 6 inches (152 mm) in diameter, has a/h = 100, L/2a = 1, and is the ninth shell of this type that was tested. Peak loading pressure was $P_m$ = 180 psi (1.24 MPa), impulse was I = 15.7 ktaps (1570 Pa-s), and peak deformation was w/a = 23%.

Table 3.7 summarizes the results of the long-duration tests and static uniform radial pressure tests performed with a hydraulic chamber surrounding the lateral surface of each test shell. Pressures $P_{2A}$ and $P_{2S}$ are the asymmetric and symmetric blast pressures, respectively, that produce buckle deformation in the range from 0+ to 5%. Pressures $P_{0A}$ and $P_{0S}$ are the highest pressures at which tests were performed and no permanent deformation was observed. Thus, the threshold pressures for deformation are between $P_{0A}$ and $P_{2A}$ for asymmetric loading and between $P_{0S}$ and $P_{2S}$ for symmetric loading. Under static loading the recorded pressure is the average of two or three tests for each shell type. In these tests the critical pressure is the value at which the hydraulic pressure dropped because of buckling. There was very little scatter in the static tests. The final column under symmetric loading gives the theoretical static buckling pressure under static radial pressure.

The buckling pressures for symmetric loading show that the measured static pressure is essentially the same as the measured pressure for long duration shock loading. This result supports our theoretical procedure of neglecting autoparametric vibration buckling that arises from oscillations of the hoop mode because of the step pressure. Also, the theoretical static buckling pressure agrees reasonably well with both the shock loading and static buckling pressures.

The buckling pressures under asymmetric loading are generally larger than for symmetric loading. This result is expected, since with a $\cos^2\theta$ distribution in pres-

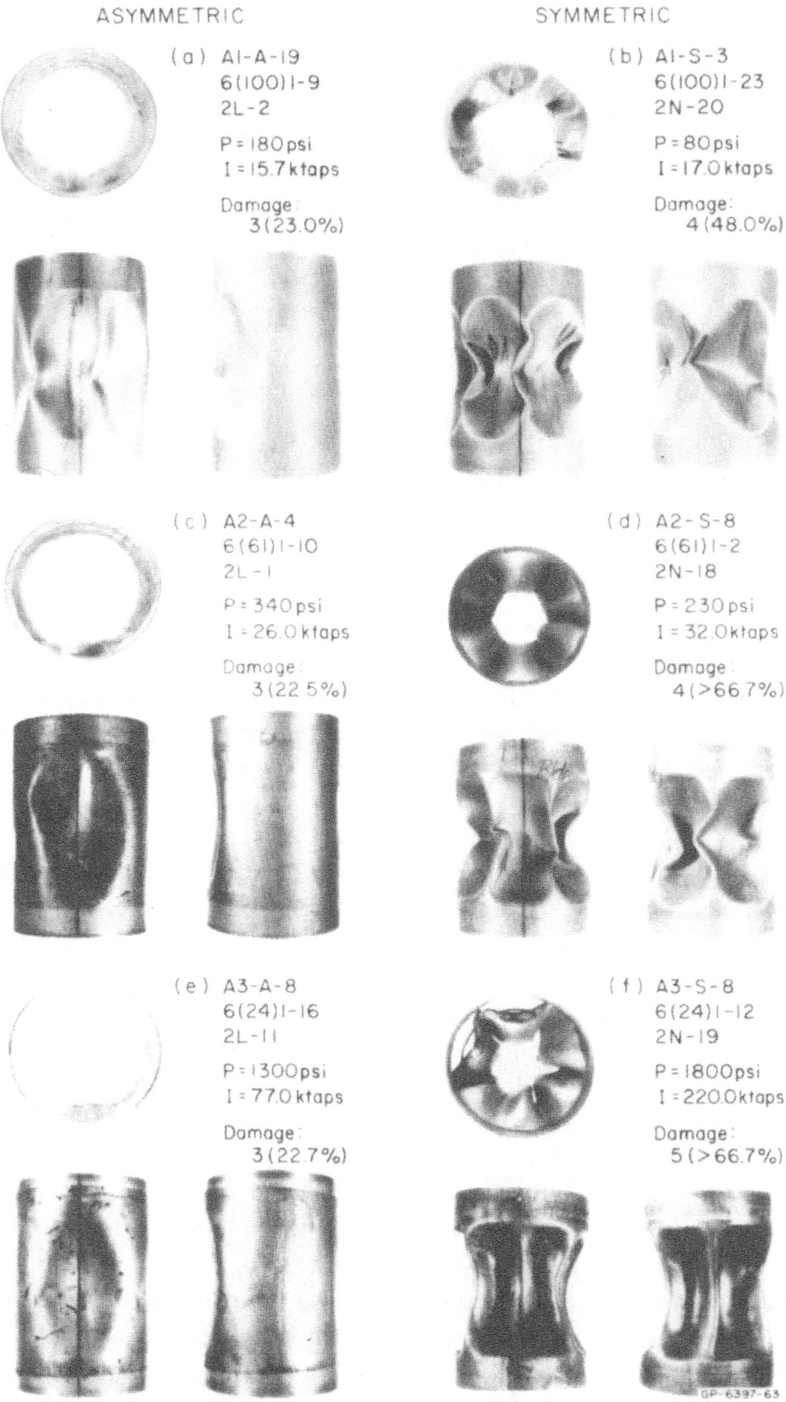


FIGURE 3.53 BUCKLING IN 6061-T6 ALUMINUM SHELLS WITH L/a = 2 UNDER QUASI-STATIC ASYMMETRIC AND SYMMETRIC LOADS (psi x 6.9 = kPa, ktaps x 100 = Pa-s)

Table 3.7

CRITICAL LOADS UNDER LONG DURATION AND STATIC LOADS*

| | | Asymmetric | | Symmetric | | | | |
|---|---|---|---|---|---|---|---|---|
| L/a | a/h | $P_{0A}$ | $P_{2A}$ | $P_{0S}$ | $P_{2S}$ | $P_{static}$ | Static Theory | Ratio $P_{2S}/P_{2A}$ |
| 1 | 100 | 110 | 115 | 40 | 45 | 62 | 46 | 0.4 |
| 1 | 61 | 270 | 280 | 180 | 190 | 190 | 158 | 0.7 |
| 1 | 24 | 650 | 1100 | 1300 | 1500 | 1500 | 1630 | 1.4 |
| 2 | 61 | 130 | 140 | 90 | 95 | — | 79 | 0.7 |

* Pressure entries are in psi; to convert to kPa, multiply by 6.9.

sure the hoop stress is near the maximum only over about one buckle wavelength. Hence, the bending resistance from the adjacent region under lower load tends to inhibit buckling until the overall level of pressure is increased. The exception is the shell with a/h = 24. In this shell the symmetric buckling pressure is near the yield pressure. The additional beam bending stresses under lateral loading probably causes local yielding that triggers buckling at a lower pressure than for symmetric loading.

Figure 3.54 gives results and photographs for quasi-impulsive symmetric loads applied to shells with a/h = 61 and L/a = 2. Loading is increased from one shell to the next with approximately constant pulse duration. First measurable deformation (1.2%) occurs at $P_m$ = 500 psi (3.45 MPa). At $P_m$ = 590 psi (4.07 MPa) buckling is well developed with permanent deformation in two lobes (w/a = 9.5%) having wavelengths corresponding to N = 8 waves around the complete circumference. At $P_m$ = 800 psi (5.52 MPa), severe permanent buckles are formed completely round the shell, again with N = 8. At $P_m$ = 1000 psi (6.90 MPa), buckling is very severe, with N = 13.

The mode number N = 8 is higher than the theoretical threshold number N = 6 for elastic quasi-impulsive response, given for a/h = 61 in Figure 3.49. This curve is presented in Figure 3.55b along with the experimental data. The difference in mode number is attributed to the need for the experimental load to be high enough to cause permanent buckles; no motion picture photographs were taken of the shells to observe purely elastic buckling. At the actual threshold of buckling, with pressure near $P_m$ = 400 psi (2.76 MPa), the shells probably buckled into N = 6 or 7 but then returned to their undeformed shape. At $P_m$ = 230 psi (1.59 MPa) with a longer pulse duration, permanent buckles were formed with N = 6 complete

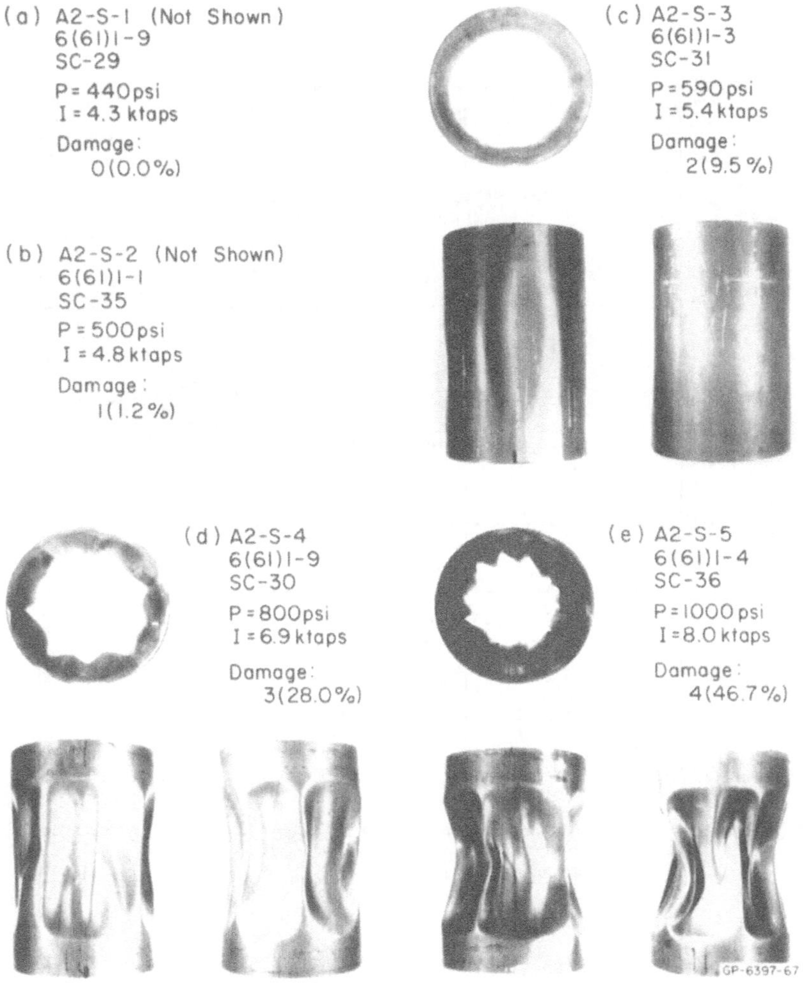


FIGURE 3.54 INCREASING DAMAGE TO SHELLS WITH a/h = 61, L/a = 2 SUBJECTED TO SYMMETRIC QUASI-IMPULSIVE LOADS (psi x 6.9 = kPa, ktaps x 100 = Pa-s)

lobes around the circumference (Figure 3.53). An identical shell tested at the same load buckled into N = 5, just as predicted by the theoretical results in Figure 3.49 for long-duration loading.

The increase in mode number to N = 13 at $P_m$ = 1000 psi (6.90 MPa) follows the trend of the theory, but results in hoop stresses beyond the range of the elastic theory. This load is precisely in the region of the cusp in Figures 3.49 and 3.55b, where neither theoretical model is valid. The sudden jump in mode number from 8 to 13, which is a value more than twice that of the elastic theory, strongly

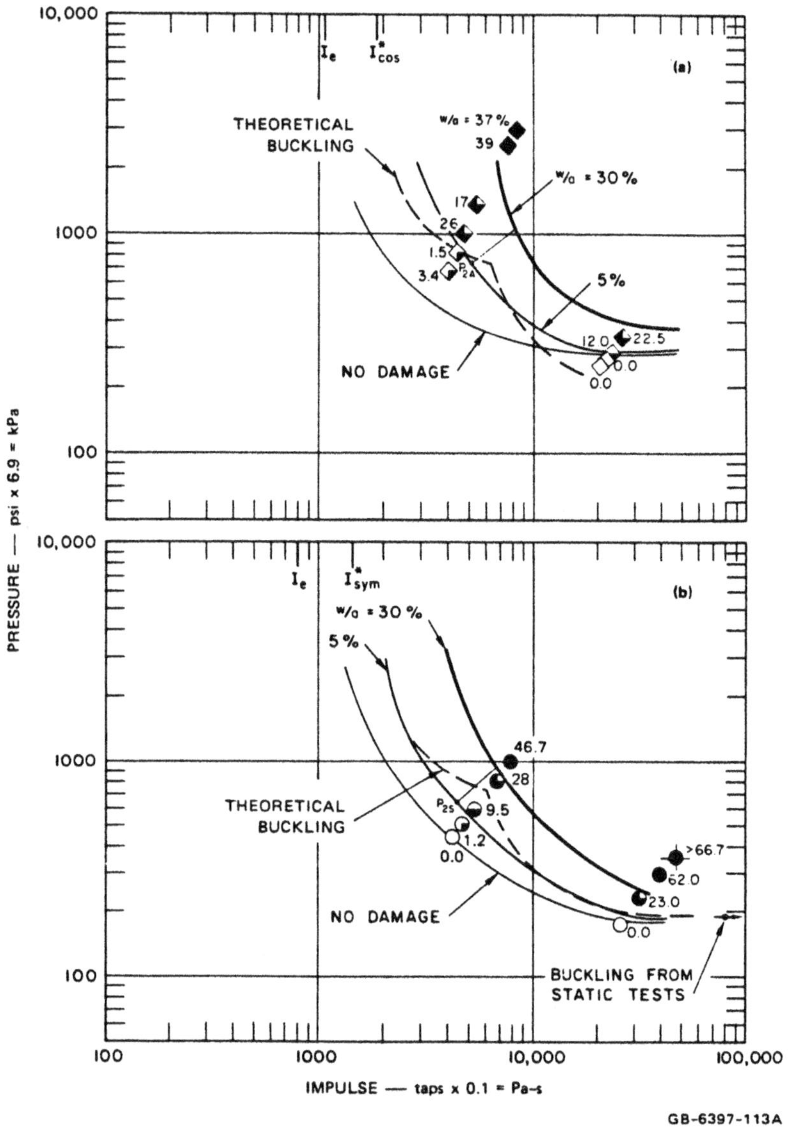


FIGURE 3.55 PRESSURE-IMPULSE DAMAGE CURVES FOR SHELLS WITH a/h = 61, L/a = 2 UNDER (a) ASYMMETRIC LOADS AND (b) SYMMETRIC LOADS

suggests that the N = 13 buckling was influenced by plastic flow in the hoop mode. However, this value is still much smaller than N ≈ 50 as predicted by the tangent modulus theory. We conclude that strain rate reversal and complex elastic-plastic interaction with the end supports also had a strong influence on the buckle motion. Nevertheless, Figure 3.55 shows that critical loads for buckling are reasonably well predicted by extending a smooth curve below the cusp to connect the valid regions of the tangent modulus and elastic theories.

Figure 3.56 gives load-damage results and photographs of another group of a/h = 61, L/a = 2 shells, but tested under asymmetric rather than symmetric loads. The results and conclusions are similar to those for symmetric loading, but the critical loads are higher and damage is limited to the loaded side of the shell. Corresponding peak pressure-impulse data are plotted in Figure 3.55a. Inclusion of the symmetric load buckling theoretical curve shows that the symmetric theory can be used to obtain rough estimates of critical loads for asymmetric loading.

Figures 3.57 and 3.58 give similar photographs and pressure-impulse plots for shells with a/h = 61 and L/a = 4, twice as long as in the previous examples. Results and comparisons with the theory are analogous to those for the shorter shells, giving support to the theoretical treatment of the effect of L/a in the preceding sections.

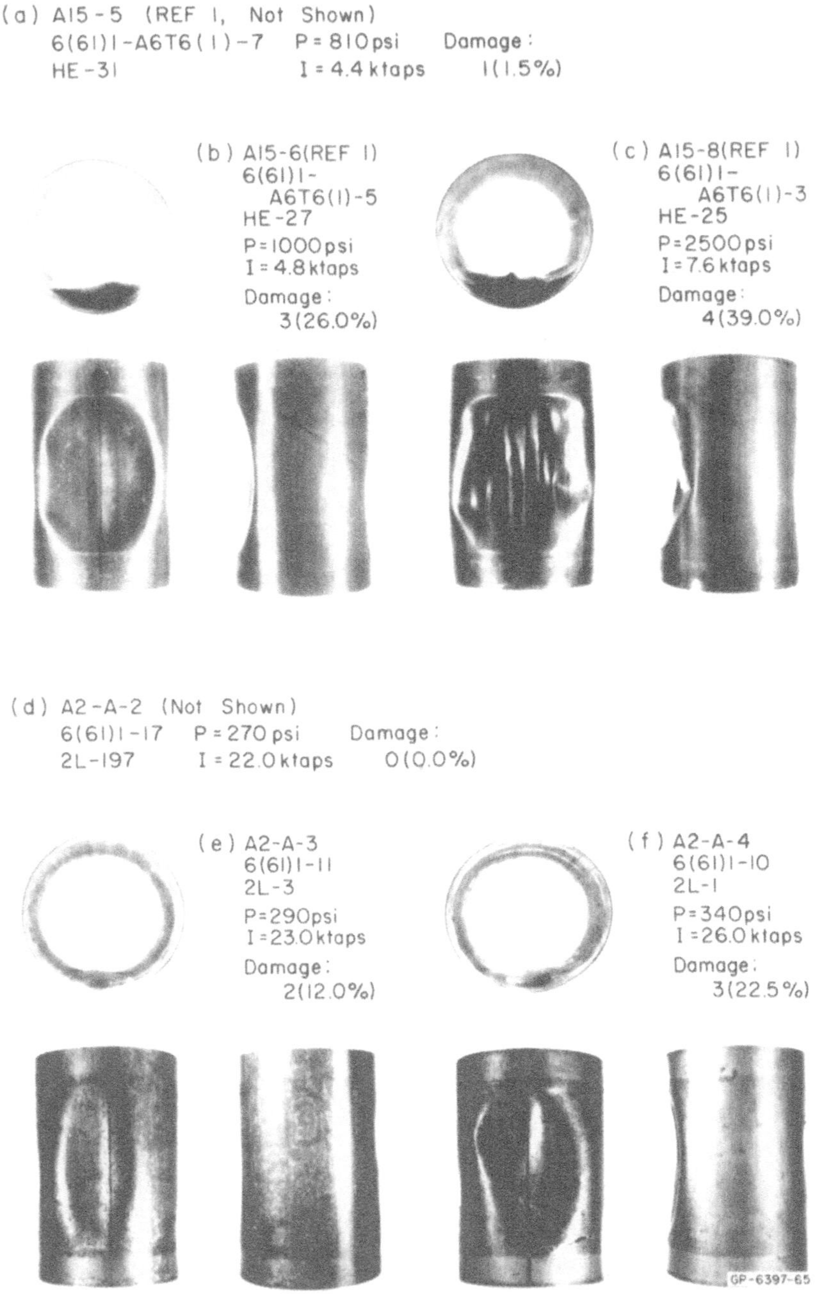


FIGURE 3.56 INCREASING DAMAGE TO SHELLS WITH a/h = 61, L/a = 2 SUBJECTED TO ASYMMETRIC LOADS. (a-c) QUASI-IMPULSIVE AND (d-f) QUASI-STATIC (psi x 6.9 = kPa, ktaps x 100 = Pa-s)

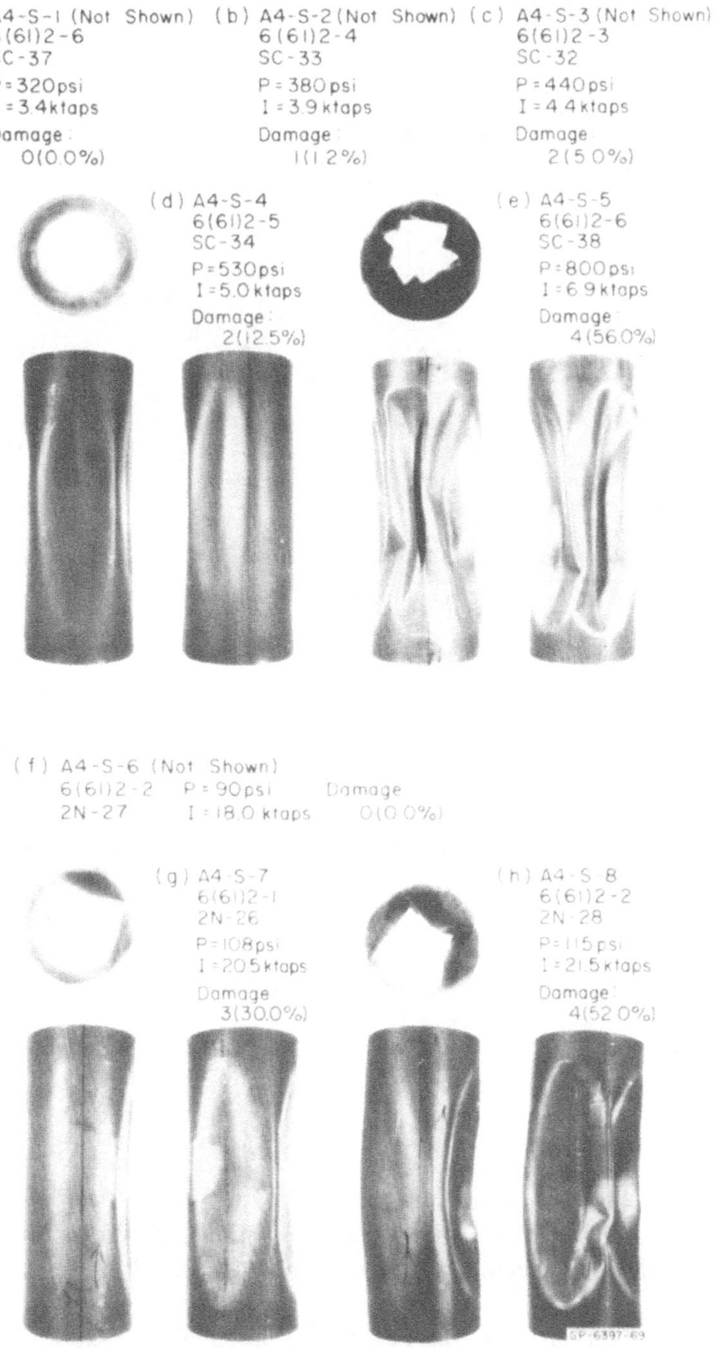


FIGURE 3.58 INCREASING DAMAGE TO SHELLS WITH a/h = 61, L/a = 4 SUBJECTED TO SYMMETRIC LOADS. (a-e) QUASI-IMPULSIVE AND (f-h) QUASI-STATIC (psi x 6.9 = kPa, ktaps x 100 = Pa-s)

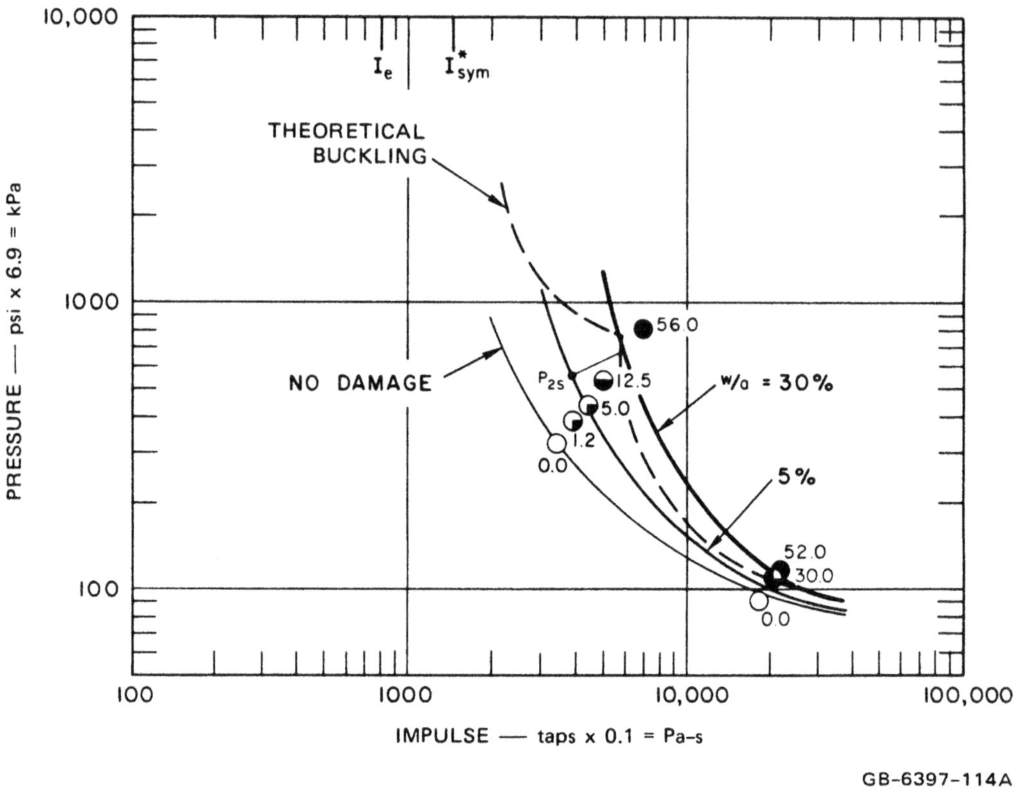


FIGURE 3.59 PRESSURE-IMPULSE DAMAGE CURVES FOR SHELLS WITH a/h = 61, L/a = 4 UNDER SYMMETRIC LOADS

## REFERENCES


1. G.R. Abrahamson and J.N. Goodier, "Dynamic Plastic Flow Buckling of a Cylindrical Shell from Uniform Radial Impulse," Proc. Fourth U.S. Nat. Congress of Appl. Mech., Berkeley, California, pp. 939-950, June 1962.

2. A. Nadai, *Theory of Flow and Fracture of Solids*, Vol. 1, 2nd ed., McGraw-Hill Book Co., Inc., New York, p. 313, 1950.

3. H.E. Lindberg, "Buckling of a Very Thin Cylindrical Shell Due to an Impulsive Pressure," ASME Trans., **86**, Series E, J. Appl. Mech., **31**, pp. 267-272, June 1964.

4. J.N. Goodier and I.K. McIvor, "The Elastic Cylindrical Shell Under Uniform Radial Impuse," ASME Trans., **86**, Series E, J. Appl. Mech., **31**, pp. 259-266, June 1964.

5. N.W. McLachlan, *Theory and Application of Mathieu Functions,* University Press, Oxford, 1941.

6. H.E. Lindberg, "Stress Amplification in a Ring Caused by Dynamic Instability," J. Appl. Mech., **41**, pp. 392-400, June 1974.

7. H.E. Lindberg, "Dynamic Plastic Buckling of a Thin Cylindrical Shell Containing an Elastic Core," ASME Trans., **87**, Series E, J. Appl. Mech., **32**, 4, pp. 803-812, December 1965.

8. D.L. Anderson and H.E. Lindberg, "Dynamic Pulse Buckling of Cylindrical Shells Under Transient Radial Pressures," AIAA J. **6**, pp 589-598, April 1968.

9. H.E. Lindberg and G.E. Sliter, "Response of Reentry-Vehicle-Type Shells to Transient Surface Pressures", AFWL-TR-68-56, Report by SRI International to Air Force Weapons Laboratory, June 1969.

10. R.G. Payton, "Dynamic Membrane Stress in a Circular Elastic Shell," ASME Trans., **83**, Series E, J. Appl. Mech., **28**, 3, pp. 417-420, September 1961.

11. H.E. Lindberg and T.C. Kennedy, "Dynamic Plastic Pulse Buckling Beyond Strain-Rate Reversal," J. Appl. Mech., **42**, 2, pp. 411-416, June 1975.

# 4. FLOW BUCKLING OF CYLINDRICAL SHELLS FROM UNIFORM RADIAL IMPULSE

## 4.1 PLASTIC FLOW BUCKLING WITH HARDENING AND DIRECTIONAL MOMENTS

In our treatment of plastic flow buckling of cylindrical shells contained in Section 3.2 we represent the shell by a ring, and the moment resisting buckling is assumed to be entirely due to strain hardening. Here, we take into account the length of the shell so that axial stresses and strains are brought into the theory. As we shall see, a consequence of yielding in a plastic biaxial state is the introduction of what we shall call a 'directional' moment contribution to the resistive moment.

The directional moment was first introduced by Goodier[1] in his work on plastic flow buckling of plates. The theory shows that the relative magnitudes of the hardening and directional moments depend on the length of the shell, with the hardening moment dominating for long shells and the directional moment dominating for short shells. This partitioning influences the predicted preferred mode of buckling, the number of buckles increasing with shell length. The predicted mode numbers agree with the experimental results presented in Section 4.1.15 for 6061-T6 aluminum alloy shells of various lengths and radius-to-thickness ratios.

### 4.1.1 Theory of Plastic Cylindrical Shells

The shell material is assumed to be rigid-plastic, yielding according to the von Mises conditions and the associated Levy-Mises flow law of incremental plasticity. The strain hardening behavior is approximated by an isotropic linear relationship. Figure 4.1 shows the coordinate system and shell dimensions.

With the strain rates in the longitudinal, circumferential, and radial directions denoted by $\dot{\epsilon}_x$, $\dot{\epsilon}_\theta$, and $\dot{\epsilon}_z$, the incompressibility condition is

$$\dot{\epsilon}_x + \dot{\epsilon}_\theta + \dot{\epsilon}_z = 0 \tag{4.1.1}$$

The generalized strain rate $\dot{\epsilon}$ and the generalized stress $\sigma$ are defined by

$$\dot{\epsilon}^2 = 2(\dot{\epsilon}_x^2 + \dot{\epsilon}_\theta^2 + \dot{\epsilon}_z^2)/3 \tag{4.1.2}$$

and

$$\sigma^2 = \left[(\sigma_x - \sigma_\theta)^2 + (\sigma_\theta - \sigma_z)^2 + (\sigma_z - \sigma_x)^2\right]/2$$

$$= 3(\sigma'^2_x + \sigma'^2_\theta + \sigma'^2_z)/2 \tag{4.1.3}$$

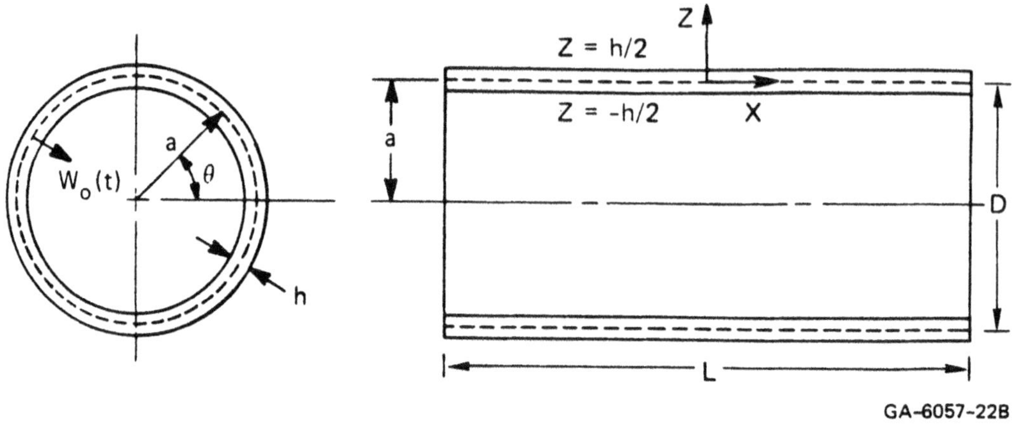


FIGURE 4.1 CYLINDRICAL SHELL — COORDINATES AND DIMENSIONS

where $\sigma_x'$, $\sigma_\theta'$, and $\sigma_z'$ are the deviatoric stress components related to the actual stress components $\sigma_x$, $\sigma_\theta$, and $\sigma_z$ by

$$\sigma'_x = \sigma_x + p \qquad \sigma'_\theta = \sigma_\theta + p \qquad \sigma'_z = \sigma_z + p \tag{4.1.4}$$

with $p = -(\sigma_x + \sigma_\theta + \sigma_z)/3$. In shell theory, the stress component perpendicular to the midsurface is assumed to be negligible, so setting $\sigma_z = 0$ in Equations (4.1.4) gives

$$\sigma_x = 2\sigma'_x + \sigma'_\theta \qquad \sigma_\theta = 2\sigma'_\theta + \sigma'_x \tag{4.1.5}$$

Using the incompressibility condition (4.1.1), we can write the generalized strain rate formula as

$$\dot\epsilon^2 = 4(\dot\epsilon_x^2 + \dot\epsilon_x\dot\epsilon_\theta + \dot\epsilon_\theta^2)/3 \tag{4.1.6}$$

The Levy-Mises flow law is

$$\dot\epsilon_x/\sigma'_x = \dot\epsilon_\theta/\sigma'_\theta = \dot\epsilon_z/\sigma'_z = \dot\lambda \tag{4.1.7}$$

where $\dot\lambda$ is the proportionality factor. In terms of generalized stress and strain rate this factor, by (4.1.2), (4.1.3), and (4.1.7) is

$$\dot\lambda = 3\dot\epsilon/2\sigma \tag{4.1.8}$$

Formulas (4.1.7) and (4.1.8) now allow Equations (4.1.5) to be written as stress-strain rate relationships

$$\sigma_x = 2(2\dot\epsilon_x + \dot\epsilon_\theta)\sigma/3\dot\epsilon \qquad \sigma_\theta = 2(2\dot\epsilon_\theta + \dot\epsilon_x)\sigma/3\dot\epsilon \tag{4.1.9}$$

Finally, the assumption of linear strain hardening gives

$$\dot{\sigma} = E_h \dot{\epsilon} \qquad \sigma = \sigma_0 + E_h \epsilon \tag{4.1.10}$$

where $E_h$ is the slope of the straight line approximating the strain-hardening portion of the stress-strain curve obtained from a simple tension or compression test, and $\sigma_0$ is the stress at zero plastic strain according to the straight line. The value of $\sigma_0$ fixes the initial size of the von Mises yield ellipse in the $\sigma_x, \sigma_\theta$ plane, and the value of $E_h$ governs the isotropic expansion of the ellipse.

### 4.1.2 Effect of Shell Length on Strain Rates

Short shells with no axial restraint from supports extend axially during the radially inward motion. The midsurface axial stress is therefore negligible and we may set $\sigma_x = 0$, which, along with the shell assumption $\sigma_z = 0$, leads to the midsurface results:

$$\sigma_\theta = -\sigma \ , \quad p = \sigma/3 \ , \quad \sigma_x' = \sigma_z' = \sigma/3 \ , \quad \text{and} \quad \sigma_\theta' = -2\sigma/3.$$

Hence the flow law (4.1.7) provides the strain-rate relationship

$$\dot{\epsilon}_x = \dot{\epsilon}_z = -\dot{\epsilon}_\theta/2$$

Shells with axial restraints from fixed supports have no axial strain, so $\dot{\epsilon}_x = 0$ and, by Equations (4.1.9), $\sigma_x = \sigma_\theta/2$.

Long shells, even without axial restraints from supports, have axial stresses induced by axial inertial forces; for the shell to extend, material has to be displaced in the axial directions away from the central cross section. For sufficiently long shells, zero axial strain conditions are established near the central cross section, whereas at each free end zero axial stress conditions prevail.

An experimental observation on imploded cylindrical shells is that the generators remain straight except in a narrow region at each end where the shell flares outwards slightly. Consequently, it is assumed that the axial strain is constant through the shell thickness; in other words, it is independent of the z coordinate. Another observation is that, even in a severly buckled shell, plane sections perpendicular to the shell axis remain plane; thus the axial strain is taken to be independent of the $\theta$ coordinate.

We assume that the effect of shell length on strain rates can be expressed by the midsurface relationship.

$$\dot{\epsilon}_x = -k(x)\dot{\epsilon}_\theta \qquad 0 < k < 1/2 \qquad z = 0 \tag{4.1.11}$$

According to Equation (4.1.11), each cross section of a given cylindrical shell has a value of k that is fixed throughout the motion and represents a measure of the axial restraint. Here we are concerned only with axial restraint caused by axial inertia. At the free ends of the shell, $x = \pm L/2$, we have $k(\pm L/2) = 0$. At the central cross section, $x = 0$, the value of $k(0)$ depends on the shell length but approaches $k(0) = 0$ for very long shells. A formula for $k(x)$ is derived below by analyzing the unperturbed motion.

### 4.1.3 The Unperturbed Motion

Shell motion consists of a dominant uniform radial motion with small flexural motions that increase in amplitude with time. During the entire uniform inward motion plastic loading occurs. The hoop motion dominates sufficiently to prevent unloading of any surface element by flexural motion. Thus the theory is restricted to describing small flexural motion amplitudes, that is, to threshold buckling.

Let the current inward displacement of a uniform shell caused by a uniform impulse be $w_0(t)$. The circumferential component of strain rate is then

$$\dot{\epsilon}_\theta = -(1 - z/a)(\dot{w}_0/a) \tag{4.1.12}$$

Experiments show that the shell generators remain straight except for slight outward flaring near the ends. Consequently, we assume that plane sections perpendicular to the axis remain plane; that is, $\dot{\epsilon}_x$ is independent of z. This assumption along with Equation (4.1.11) means that the axial strain rate is

$$\dot{\epsilon}_x = k(x)\,\dot{w}_0/a \tag{4.1.13}$$

To simplify expressions to be derived, we let

$$K_1 = 2(1 - k + k^2) \qquad K_2 = (3K_1/2)^{1/2} \qquad K_3 = (2 - k)/K_1 \tag{4.1.14}$$

We substitute the strain rates (4.1.12) and (4.1.13) into Equation (4.1.6) to give the generalized strain rate

$$\dot{\epsilon} = (2K_2/3)(1 - K_3 z/a)(\dot{w}_0/a) \tag{4.1.15}$$

We substitute the strain rates (4.1.12), (4.1.13), and (4.1.15) into Equations (4.1.9) and (4.1.10) to give the stresses

$$\sigma_x = -(1 - 2k)\,\sigma^0/k_2 + (z/a)\,K_3\left[3k\sigma^0/(2 - k)\,K_2 + 2(1 - 2k)\,E_h w_0/3a\right] \tag{4.1.16}$$

$$\sigma_\theta = -(2 - k)\,\sigma^0/K_2 + (z/a)\,K_3\left[3k^2\sigma^0/(2 - k)\,K_2 + 2(2 - k)\,E_h w_0/3a\right] \tag{4.1.17}$$

$$\sigma = \sigma_0 + (2E_h K_2/3)(1 - K_3 z/a)(w_0/a) \tag{4.1.18}$$

$$\sigma^0 = \sigma_0 + 2E_h K_2 w_0/3a \tag{4.1.19}$$

By comparing Equations (4.1.18) and (4.1.19), we see that $\sigma^0$ required in Equations (4.1.16) and (4.1.17) is the current generalized stress at the midsurface.

In the notation of Figure 4.2, the equation of motion of an element of shell subtending an angle $d\theta$ and of unit axial length is

$$N_\theta = a\rho h\ddot{w}_0 \tag{4.1.20}$$

Integration of the circumferential component of stress through the shell thickness, h, gives

$$N_\theta = -(2 - k)\,\sigma^0 h/K_2 \tag{4.1.21}$$

Hence substitution of Equation (4.1.19) in (4.1.21) and Equation (4.1.21) in (4.1.20) yields

$$\ddot{w}_0 + 2(2 - k)\,E_h w_0/3a^2\rho = -(2 - k)\,\sigma_0/K_2 a\rho \tag{4.1.22}$$

The solution of (4.1.22) with the initial conditions for the impulsive loading

$$w_0(0) = 0 \qquad \dot{w}_0(0) = V_0 \tag{4.1.23}$$

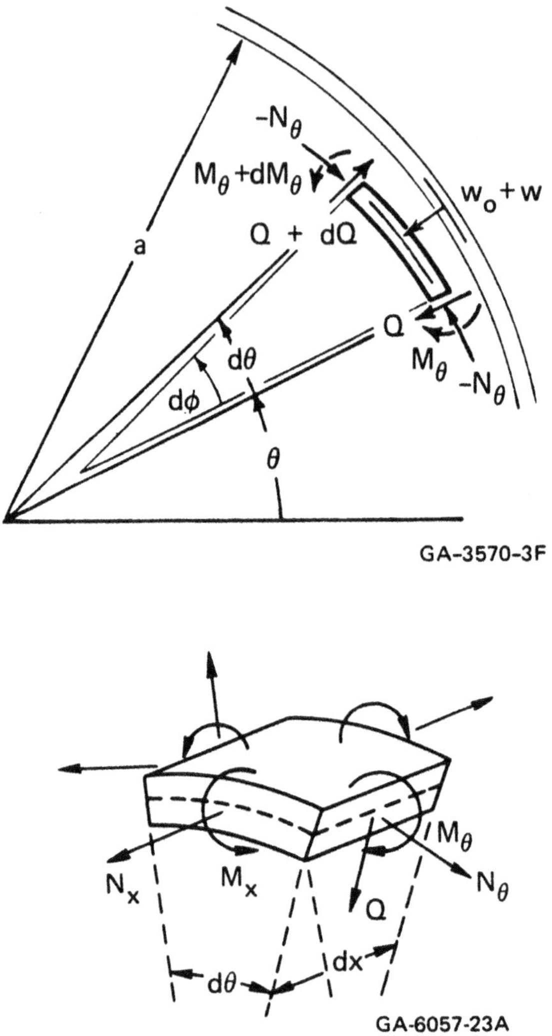


FIGURE 4.2 SHELL ELEMENT NOTATION

is

$$w_0/a = (V_0/\lambda c_h)\sin(\lambda c_h t/a) - (3\sigma_0/2K_2E_h)\left[1 - \cos(\lambda c_h t/a)\right] \quad (4.1.24)$$

where

$$\lambda^2 = 2(2 - k)/3 \qquad c_h^2 = E_h/\rho$$

For many practical materials the variable work-hardening term in Equation (4.1.19) is small compared with $\sigma_0$, the initial value of the generalized stress. Hence, as an analytical simplification, a constant mean value $\overline{\sigma}$ is taken instead of $\sigma^0$. For shells of interest, buckling impulses produce final generalized strains of about 5%, so $\overline{\sigma}$ is estimated from the stress-strain curve over this range of strain. This simplification reduces the equation of motion (4.1.22) to

$$\ddot{w}_0 = -(2 - k)\overline{\sigma}/K_2 a\rho \qquad (4.1.25)$$

and the solution (4.1.24) becomes

$$w_0/a = \tau(2 - \tau/\tau_f) \qquad (4.1.26)$$

where

$$\tau = V_0 t/2a \qquad (4.1.27)$$

and

$$\tau_f = \rho V_0^2 K_2/2(2 - k)\overline{\sigma} \qquad (4.1.28)$$

Because solution (4.1.26) approximately describes the unperturbed motion and allows the derivation of an analytical solution of the perturbed motion, we use this solution instead of Equation (4.1.24). The symbol $\tau_f$ is the maximum midsurface circumferential strain to which the shell is driven by the impulsive load; from Equation (4.1.26), when $\tau = \tau_f$, we have $\dot{w}_0 = 0$ and $w_0/a = \tau_f$.

### 4.1.4 Axial Strain Distribution

We have assumed that the axial and circumferential strain rates are related by Equation (4.1.11). We now use the unperturbed motion to derive an approximate formula for k(x). Let $u_0$ be the axial displacement during the unperturbed motion. The equation governing motion in the axial direction is

$$N_x' = \rho h \ddot{u}_0 \qquad (4.1.29)$$

where primes denote differentiation with respect to x. The kinematic relations $\dot{u}_0' = \dot{\epsilon}_x$ and $\dot{\epsilon}_\theta = -\dot{w}_0/a$, when substituted in $\dot{\epsilon}_x = -k\dot{\epsilon}_\theta$, give $\dot{u}_0' = k\dot{w}_0/a$. Hence, the equations of motion (4.1.20) and (4.1.29) can be combined to give

$$N_x' + \left(k/a^2\right) N_\theta = 0 \qquad (4.1.30)$$

We now substitute in Equation (4.1.30) the membrane force expressions

$$N_x = (2k - 1)\sigma^0 h/K_2 \qquad N_\theta = -(2 - k)\sigma^0 h/K_2 \tag{4.1.31}$$

where $\sigma^0$ is given by Equation (4.1.19), which leads to the equation

$$k'' + \left[9(1 - 2k)\,\Omega^2/K_1^2(2 - k)\right]k'^2 - (\Omega L/D)^2 k = 0 \tag{4.1.32}$$

where primes now denote differentiation with respect to $\xi = 2x/L$. In Equation (4.1.32) $D = 2a$ is the shell diameter, $L$ is the shell length, and

$$\Omega^2 = (2 - k)K_1/\left[K_1 - (1 - 2k)^2\right] \tag{4.1.33}$$

As a simplification, the approximation $\sigma^0 \approx \sigma_0$ was made in the derivation of Equation (4.1.32), which assumes again that the variable work-hardening term in Equation (4.1.19) is small compared with $\sigma_0$. A further simplification is possible if we observe that for $0 < k < \frac{1}{2}$ formula (4.1.33) gives $3/4 < \Omega^2 < 4/3$ and the coefficient of $k'^2$ lies in the range $(0, 3/2)$. If, in addition, $k'^2$ is small enough to allow us to neglect the middle term in Equation (4.1.32) the equation for k is

$$k'' - (\Omega L/D)^2 k = 0 \tag{4.1.34}$$

The solution of Equation (4.1.34) satisfying $k'(0) = 0$ and $k(1) = \frac{1}{2}$ is

$$k(x) = \cosh(2\Omega x/D)\,/\,2\cosh(\Omega L/D) \qquad -L/2 < x < L/2 \tag{4.1.35}$$

where $\sqrt{3}/2 < \Omega < 2/\sqrt{3}$. Values of $k(0)$ from Equation (4.1.35) are compared in Section 4.1.16 with values obtained from experiments on impulsively loaded shells of various length-to-diameter ratios.

### 4.1.5 Perturbed Motion

Because of imperfections, the shell cross section is not exactly circular, so the uniform radial motion is perturbed. Let $w(\theta, t)$ be the additional deflection caused by imperfections. Then the total deflection is $w_0 + w$. In this theory, $w(\theta, t)$ is inextensional and small compared with $w_0(t)$.

In the nomenclature of Figure 4.2, the curvature of the deformed shell is

$$\partial\phi/\partial\lambda = \kappa = 1/a + (w_0 + w)/a^2 + w''/a^2 \tag{4.1.36}$$

where primes denote partial differentiation with respect to $\theta$. The circumferential strain rate is then

$$\dot{\epsilon}_\theta = -(1 - z/a)\ (w_0 + w)^{\cdot}/a + (z/a)(\dot{w}''/a) \tag{4.1.37}$$

We assume that the axial strain is unaffected by the perturbations and consequently may continue to be described by Equation (4.1.13). During the process of deriving expressions for the generalized strain and stress and for the components of stress, we neglect all terms with products of perturbation quantities and, as before, all terms in z/a of higher order than linear. A further simplification in the analysis is that the perturbation displacement $w(\theta, t)$ is small relative to the unperturbed displacement $w_0(t)$. This assertion is based on the requirement that strain rate reversal should not occur and on our objective of determining threshold buckling impulses. Thus we continue our formulation with

$$\dot{\epsilon}_\theta = -(1 - z/a)(\dot{w}_0/a) + (z/a)(\dot{w}''/a) \tag{4.1.38}$$

instead of Equation (4.1.37).

Substitution of the strain rates (4.1.13) and (4.1.38) into Equation (4.1.6) gives the generalized strain rate

$$\dot{\epsilon} = (2K_2/3)\left[\dot{w}_0/a - K_3(z/a)(\dot{w}_0 + \dot{w}'')/a\right] \tag{4.1.39}$$

Temporal integration of Equation (4.1.39) gives the generalized strain

$$\epsilon = (2K_2/3)\left[w_0/a - K_3(z/a)(w_0 + w'' - \tilde{w}'')/a\right] \tag{4.1.40}$$

where $\tilde{w}(\theta,0)$ is the initial value of the displacement perturbation. Substitution of Equation (4.1.40) in (4.1.10) gives the generalized stress

$$\sigma = \sigma^0 - (2K_2K_3E_h/3)\ (z/a)\ (w_0 + w'' - \tilde{w}'')/a \tag{4.1.41}$$

A two-term binomial approximation for $\dot{\epsilon}^{-1}$ from Equation (4.1.39), strain rates (4.1.38) and (4.1.13), and the generalized stress Equation (4.1.41) substituted in the stress formulas (4.1.9) lead to

$$\sigma_x = -(1 - 2k)\,\sigma^0/K_2 + (z/a)\,K_3\Big[3k\sigma^0(1 + \dot{w}''/\dot{w}_0)/(2 - k)\,K_2$$

$$+2(1 - 2k)\,E_h(w_0 + w'' - \tilde{w}'')/3a\Big] \tag{4.1.42}$$

$$\sigma_\theta = -(2-k)\sigma^0/K_2 + (z/a)K_3\Big[3k^2\sigma^0(1+\dot{w}''/\dot{w}_0)/(2-k)K_2$$

$$+2(2-k)E_h(w_0+w''-\tilde{w}'')/3a\Big] \qquad (4.1.43)$$

The stress distributions (4.1.42) and (4.1.43) have the following membrane forces and bending moments:

$$\begin{aligned} N_x &= -h(1-2k)\bar{\sigma}/K_2 \\ N_\theta &= -h(2-k)\bar{\sigma}/K_2 \end{aligned} \qquad (4.1.44)$$

$$M_x = \Big[3k\bar{\sigma}(1+\dot{w}''/\dot{w}_0)/(2-k)K_2$$

$$+2(1-2k)E_h(w_0+w''-\tilde{w}'')/3a\Big]K_3(h^3/12a)$$

$$M_\theta = \Big[3k^2\bar{\sigma}(1+\dot{w}''/\dot{w}_0)/(2-k)K_2$$

$$+2(2-k)E_h(w_0+w''-\tilde{w}'')/3a\Big]K_3(h^3/12a) \qquad (4.1.45)$$

where we have again replaced the generalized strain at the midsurface with the constant stress $\bar{\sigma}$ approximating its mean value during the motion. Formulas (4.1.44) and (4.1.46) for the circumferential membrane force and bending moment are used in Section 4.1.7 in the derivation of the equation governing the dynamic buckling.

### 4.1.6 Directional Moments

The circumferential bending moment consists of two parts. As seen in formula (4.1.45), the second term is the hardening moment, which is proportional to the hardening modulus $E_h$. The first term ensures that a bending moment exists even with no strain hardening present, providing $k \neq 0$ ($k = 0$ corresponds to infinitely long shells). This contribution is called the directional moment because the strain vector at each location z through a shell cross section has a different direction in the strain-rate plane, a state that corresponds to a stress distribution through the thickness according to the von Mises yield criterion and the associated flow rule.

To simplify the discussion of the directional moment, let the hardening modulus be zero. Then, according to (4.1.42) and (4.1.43), the stresses are

$$\sigma_x = -(1 - 2k)\sigma_0/K_2 + (z/a)\,3kK_3\sigma_0(1 + \dot{w}''/\dot{w}_0)/(2 - k)K_2 \quad (4.1.46)$$

$$\sigma_\theta = -(2 - k)\sigma_0/K_2 + (z/a)3k^2K_3\sigma_0(1 + \dot{w}''/\dot{w}_0)/(2 - k)K_2 \quad (4.1.47)$$

At the midsurface

$$\sigma_x/\sigma_\theta = (1 - 2k)/(2 - k) \quad (4.1.48)$$

and, since k lies in the range $0 < k < 1/2$, we have

$$0 < \sigma_x/\sigma_\theta < 1/2$$

for the midsurface values. This stress state is represented by point M which lies on the thickened portion of the yield ellipse as shown in Figure 4.3. From (4.1.46) and (4.1.47), the difference between the outer fiber stresses ($z = h/2$) and the midsurface stresses are

$$\Delta\sigma_x = (h/2a)3kK_3\sigma_0(1 + \dot{w}''/\dot{w}_0)/(2 - k)K_2 \quad (4.1.49)$$

$$\Delta\sigma_\theta = k\Delta\sigma_x \quad (4.1.50)$$

which, if $\dot{w}'' + \dot{w}_0 > 0$, means that the outer fiber stress state is represented by some adjacent point O to the right of M. Similarly, for $\dot{w}'' + \dot{w}_0 > 0$, the inner fiber ($z = -h/2$) stress state is represented by an adjacent point I to the left of M. If $\dot{w}'' + \dot{w}_0 < 0$, the points I and O lie to the right and left of M. The yield ellipse has the equation

$$\sigma_x^2 - \sigma_x\sigma_\theta + \sigma_\theta^2 = \sigma_0^2 \quad (4.1.51)$$

Differentiation of Equation (4.1.51) with respect to z and substitution of the ratio (4.1.48) lead to the result of Equation (4.1.50) for the tangent condition at M.

The associated flow rule provides strain rate vectors that are outward pointing normals to the yield ellipse. Figure 4.3 shows the vectors at I, M, and O. Each vector has the same component $\dot{\epsilon}_x = k\dot{w}_0/a$, and by Equation (4.1.38), has the component

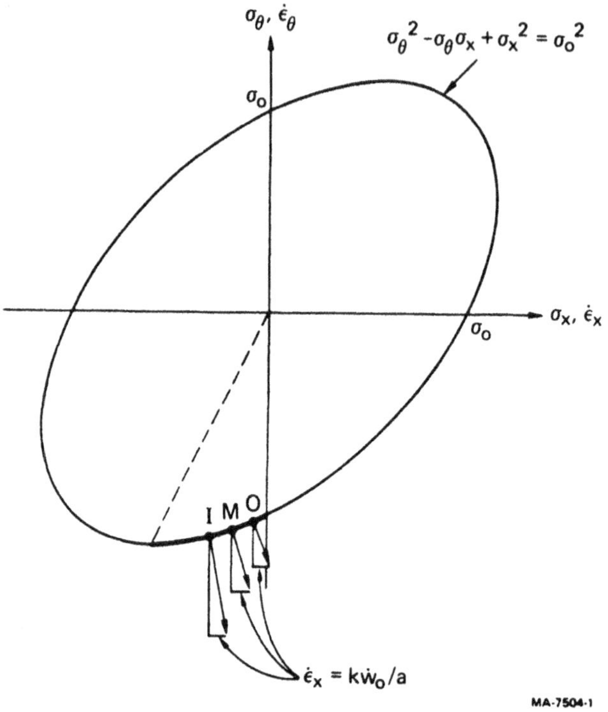


FIGURE 4.3 PLASTIC REGIMES FOR INNER, MIDDLE, AND OUTER FIBERS FOR NO STRAIN HARDENING

$$\dot\epsilon_\theta = -(\dot w_0/a)\left[1 - (z/a)(1 + \dot w''/\dot w_0)\right] \tag{4.1.38a}$$

with the appropriate value of z. At M, $\dot\epsilon_\theta/\dot\epsilon_x = -1/k$ in agreement with the normality condition (because $d\sigma_\theta/d\sigma_x = k$ at M). At the outer fiber where $z = h/2$, Equation (4.1.38a) shows that the magnitude of $\dot\epsilon_\theta$ is less than that at the midsurface if $\dot w'' + \dot w_0 > 0$, and since $\dot\epsilon_x$ remains the same, the point O lies to the right of M. Similarly, I representing the inner fiber, lies to the left of M whenever $\dot w'' + \dot w_0 > 0$.

The quantity $(\dot w'' + \dot w_0)/a^2$ is the rate of change of curvature, $\dot\kappa$, so that $\dot\kappa > 0$ means that the plastic regimes shown in Figure 4.3 are for sections such as $\theta_2$ in Figure 4.4 where the buckling is outward relative to the circle of the unperturbed motion. The relative compressive stresses at I, M, and O show that the resulting moment is resistive. Section $\theta_1$ in Figure 4.4 corresponds to a case of $\dot\kappa < 0$.

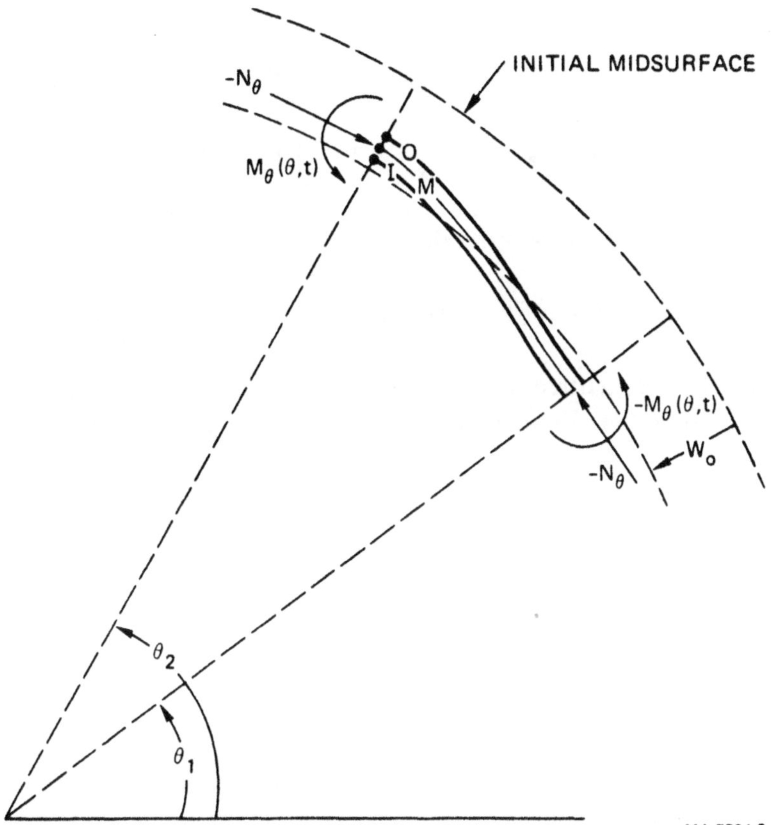


FIGURE 4.4 SHELL SEGMENT

### 4.1.7 Governing Equation

With the aid of the lower part of Figure 4.2, which shows a unit element of shell with its attendant forces and moments, the following two equilibrium equations can be derived:

$$Q = \frac{\partial M_\theta}{\partial \lambda} \quad , \qquad \frac{\partial Q}{\partial \lambda} - N_\theta \frac{\partial \phi}{\partial \lambda} = -\rho h \frac{\partial^2}{\partial t^2}(w_0 + w)$$

The curvature $\kappa = \partial\phi/\partial\lambda$ is given by Equation (4.1.36). Eliminating Q and $\phi$ leads to

$$M_\theta''/a^2 - N_\theta(1 + w_0/a + w''/a)/a + \rho h(\ddot{w}_0 + \ddot{w}) = 0 \qquad (4.1.52)$$

Using the more accurate version of Equation (4.1.20), specifically $N_\theta(1 + w_0/a) = \rho h \ddot{w}_0$, and substituting formulas (4.1.44) and (4.1.45) for $N_\theta$

and $M_\theta$, we find that (4.1.52) becomes

$$\rho h\ddot{w} + h\bar{\sigma}(2 - k)(w''/a)/K_2 a + K_3\Big[3k^2\bar{\sigma}(\dot{w}''/\dot{w}_0)/(2 - k)K_2$$

$$+2(2 - k)E_h(w'' - \tilde{w}'')/3a\Big]''(h^3/12a^3) = 0 \qquad (4.1.53)$$

Introducing the dimensionless variables

$$u = w/a \ , \ \tilde{u} = \tilde{w}/a \ , \ u_0 = w_o/a \ , \ \text{and} \ \ \xi = 1 - \tau/\tau_f \qquad (4.1.54)$$

where $\tau$ and $\tau_f$ are defined by (4.1.27) and (4.1.28), and integrating Equation (4.1.25) to obtain $\dot{w}_0 = V\xi$ enables Equation (4.1.53) to be written in the form

$$\ddot{u} + 2\tau_f u'' - \alpha^2\Big[3k^2K_3(\dot{u}''/\xi)/(2 - k)^2$$

$$-2(2 - k)\tau_f\beta(u'' - \tilde{u}'')/K_2\Big]'' = 0 \qquad (4.1.55)$$

where

$$\alpha^2 = h^2/12a^2 \qquad \beta = E_h/\bar{\sigma} \qquad (4.1.56)$$

and now the dots denote differentiation with respect to $\xi$.

### 4.1.8 Modal Solution

General displacement perturbations are represented by the sum of a Fourier series in $\sin n\theta$ or $\cos n\theta$, but because the differential equation governing the time variation of the coefficients $u_n(\xi)$ for the cosine series is the same as that for the sine series, we consider only the displacement perturbation

$$u(\theta, \xi) = \sum_1^\infty u_n(\xi) \sin n\theta \qquad (4.1.57)$$

having as initial displacement perturbations

$$\tilde{u}(\theta, 1) = \sum_1^\infty a_n \sin \theta \qquad (4.1.58)$$

When we substitute Equations (4.1.57) and (4.1.58) in the governing Equation (4.1.55) we obtain for each amplitude $u_n$ the differential equation

$$\ddot{u}_n - Q_n \dot{u}_n/\xi - R_n^2 u_n = S_n a_n \tag{4.1.59}$$

where

$$Q_n = 3\alpha^2 k^2 K_3 n^4/(2-k)^2 \tag{4.1.60}$$

$$R_n^2 = 2\tau_f n^2\left[1 - (2-k)\alpha^2\beta n^2/K_2\right] \tag{4.1.61}$$

$$S_n = 2\tau_f(2-k)\alpha^2\beta n^4/K_2 \tag{4.1.62}$$

The representation

$$V = V_0\left(1 + \sum_1^\infty b_n \sin n\theta\right) \tag{4.1.63}$$

for initial velocity imperfections and Equation (4.1.58) for initial displacement imperfections means that the coefficients $u_n(\xi)$ in Equation (4.1.58) must satisfy the initial conditions

$$u_n(1) = a_n \qquad \dot{u}_n(1) = -2\tau_f b_n \tag{4.1.64}$$

Whenever $R_n^2 > 0$, the solution of Equation (4.1.38) satisfying initial conditions (4.1.64) is

$$u_n(\xi) = A_n(\xi)a_n + B_n(\xi)b_n \tag{4.1.65}$$

where $A_n$ and $B_n$ are the amplification functions

$$A_n(\xi) = 2\tau_f\xi^\nu\left[I_\nu(R_n\xi)K_{\nu-1}(R_n) + K_\nu(R_n\xi)I_{\nu-1}(R_n)\right]n^2/R_n - S_n/R_n^2 \tag{4.1.66}$$

$$B_n(\xi) = 2\tau_f\xi^\nu\left[-I_\nu(R_n\xi)K_\nu(R_n) + K_\nu(R_n\xi)I_\nu(R_n)\right] \tag{4.1.67}$$

in which

$$\nu = (1 + Q_n)/2 \tag{4.1.68}$$

In Equations (4.1.66) and (4.1.67), $I_\nu$ and $K_\nu$ are modified Bessel functions[2] of the first and second kinds of order $\nu$. For the case $R_n^2 < 0$, expressions similar to (4.1.66) and (4.1.67) arise, but involving $J_\nu$ and $Y_\nu$, which are Bessel functions of the first and second kinds of order $\nu$; the solution is therefore of a damped oscillatory nature with $\xi^\nu J_\nu$ and $\xi^\nu Y_\nu$ approaching zero as $\xi$ approaches zero (i.e., as motion terminates). Thus the growth of modes is possible only when $R_n^2 > 0$ or, from Equation (4.1.61), when $n^2 < K_2/(2 - k)\alpha^2\beta$. This transition is analogous to that occurring in very long shells ($k = 0$), in which only strain-hardening moments are present, where hyperbolic and trigonometric functions occur instead of modified Bessel and Bessel functions.

### 4.1.9 Amplification Functions

We illustrate the nature of amplification functions $A_n$ and $B_n$ by means of an example that corresponds to one of the experiments described in Section 4.1.16 on 6061-T6 aluminum alloy shells. The data for this example are given below

| | | | |
|---|---|---|---|
| shell radius | $a$ | = | 3.734 cm (1.47 inch) |
| shell thickness | $h$ | = | 0.165 cm (0.065 inch) |
| shell length | $L$ | = | 10.160 cm (4.0 inch) |
| flow stress (average) | $\bar{\sigma}$ | = | 290 MPa (42,000 psi) |
| hardening modulus | $E_h$ | = | 2275 MPa (330,000 psi) |
| density | $\rho$ | = | 2.7 g/cm$^3$ (0.097 lb/in$^3$) |
| impulse | $I$ | = | 640 N-s/m$^2$ (0.095 psi-s) |

According to Equation (4.1.35) with the choice $\Omega = \sqrt{3}/2$, the value of k at $x = 0$ is $k = 0.28$. Figures 4.5 and 4.6 show the amplification functions $A_n(\xi)$ and $B_n(\xi)$ from formulas (4.1.66) and (4.1.67) for n between 1 and 26 that give $R_n^2 > 0$ and at time intervals corresponding to $\xi = 0.2$. The curves show that the amplification spectra develop a strong preference for a narrow band of harmonics. It is reasonable to select the most amplified harmonic to represent the predicted mode of buckling. The variation of the preferred harmonic with $\xi$ is shown by the dashed line through the peaks. After $\xi = 0.4$ ($\tau = 0.6\tau_f$), the preferred mode does not change appreciably. At $\xi = 0$ ($\tau = \tau_f$), the preferred mode numbers stemming from initial displacement and velocity perturbations are 16 and 14. The amplification curves for the other shells in the experiments exhibit the same characteristics as those in Figure 4.5.

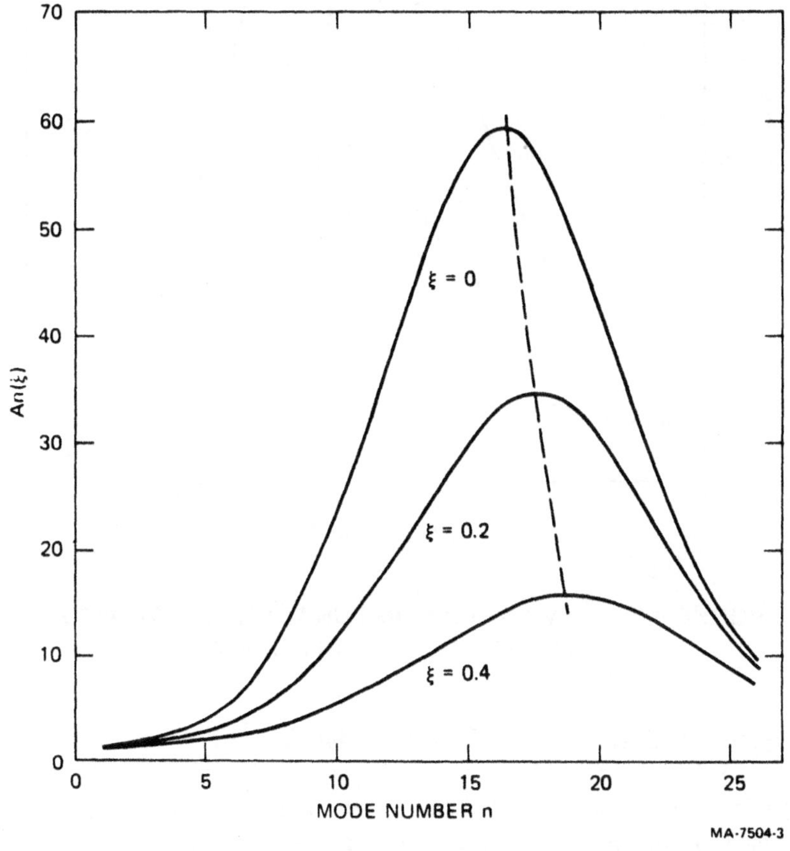


FIGURE 4.5 AMPLIFICATION FUNCTIONS FROM INITIAL DISPLACEMENTS (Shell 1 in Table 4.1)

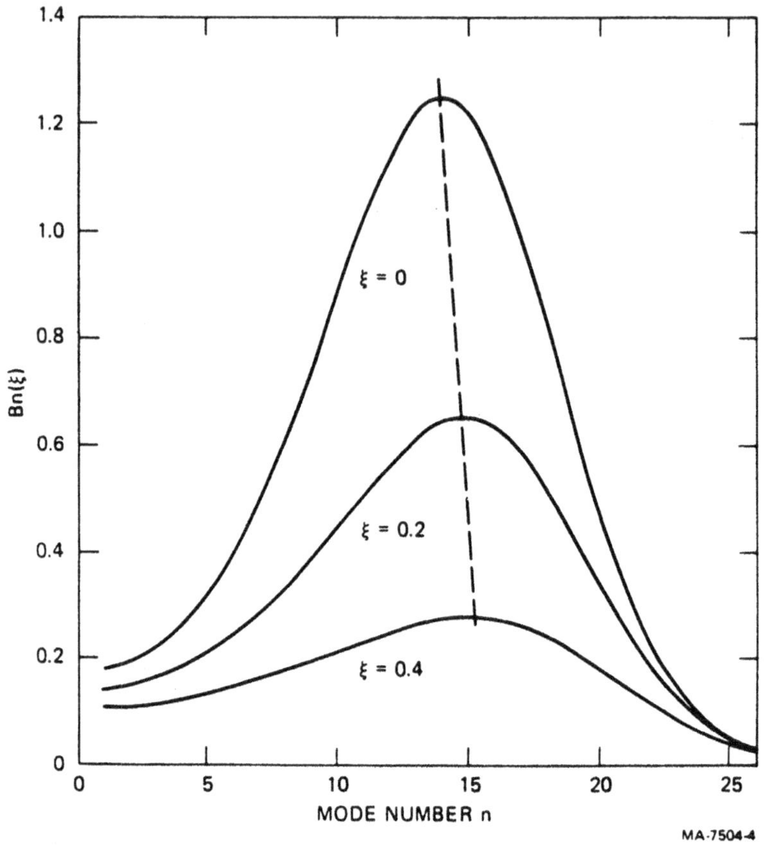


FIGURE 4.6 AMPLIFICATION FUNCTIONS FROM INITIAL VELOCITIES (Shell 1 in Table 4.1)

### 4.1.10 Asymptotic Solutions for Terminal Motion

We can derive simple formulas that approximate the amplification functions at the end of the motion, that is, when $\xi = 0$. It may be shown that

$$\mathop{\mathrm{Lt}}_{\xi \to 0} \xi^\nu I_\nu(R_n \xi) = 0 \qquad \mathop{\mathrm{Lt}}_{\xi \to 0} \xi^\nu K_\nu(R_n \xi) = 2^{\nu-1}\Gamma(\nu)/R_n^\nu$$

where $\Gamma(\nu)$ is the gamma function; from (4.1.60) and (4.1.68), $\nu > \frac{1}{2}$. Let $\bar{n}$ be the value of n above which no amplification occurs so that $R_{\bar{n}} = 0$. Then

$$\bar{n}^2 = K_2/(2 - k)\alpha^2\beta \tag{4.1.69}$$

and

$$S_n/R_n^2 = (n/\bar{n})^2/\left[1 - (n/\bar{n})^2\right]$$

In general, mode numbers in the vicinity of the most amplified mode lie well below $\bar{n}$. In fact, $n \approx \bar{n}/\sqrt{2}$ is a reasonable estimate. Consequently, the term $S_n/R_n^2$ does not contribute substantially to the amplification function $A_n(0)$. Formulas (4.1.66) and (4.1.67) thus reduce to

$$A_n(0) = 2^{\nu}\tau_f\Gamma(\nu)I_{\nu-1}(R_n)n^2/R_n^{\nu+1} \tag{4.1.70}$$

$$B_n(0) = 2^{\nu}\tau_f\Gamma(\nu)I_{\nu}(R_n)/R_n^{\nu} \tag{4.1.71}$$

Further simplification is possible for many practical cases in which $8R_n > 4\nu^2 - 1$, allowing us to use the asymptotic form

$$I_{\nu}(R_n) \approx e^{R_n}\left[1 - (4\nu^2 - 1)/8R_n\right]/(2\pi R_n)^{1/2}$$

in (4.1.70) and (4.1.71) to give

$$A_n(0) \approx 2^{\nu}\tau_f\Gamma(\nu)e^{R_n}n^2\left[1 - (4(\nu - 1)^2 - 1)/8R_n\right]/(2\pi)^{1/2}R_n^{\nu+3/2} \tag{4.1.72}$$

$$B_n(0) \approx 2^{\nu}\tau_f\Gamma(\nu)e^{R_n}\left[1 - (4\nu^2 - 1)/8R_n\right]/(2\pi)^{1/2}R^{\nu+1/2} \tag{4.1.73}$$

We can see whether we are allowed to use the result $8R_n > 4\nu^2 - 1$ by estimating the values of n near the peak of the amplification curves by $n = \bar{n}/\sqrt{2}$, where $\bar{n}$ is the cutoff value determined by (4.1.69). In the example of Section 4.1.9, the cutoff mode number is $\bar{n} = 27$ and the estimated number of the most amplified mode is $N = \bar{n}/\sqrt{2} = 19$. Thus $Q_N = 1.74$, $\nu = 1.37$, and $R_N = 5.51$, giving $8R_n > 4\nu^2 - 1$.

### 4.1.11 Strain Hardening Moments Only

A simple special case is the very long shell for which $k \approx 0$ and the directional moments are negligible. For this case, (4.1.60), (4.1.61), and (4.1.62) are

$$Q_n = 0 \qquad R_n^2 = 2\tau_f n^2(1 - 2\alpha^2\beta n^2/\sqrt{3}) \qquad S_n = 4\tau_f\alpha^2\beta n^4/\sqrt{3}$$

The solution of Equation (4.1.59) with $Q_n = 0$ that satisfies initial conditions (4.1.64) gives the amplification functions

$$A_n(\xi) = (2\tau_f n^2/R_n^2)\cosh R_n(1 - \xi) - S_n/R_n^2 \tag{4.1.74}$$

$$B_n(\xi) = (2\tau_f/R_n)\sinh R_n(1 - \xi) \tag{4.1.75}$$

which, at the end of motion ($\xi = 0$), are approximately

$$A_n(0) \approx \tau_f e^{R_n} n^2/R_n^2 \tag{4.1.76}$$

$$B_n(0) \approx \tau_f e^{R_n}/R_n \tag{4.1.77}$$

Formulas (4.1.76) and (4.1.77) are also obtained if we set $\nu = \frac{1}{2}$, corresponding to $Q_n = 0$, in formulas (4.1.72) and (4.1.73). Regarding $A_n(0)$ as a continuous function of the variable n, so that we can differentiate with respect to n, and equating the first derivative to zero yield the preferred mode number

$$N = \bar{n}/\sqrt{2} = \left(\sqrt{3}/4\alpha^2\beta\right)^{1/2} \tag{4.1.78}$$

and the preferred mode amplification

$$A_N(0) = e^{R_N} \qquad R_N^2 = \tau_f N^2 = \sqrt{3}\,\tau_f/4\alpha^2\beta \tag{4.1.79}$$

In (4.1.78), $\bar{n}$ is again the cutoff mode. Treating the other amplification function (4.1.77) in the same manner leads to the same preferred mode (4.1.78) with the amplification

$$B_N(0) = \tau_f e^{R_N}/R_N \qquad R_N^2 = \tau_f N^2 = \sqrt{3}\,\tau_f/4\alpha^2\beta \tag{4.1.80}$$

When $k = 0$, Equation (4.1.28) for $\tau_f$, the final circumferential strain, becomes

$$\tau_f = \sqrt{3}\,\rho V_0^2/4\bar{\sigma} = \sqrt{3}\,I^2/4\rho h^2\bar{\sigma} \tag{4.1.81}$$

where $I = \rho h V_0$ is the applied impulse. Thus (4.1.79) and (4.1.81) provide the relationship between the impulse and the preferred mode amplification

$$I = \sqrt{\rho E_h}\,(2h^2/3a)\,\ell n A_N \tag{4.1.82}$$

An implicit expression is found for I if $B_n$ is used. Formula (4.1.82) is useful for determining the threshold impulse.

In the example of Section 4.1.9 with $k = 0$, the preferred mode number is $N = 18$, the displacement amplification is $A_n = 228$, and the impulse is $I = 6500$ dyne-s/cm$^2$ (650 N-s/m$^2$).

### 4.1.12 Directional Moments Only

The special case of negligible strain hardening is obtained by assuming that $\beta$ is small enough to give $S_n \approx 0$ and $R_n^2 \approx 2\tau_f n^2$; $Q_n$ remains the same as (4.1.60). Note that $R_n^2$ is always positive, so all modes are amplified. However, if n is large enough to make $4\nu > R_n^2$, then $A_n(0)$ in (4.1.70) reduces to

$$A_n(0) = 1 + \frac{(R_n/2)^2}{1.\nu} + \frac{(R_n/2)^4}{1.2.\nu(\nu + 1)} + \ldots \qquad (4.1.83)$$

which shows that substantial amplification is possible only for mode numbers satisfying $4\nu < R_n^2$, that is, for

$$n^2 < (2 - k)K_1\tau_f/3k^2\alpha^2 \qquad (4.1.84)$$

If k is a small number corresponding to a long shell, inequality (4.1.84) implies that all modes are capable of being amplified substantially; let $k \to 0$ in (4.1.84) $(K_1 \to 2)$. If k is small, the restoring directional moment is small and the buckling is highly unstable. If $k \approx ½$, corresponding to a short shell, inequality (4.1.84) becomes

$$n^2 < 3\tau_f/\alpha^2 \qquad (4.1.85)$$

For the example in Section 4.1.9, inequality (4.1.85) gives amplification of modes with $n < 40$. With strain-hardening moments only, (4.1.78) gives the cutoff mode number $\bar{n} = 26$.

The amplification functions are again (4.1.70) and (4.1.71), which for cases with $8R_n > 4\nu^2 - 1$, become (4.1.72) and (4.1.73). With $R_n^2 = 2\tau_f n^2$, differentiation of (4.1.72) with respect to n shows that the maximum value of $A_n(0)$ occurs when n is a solution of

$$R_n - 2Q_n \ln(R_n/2) - Q_n/2 + 2Q_n\psi - Q_n(7Q_n - 6)/8R_n = 0 \qquad (4.1.86)$$

where $\psi = \psi(\nu)$ is the Psi or Digamma function.[2] Similarly the maximum value of $B_n(0)$ occurs when n is a solution of

$$R_n - 2Q_n \ln(R_n/2) - Q_n/2 - 1 + 2Q_n\psi - Q_n(7Q_n + 6)/8R_n = 0 \qquad (4.1.87)$$

The preferred modes determined by (4.1.86) and (4.1.87) can have substantial amplification when $2(Q_n + 1) < R_n^2$ and $2(Q_n + 3) < R_n^2$, respectively. Also, (4.1.86) and (4.1.87) are valid approximate equations for the preferred modes when $Q_n(Q_n - 2) < 8R_n$ and $Q_n(Q_n + 2) < 8R_n$, respectively; otherwise, n must be found to maximize $A_n(0)$ and $B_n(0)$ in the forms (4.1.70) and (4.1.71).

Figure 4.7 shows the two curves for the real values of $Q_n$ and $R_n$ that satisfy (4.1.86) and (4.1.87). For numerical convenience, these curves may be approximated by the straight lines

$$2Q_n = 2R_n/3 + 1 \tag{4.1.88}$$

$$16Q_n = 5(R_n - 1) \tag{4.1.89}$$

These equations are of the form

$$n^4 - a_1 n + a_0 = 0 \qquad a_1 > 0 \tag{4.1.90}$$

where the constants $a_0$ and $a_1$ are readily determined from (4.1.60) and (4.1.61), with $\beta = 0$, for $Q_n$ and $R_n$.

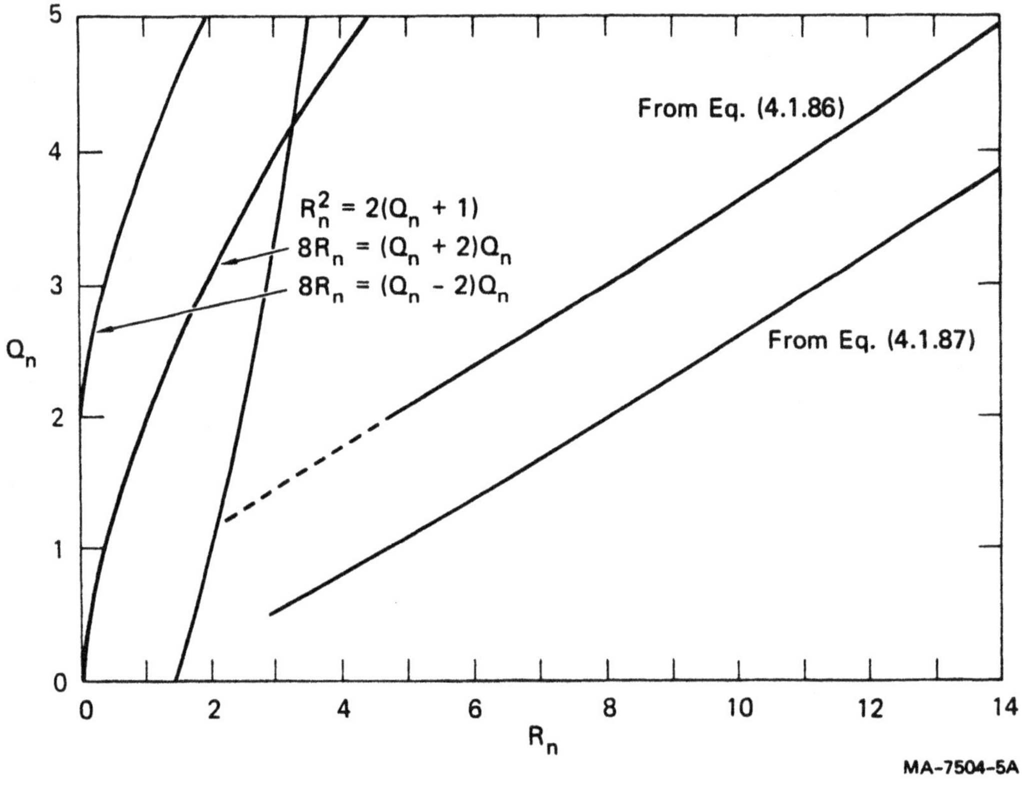


FIGURE 4.7 $Q_n$ *vs* $R_n$ FOR PREFERRED MODE NUMBER (No Strain Hardening)

An adequate and simpler approximation for the preferred mode number is obtained by letting the straight lines pass through the origin. Then, instead of (4.1.88) and (4.1.89) we have

$$Q_n = 5R_n/14 \qquad Q_n = 5R_n/17 \tag{4.1.91}$$

or

$$N^3 = c\sqrt{2\tau_f}\,(2 - k)K_1/3k^2\alpha^2 \tag{4.1.92}$$

where

$c = 5/14$ for displacement imperfections
$c = 5/17$ for velocity imperfections.

In the example of Section 4.1.9, formula (4.1.92) predicts preferred mode numbers of $N = 22$ and 21 for amplification of displacement and velocity imperfections. For $N = 22$, we have $Q_n = 3.3$ and $R_n = 9.2$.

The above approximations for obtaining the explicit formula (4.1.92) for the preferred mode allow us to derive an explicit relationship between the applied impulse and the amplification of the preferred mode stemming from initial lack of circularity. The procedure outlined below for obtaining this explicit relationship leads to an implicit relationship between the impulse and the amplification of the preferred mode stemming from initial velocity imperfections.

In (4.1.72) we shall first neglect the term $[4(\nu - 1)^2 - 2]/8R_n$. Then, since $R_n^2 = 2\tau_f n^2$, formula (4.1.72) becomes

$$A_n(0) = 2^{\nu-1}\Gamma(\nu)e^{R_n}/(2\pi)^{1/2}R_n^{\nu-1/2} \tag{4.1.93}$$

With $\nu > 1/2$, an approximation adequate for our derivation is

$$\Gamma(\nu) \approx e^{-\nu}\nu^{\nu-1/2}(2\pi)^{1/2} \tag{4.1.94}$$

which reduces (4.1.93) to

$$\sqrt{2e}\;A_n(0) = e^{R_n}(2\nu/eR_n)^{\nu-1/2} \tag{4.1.95}$$

After taking the logarithm of (4.1.95) and substituting the relationship $Q_n = CR_n$ from (4.1.91) with $c = 5/14$, we obtain

$$\ln\left(\sqrt{2e}\;A_n\right) = R_n\left[1 - (c/2)\ln\left\{eQ_n/c(Q_n + 1)\right\}\right] \tag{4.1.96}$$

We now make use of the observation that the second term in the square brackets in (4.1.96) is a slowly varying function of $Q_n$. As $Q_n$ increases from 1 to $\infty$, this function increases from 0.24 to 0.36. A reasonable value is 0.3, corresponding to value of $Q_n = 2.5$, which is in a practical range of $Q_n$ values. This approximation reduces (4.1.96) to

$$\ell n\left(\sqrt{2e}\, A_n\right) = c_0 R_n \qquad c_0 = 0.7 \tag{4.1.97}$$

For the preferred mode, $R_n^2 = 2\tau_f N^2$, where N is given by (4.1.92). The dimensionless duration of motion or final hoop strain $\tau_f$ is related to the initial uniform velocity $V_0$ and hence the impulse by (4.1.28). Consequently, (4.1.97) may be manipulated into the form

$$I = \left(k^2K_3/4c_0^3cK_2^2\right)^{1/4}\sqrt{\rho\bar{\sigma}}\; a(h/a)^{3/2}\left[\ell n\left(\sqrt{2e}\, A_n\right)\right]^{3/4} \tag{4.1.98}$$

In the example of Section 4.1.9, formula (4.1.97) predicts an amplification of $A_n = 270$ and formula (4.1.98) predicts an impulse of $I = 6320$ dyne-s/cm$^2$ (632 N-s/m$^2$).

### 4.1.13 Directional and Hardening Moments

In the general case that includes directional and hardening moments, we must retain the general forms (4.1.60), (4.1.61), and (4.1.62) for the coefficients $Q_n$, $R_n$, and $S_n$ in the governing differential Equation (4.1.59). We recall that mode growth is possible only if $R_n^2 > 0$, that is, if $n < \bar{n}$ where $\bar{n}$ is a cutoff mode number given by (4.1.69). In this section we continue our discussion of the function (4.1.79) that amplifies the initial displacement imperfections. This function reduces to the series (4.1.83) if n is large enough to make $R_n^2 < 4\nu = 2(Q_n + 1)$, so substantial amplification occurs only if $R_n^2 > 2(Q_n + 1)$. The amplification function may be approximated by (4.1.72) if $8R_n > 4(\nu - 1)^2 - 1 = (Q_n - 2)Q_n$.

If we substitute expressions (4.1.60) and (4.1.61) for $Q_n$ and $R_n$ in the inequality $R_n^2 > 2(Q_n + 1)$, we obtain an upper bound that is below $\bar{n}$ for the numbers of the modes that are amplified. This substitution gives

$$n^2 < \tilde{n}^2 = \bar{n}^2/(2q + 1) \tag{4.1.99}$$

where

$$q = 3k^2K_2/2(2 - k)^2K_1\tau_f\beta \tag{4.1.100}$$

The value of n that maximizes $A_n(0)$ given by (4.1.72) is the solution of

$$R_n - 2Q_n \ell n(R_n/2) - Q_n/2 + 2Q_n\psi$$

$$- Q_n(7Q_n - 6)/8R_n = (R_n - Q_n/2 - 2)Q_n/qR_n^2 \quad (4.1.101)$$

where q is defined by (4.1.100) and $\psi = \psi(\nu)$ is again the Psi function.[2] When hardening is negligible, Equation (4.1.101) is the same as Equation (4.1.87). Figure 4.8 shows solution curves that relate $Q_n$ and $R_n$ for several values of the parameter q. These curves may be approximated by straight lines through the origin so that

$$Q_n = C(q)R_n \quad (4.1.102)$$

The function C(q) is satisfactorily represented by

$$C(q) = c\left(1 - c_1 e^{-q/q_0}\right) \quad (4.1.103)$$

where $c = 0.36$, $c_1 = 0.85$, $q_0 = 0.10$, and $0.01 < q$.

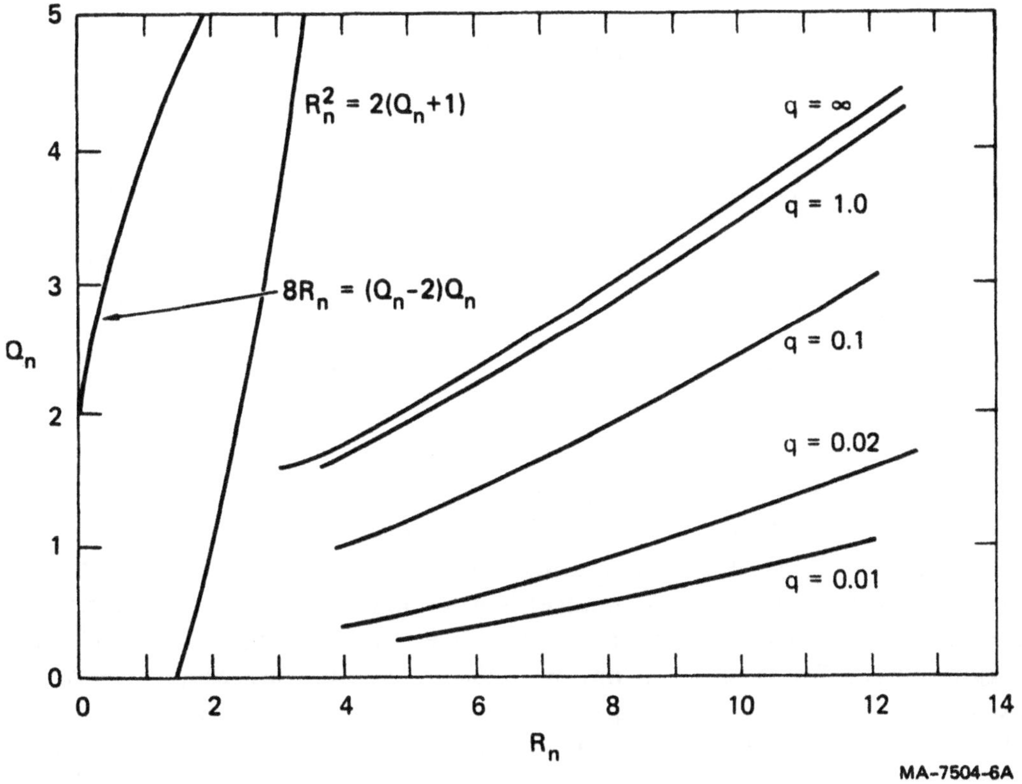


FIGURE 4.8 $Q_n$ *vs* $R_n$ FOR PREFERRED MODE NUMBER
(No Strain Hardening Moment: q = ∞; No Direction Moment: q = 0)

If we substitute expressions (4.1.60) and (4.1.61) for $Q_n$ and $R_n$ in (4.1.102) we obtain for the preferred mode number the formula

$$N^3 = C(q)\sqrt{2\tau_f}\,(2 - k)K_1 f\left(\frac{N}{\bar{n}}\right)/3k^2\alpha^2 \qquad 0.01 < q \qquad (4.1.104)$$

where

$$f\left(\frac{N}{\bar{n}}\right) = \left[1 - \left(\frac{N}{\bar{n}}\right)^2\right]^{1/2} \qquad (4.1.105)$$

and $C(q)$ is determined by (4.1.103), q being the parameter in (4.1.100). Negligible strain hardening ($\beta \approx 0$) corresponds to large values of q for which $C(q) \approx c$ and $N/\bar{n} \approx 0$. In these cases, (4.1.105) reduces to (4.1.92). Negligible directional moments ($k \approx 0$) corresponds to small values of q for which $C(q) \approx c(1 - c_1)$. In these cases, (4.1.105) does not reduce to the predictive formula (4.1.78). However, if curves of the type shown in Figure 4.8 are approximated by parabolas whenever $q < 0.01$, that is, if

$$Q_n = D(q)R_n^2 + d(q) \qquad q < 0.01 \qquad (4.1.106)$$

Equation (4.1.78) can be obtained in the limit as $q \rightarrow 0$. Suitable functions that allow a fit of the curves in Figure 4.8 and ensure the correct limiting prediction of N as $q \rightarrow 0$ are

$$D(q) = q(1 - 44q) \qquad d(q) = q^2(1.08 - 80q)10^4 \qquad (4.1.107)$$

If we substitute expressions (4.1.60) and (4.1.61) for $Q_n$ and $R_n$ in Equation (4.1.106), we obtain a quadratic equation for $N^2$, the solution of which may be expressed in the form

$$(N/\bar{n})^2 = \left[1 + \{1 + 2d(q/D + 1)/\tau_f\bar{n}^2 D\}^{1/2}\right]/2(q/D + 1) \qquad q < 0.01 \quad (4.1.108)$$

Formulas (4.1.107) ensure that $(N/\bar{n})^2 \rightarrow 2$ as $q \rightarrow 0$, in agreement with (4.1.78) for the case of vanishing directional moments. The factor in (4.1.108) that strongly influences N is $\bar{n}$, which according to its definition (4.1.69), is primarily a function of $\beta$. Thus the preferred mode number is determined by the hardening parameter $\beta$ whenever $q < 0.01$. In the upper part of the range of q, the slopes $C(q)$ from Figure 4.8 and from the approximating function (4.1.103) are within 20% of the asymptotic value of $c = 0.36$ at $q = \infty$ whenever $q > 0.2$. Inspection of (4.1.104) shows that whenever $q > 0.2$, the preferred mode number is weakly dependent on $\beta$ through the function (4.1.105) and hence N is predominantly influenced by the directional moments. In the range $0.01 < q < 0.2$, the major influence on N changes from hardening to directional as q increases.

Figure 4.9 shows how the preferred mode number depends on the hardening and directional parameters $\beta$ and k for the shell of Section 4.1.9 with a/h = 22.6 and $\tau_f = 0.05$. The full lines were obtained from (4.1.104) for $0.01 < q$, and the dashed lines were obtained from (4.1.108) for $q < 0.01$. The main feature is that the mode number decreases as $\beta$ or k increases, demonstrating the increase of wavelength with resistive moment. For very short shells having $0.4 < k \leqslant 0.5$, directional moments predominate. Even for very long shells having $0 < k \leqslant 0.1$, directional moments have a strong influence if $\beta$ is small enough (say $\beta < 5$ in Figure 4.9).

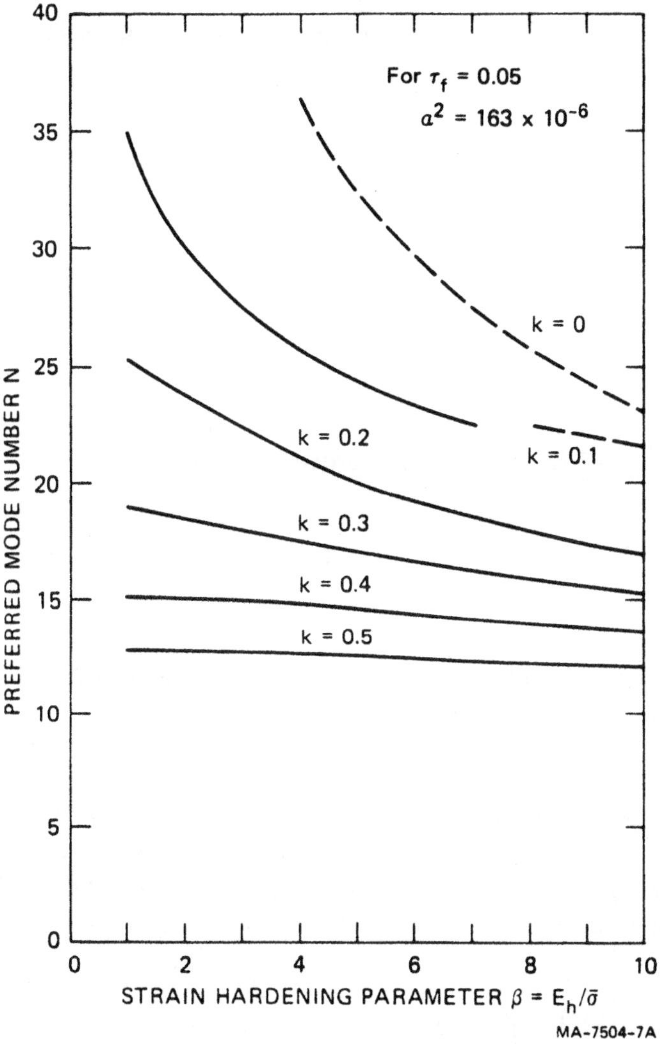


FIGURE 4.9 INFLUENCE OF DIRECTIONAL AND HARDENING MOMENTS ON PREFERRED MODE NUMBER

We now use (4.1.104) and (4.1.108) to obtain formulas relating the final amplification factor of the preferred mode to the applied impulse. In (4.1.72), we neglect the term $[4(\nu-1)^2-1]/8R_n$ and make use of the approximation (4.1.94) to give

$$\sqrt{2e}\;A_n(0) = (1 + S_n/R_n^2)e^{R_n}(2\nu/eR_n)^{\nu-1/2} \tag{4.1.109}$$

Formula (4.1.109) reduces to (4.1.95) when $\beta = 0$. If we introduce (4.1.102) for $0.01 < q$ and (4.1.106) for $0 < q < 0.01$, (4.1.109) becomes

$$\ell n(\sqrt{2e}\;A_n) = C_0(q)R_n + \ell n(1 + S_n/R_n^2) \tag{4.1.110}$$

where

$$C_0(q) = \begin{cases} 1 - \left(\dfrac{C}{2}\right)\ell n\left(\dfrac{eR_n}{CR_n+1}\right) & 0.01 < q \\ 1 - \left(\dfrac{DR_n^2+d}{2R_n}\right)\ell n\left(\dfrac{eR_n}{DR_n^2+d+1}\right) & q < 0.01 \end{cases} \tag{4.1.111}$$

When $0.01 < q$, $C_0(q)$ does not depend strongly on $R_n$. When $q < 0.01$, $C_0(q) \approx 1$. In fact, $C_0(q)$ can be represented adequately by

$$C_0(q) = \begin{cases} c_0 + (1-c_0)e^{-q/q_0} & 0.01 < q \qquad c_0 = 0.71 \qquad q_0 = 0.06 \\ 1 & 0 < q < 0.01 \end{cases} \tag{4.1.112}$$

Again, the term in (4.1.110) that includes $S_n$ does not contribute substantially to the amplification. The term is retained only for $0 < q < 0.01$ but replaced by $\ell n\sqrt{2e}$.

Formulas (4.1.102) and (4.1.103) relating $Q_n$ and $R_n$ for $0.01 < q$, formulas (4.1.106) and (4.1.107) relating $Q_n$ and $R_n$ for $q < 0.01$, formulas (4.1.104) and (4.1.108) for the preferred mode number, and (4.1.28) relating the final strain $\tau_f$ to the impulse are all used in (4.1.110) to derive the expressions

$$I = \begin{cases} \left(\dfrac{k^2K_3}{4K_2^2C_0^3C}\right)^{1/4} f\left(\dfrac{N}{\bar{n}}\right)\sqrt{\rho\bar{\sigma}}\,a\left(\dfrac{h}{a}\right)^{3/2}\left[\ell n\sqrt{2e}\;A_n\right]^{3/4} & 0.01 < q \qquad (4.1.113) \\ \left(\dfrac{2-k}{2\sqrt{3}K_2}\right) F\left(\dfrac{N}{\bar{n}}\right)\sqrt{\rho E_h}\,a\left(\dfrac{h}{a}\right)^2 \ell n\,A_n & q < 0.01 \qquad (4.1.114) \end{cases}$$

where

$$f\left(\frac{N}{\bar{n}}\right) = \frac{N}{\bar{n}}F\left(\frac{N}{\bar{n}}\right) = \left[1 - \left(\frac{N}{\bar{n}}\right)^2\right]^{-1/2} \quad (4.1.115)$$

The impulse formula (4.1.114) approaches (4.1.82) for vanishing directional moments, as k (and hence q) approaches zero.

Figure 4.10 shows the variation with k and $\beta$ of the coefficient in the impulse-amplification formula (4.1.113); for a given shell, Figure 4.10 therefore shows how the impulse-amplification relationship depends on directional and hardening moments, represented by parameters k and $\beta$. For $k > 0.2$ (shell length less than two diameters), the impulse is not sensitive to strain hardening; the

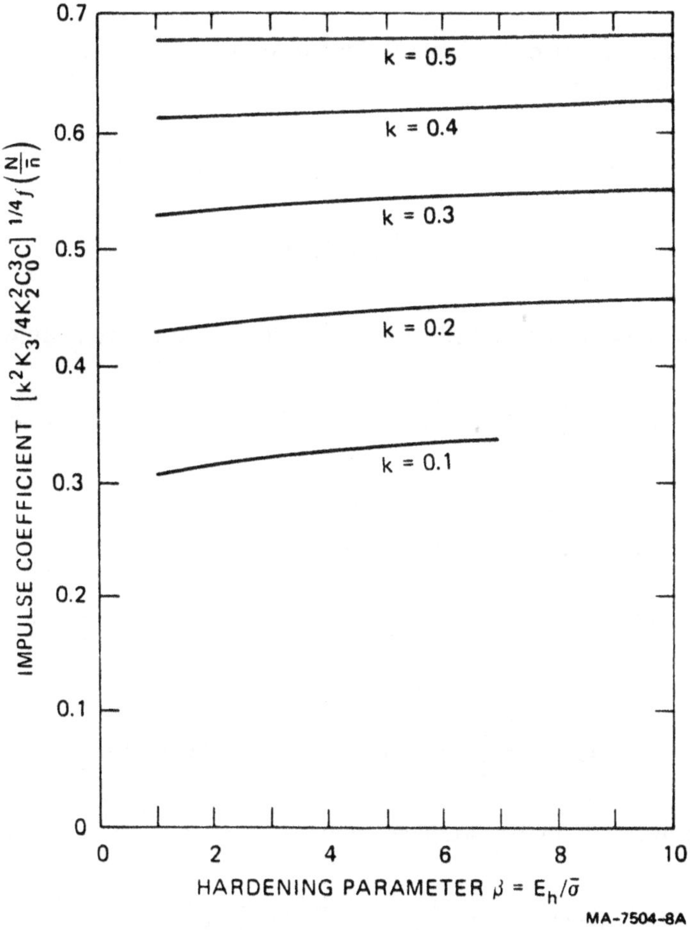


FIGURE 4.10 INFLUENCE OF DIRECTIONAL AND HARDENING MOMENTS ON THRESHOLD IMPULSE

coefficient increases from about 0.45 and 0.70 as k increases from 0.2 to 0.5. Thus, for a specified amplification $A_n$, an increase in k from 0.2 to 0.5 increases the impulse required to produce this amplification by about 50%.

We have seen how hardening and directional moments influence the predicted preferred mode number and the impulse-amplification relationship. For the remainder of this section, we examine the moment components themselves.

From formula (4.1.45) the directional and hardening moments are

$$M_d = \left[3k^2K_3/(2 - k)K_2\right]\bar{\sigma}(h^3/12a)(1 + \dot{w}''/\dot{w}_0) \qquad (4.1.116)$$

$$M_h = \left[2(2 - k)K_3/3\right]E_h(h^3/12a^2)(w_0 + w' - \tilde{w}'') \qquad (4.1.117)$$

In terms of the dimensionless quantities introduced in Section 4.1.7, Equations (4.1.116) and (4.1.117) are

$$M_d = -\left[3k^2K_3/2(2 - k)K_2\right](\bar{\sigma}ha\alpha^2/\tau_f)(\dot{u}''/\xi) \qquad (4.1.118)$$

$$M_h = \left[2(2 - k)K_3/3\right](\bar{\sigma}ha\alpha^2\beta)(u - \tilde{u})'' \qquad (4.1.119)$$

Substitution in Equations (4.1.118) and (4.1.119) of the Fourier representations

$$M_d = \sum_1^\infty M_d^{(n)}(\xi)\sin n\theta \qquad M_h = \sum_1^\infty M_h^{(n)}(\xi)\sin n\theta \qquad (4.1.120)$$

along with the similar representations of (4.1.57) and (4.1.58) for the current and initial displacement perturbations provides the coefficients

$$M_d^{(n)}(\xi) = \left[3k^2K_3/2(2 - k)K_2\right](\bar{\sigma}ha\alpha^2/\tau_f)n^2\dot{u}_n/\xi \qquad (4.1.121)$$

$$M_h^{(n)}(\xi) = -\left[2(2 - k)K_3/3\right](\bar{\sigma}ha\alpha^2\beta)n^2(u_n - a_n) \qquad (4.1.122)$$

Displacement imperfections alone lead to

$$\dot{u}_n = \dot{A}_n(\xi)a_n \qquad u_n - a_n = \left[A_n(\xi) - 1\right]a_n$$

where the amplification function is determined by (4.1.66). A measure of the relative magnitude of the directional and hardening moments may be obtained by the ratio

$$\Lambda_a^{(n)} = \frac{\int_1^0 M_d^{(n)} d\xi}{\int_1^0 M_d^{(n)} d\xi} = \frac{-q\int_0^1 \dot{A}_n d\xi/\xi}{\int_0^1 (A_n - 1)d\xi} \quad (4.1.123)$$

where q is given by (4.1.100). Substitution of $A_n$ from (4.1.66) into (4.1.123) and integration lead to

$$\Lambda_a^{(n)} = (qR_n^2/Q_n)\Big/\Big[1 - R_n^{\nu}/2^{\nu-1}\pi^{1/2}\Gamma(\nu + \tfrac{1}{2})L_{\nu-1}(R_n)\Big] \quad (4.1.124)$$

where $L_\nu$ is the modified Struve function[2] of order $\nu$. If we treat the amplification functions for the initial velocity perturbations in the same way, we can derive

$$\Lambda_b^{(n)} = (qR_n^2/Q_n)\Big/\Big[1 + R_n^{\nu-1}/2^{\nu-1}\pi^{1/2}\Gamma(\nu + \tfrac{1}{2})L_{\nu}(R_n)\Big] \quad (4.1.125)$$

As an illustration of the magnitude of the ratio (4.1.124) let us take the practical value $Q_n = 2$, which gives $\nu = 3/2$. If n is the preferred mode number, $Q_n = C(q)R_n$ where C(q) is determined by (4.1.103) for $0.01 < q$. When $\nu = 3/2$, the ratio (4.1.124) becomes

$$\Lambda_a^{(n)} = (qR_n^2/Q_n)\Big/\Big[1 - R_n^2/2(\cosh R_n - 1)\Big] \quad (4.1.126)$$

The smallest value of $\Lambda_a$ is achieved by the smallest permissible value of $R_n$, which in turn is obtained by the maximum permissible C(q). Now, by (4.1.103), max.C(q) = 0.36 whenever $q > 0.3$ and, since $Q_n = 2$, we have $R_n = 5.6$. With these values of q, $Q_n$, and $R_n$, we obtain $\Lambda_a^{(n)} = 5.3$ for the ratio of directional to hardening moments according to our definition (4.1.123). Again for $Q_n = 2$ but with q reduced to $q = 0.1$ we obtain a smaller ratio $\Lambda_a^{(n)} = 2.5$. Fixing $Q_n$ fixes k, so reducing q, given by (4.1.100), means increasing the hardening by increasing $\beta$, which accounts for a reduced moment ratio.

### 4.1.14 Displacement and Velocity Imperfections

The final shell buckling deformation $w(\theta, 0)$ stems from the growth of initial displacement of velocity imperfections. We would like to compare these initial imperfections when they give rise to equal contributions to the final shape. We simplify the procedure for obtaining this comparison if we visualize the final shape as consisting only of the preferred mode

$$u(\theta,o) = (A_N a_N + B_N b_N)\sin N\theta \tag{4.1.127}$$

The amplification functions $A_N$ and $B_N$ are given by (4.1.70) and (4.1.71). Thus the amplitude of mode (4.1.127) is

$$\max u(\theta,o) = 2^{\nu}\tau_f\Gamma(\nu)\Big[I_{\nu-1}(R_N)N^2 a_N + I_\nu(R_N)R_N b_n\Big]/R_N^{\nu+1} \tag{4.1.128}$$

For equal contributions to this amplitude stemming from the growth of initial displacement and velocity imperfections, the imperfection ratio has to be

$$b_N/a_N = N^2 I_{\nu-1}(R_N)/R_N I_\nu(R_N) \tag{4.1.129}$$

Many practical cases have $8R_N > 4\nu^2 - 1$ so that $I_{\nu-1} \approx I_\nu$, which allows the ratio (4.1.129) to be estimated by

$$b_N/a_N \approx N^2/R_N = N/(2\tau_f)^{1/2}\Big[1 - (N/\bar{n})^2\Big]^{1/2} \tag{4.1.130}$$

Because the preferred mode number N is less than the cutoff mode number $\bar{n}$, we have

$$b_N/a_N > N/(2\tau_f)^{1/2} \tag{4.1.131}$$

In the example of Section 4.1.9, the preferred mode number and the final unperturbed hoop strain are $N = 19$ and $\tau_f = 0.0865$, so (4.1.131) gives $b_N/a_N > 46$. This result indicates that for equal contributions to the final amplitude the velocity imperfection has to be a much larger percentage of the initial uniform velocity than the displacement imperfection is of the initial uniform shell radius.

### 4.1.15 Threshold Impulse

In Section 4.1.13, we derived Equations (4.1.113) and (4.1.114) for the impulse-amplification relationships. If the directional moment is negligible, the impulse formula is (4.1.82), and if the strain hardening moment is negligible, the impulse formula is (4.1.98). Figure 4.11 shows the impulse-amplification curves constructed from formulas (4.1.113) and (4.1.114) for shell 1, chosen from the experimental shells of Section 4.1.16. Each curve is associated with a fixed shell length and hence a fixed value of k. The curves show the exponential rise of the amplification with increasing impulse and, for a fixed amplification, the increase of impulse with decreasing shell length. The discontinuity in the curve L = 8 is caused by mismatch of the two approximate predictive formulas (4.1.113) and (4.1.114).

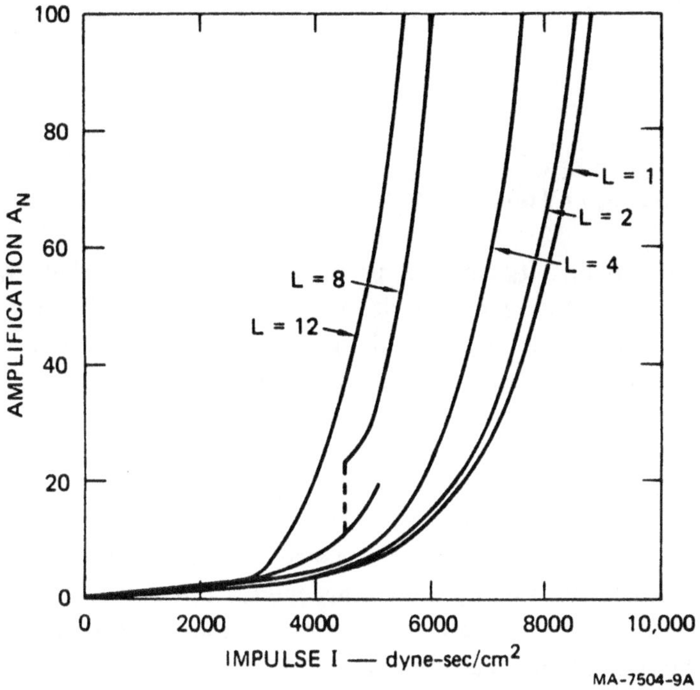


FIGURE 4.11 IMPULSE-AMPLIFICATION CURVES (Shell 1 in Table 4.1)

A reasonable and simple choice of a criterion for the threshold impulse is to take as the amplification the value $A_N = 100$. For this value, quantities required in the impulse formulas are

$$\left[\ln(\sqrt{2e}\ A_N)\right]^{3/4} = 3.57 \qquad \ln A_N = 4.61$$

For long shells with $k \approx 0$ the threshold impulse, from (4.1.82), is

$$I_c = 3.07\sqrt{\rho E_h}\ a(h/a)^2 \tag{4.1.132}$$

For short shells with $k \approx 0.5$ and a strain-hardening value such that $q > 0.5$, the threshold impulse, from (4.1.98), is

$$I_c = 2.46\sqrt{\rho\bar{\sigma}}\ a(h/a)^{3/2} \tag{4.1.133}$$

If we consider again shell 1 with lengths L = 12 inches (30.5 cm) and L = 1 inch (2.54 cm) ($k \approx 0$ and $k \approx 0.5$), formulas (4.1.132) and (4.1.133) give threshold impulses of 5600 and 7600 dyne-s/cm$^2$ (560 and 760 N-s/cm$^2$).

### 4.1.16 Comparison of Theory and Experiment

Table 4.1 contains the main experimental data. Each shell was impulsively loaded with sheet explosive as described in Section 3.2.7. Profiles were taken at ½-inch intervals along the shells, and the number of waves at each section was recorded. The mode numbers were highly reproducible and the averages are listed in the last column.

The experimental values of k were determined from the final strains $\epsilon_x$ and $\epsilon_\theta$. The axial strains $\epsilon_x$ were found by scribing fine grid lines, approximately 1-cm apart, along an outer generator. The distances between the grid lines were measured before and after loading by a traveling microscope accurate to one thousandth of a millimeter. In each shell the strain distribution was regular except for increases in the flared regions at the ends.

The circumferential strain was found by the following two methods. For the short shells in which no flaring occurred, the end profiles of the outer and inner surfaces were reproduced directly by standing the shells on a flat surface and tracing around the contact circles. A circle representing an average of the outer and inner traces was constructed to determine $\epsilon_\theta$ at the midsurface. For the larger shells, in which flaring was noticeable at the ends, the outer circumference at several positions along the shell was found by measuring the length of a strip required to pass around the shell. The midsurface strain was then found by adjustment for the shell thickness. Again, the strain distribution was regular except for a slight decrease near the flared ends. An interesting observation was the slight decrease in mode number near the ends of the shells corresponding to the local increase in k because of the adjacent free ends.

The values of the final strain ratio and the mode number given in Table 4.1 are those prevailing over the uniform region away from the ends. These values of k are plotted in Figure 4.12. Also shown for comparison is the theoretical strain ratio derived in Section 4.1.4; this function of the length-to-diameter ratio L/D is formula (4.1.35) with $x = 0$ and, for the best fit, with the parameter $\Omega = \sqrt{3}/2$. The experimental and theoretical k-values are in agreement.

Table 4.1

EXPERIMENTAL DATA FOR CYLINDRICAL SHELLS OF 6061-T6 ALUMINUM

| Shell Number | Number of Shells | Radius a (cm) | Thickness h (cm) | a/h | Length L (cm) | Impulse I (N-s/m²) | Strain $\epsilon_\theta$ (%) | $k = \frac{\epsilon x}{\epsilon_\theta}$ | Mode N |
|---|---|---|---|---|---|---|---|---|---|
| | 3 | | | | 2.54 | 650 | 9.4 | 0.43 | 15 |
| | 3 | | | | 5.08 | 650 | 9.4 | 0.37 | 15 |
| 1 | 2 | 3.729 | 0.165 | 22.6 | 10.16 | 650 | 9.4 | 0.26 | 16 |
| | 1 | | | | 20.32 | 630 | 9.4 | 0.10 | 19 |
| | 1 | | | | 30.48 | 630 | 9.4 | 0.04 | 20 |
| | 3 | | | | 2.54 | 660 | 4.8 | 0.48 | 14 |
| 2 | 3 | 3.706 | 0.211 | 17.6 | 5.08 | 660 | 4.8 | 0.38 | 15 |
| | 2 | | | | 10.16 | 660 | 4.8 | 0.31 | 15 |
| | 3 | | | | 2.54 | 690 | 5.3 | 0.45 | 11 |
| | 3 | | | | 5.08 | 690 | 5.3 | 0.45 | 11 |
| 3 | 2 | 3.691 | 0.241 | 15.3 | 10.16 | 690 | 5.3 | 0.31 | 12 |
| | 1 | | | | 30.48 | 690 | 5.2 | 0.06 | 16 |

Shell data: 6061-T6 aluminum.
$E_h$ = 2275 MPa (330,000 psi), $\bar{\sigma}$ = 290 MPa (42,000 psi),
$\rho$ = 2.7 g/cm³ (0.097 lb/in³).

Table 4.1 (concluded)

EXPERIMENTAL DATA FOR CYLINDRICAL SHELLS OF 6061-T6 ALUMINUM

| Shell Number | Number of Shells | Radius a (inch) | Thickness h (inch) | a/h | Length L (inch) | Impulse I (dyne-s/cm²) | Strain $\epsilon_\theta$ (%) | $k = \frac{\epsilon x}{\epsilon_\theta}$ | Mode N |
|---|---|---|---|---|---|---|---|---|---|
| | 3 | | | | 1 | 6500 | 9.4 | 0.43 | 15 |
| | 3 | | | | 2 | 6500 | 9.4 | 0.37 | 15 |
| 1 | 2 | 1.468 | 0.065 | 22.6 | 4 | 6500 | 9.4 | 0.26 | 16 |
| | 1 | | | | 8 | 6300 | 9.4 | 0.10 | 19 |
| | 1 | | | | 12 | 6300 | 9.4 | 0.04 | 20 |
| | 3 | | | | 1 | 6600 | 4.8 | 0.48 | 14 |
| 2 | 3 | 1.459 | 0.083 | 17.6 | 2 | 6600 | 4.8 | 0.38 | 15 |
| | 2 | | | | 4 | 6600 | 4.8 | 0.31 | 15 |
| | 3 | | | | 1 | 6900 | 5.3 | 0.45 | 11 |
| | 3 | | | | 2 | 6900 | 5.3 | 0.45 | 11 |
| 3 | 2 | 1.453 | 0.095 | 15.3 | 4 | 6900 | 5.3 | 0.31 | 12 |
| | 1 | | | | 12 | 6900 | 5.2 | 0.06 | 16 |

Shell data: 6061-T6 aluminum.
$E_h$ = 2275 MPa (330,000 psi), $\bar{\sigma}$ = 290 MPa (42,000 psi),
$\rho$ = 2.7 g/cm³ (0.097 lb/in³).

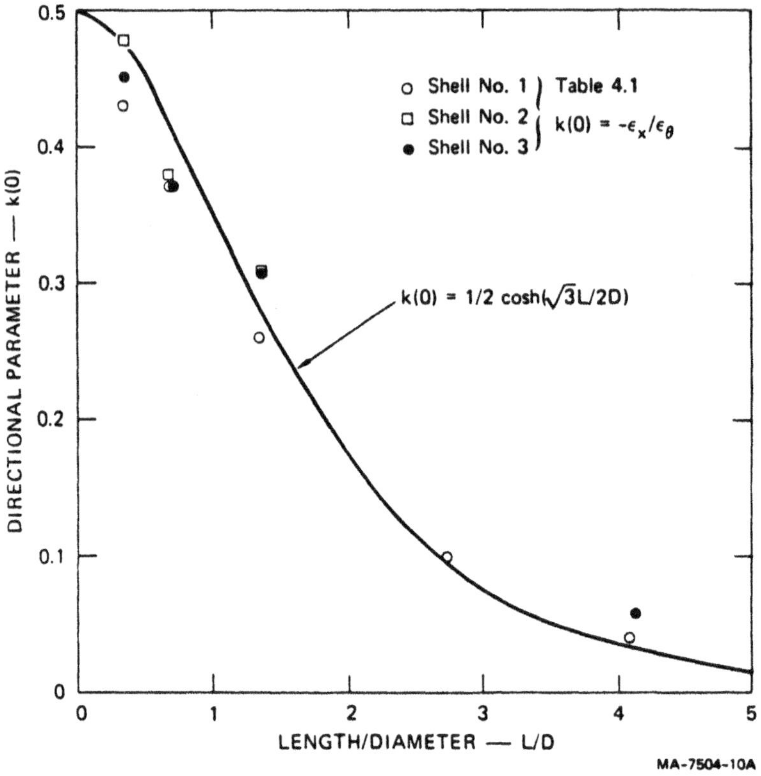


FIGURE 4.12 COMPARISON OF THEORETICAL AND EXPERIMENTAL DIRECTIONAL PARAMETERS

The mode numbers were obtained from profiles such as the two samples shown in Figure 4.13 for shell 1 of Table 4.1. As shown, the samples are for shell lengths of L = 2 inches and L = 12 inches. The average count from all profiles at ½-inch spacing, and from repeat tests in the case of the 2-inch-long shells, resulted in preferred mode numbers of N = 15 and N = 20, respectively. The experimental mode numbers for all the shells are listed in Table 4.1 and, for comparison with predictions, in Table 4.2.

The mode numbers were predicted in three ways. In the first way, each number in the column headed by $A_n(0)$ is the mode number at the peak of the amplification curve at the end of motion ($\xi = 0$, $\tau = \tau_f$), the growth of the mode amplitude stemming from initial displacement imperfections. The preferred mode number is therefore the value of N that maximizes $A_n(0)$ given by formula (4.1.70). An example of the curve $A_n(0)$ versus n is shown in Figure 4.5. In the second way, each number in the column headed by $B_n(0)$ is similarly determined, the growth now stemming from initial velocity imperfections. An example of the curve $B_n(0)$ versus n obtained from formula (4.1.71) is shown in Figure 4.6. In

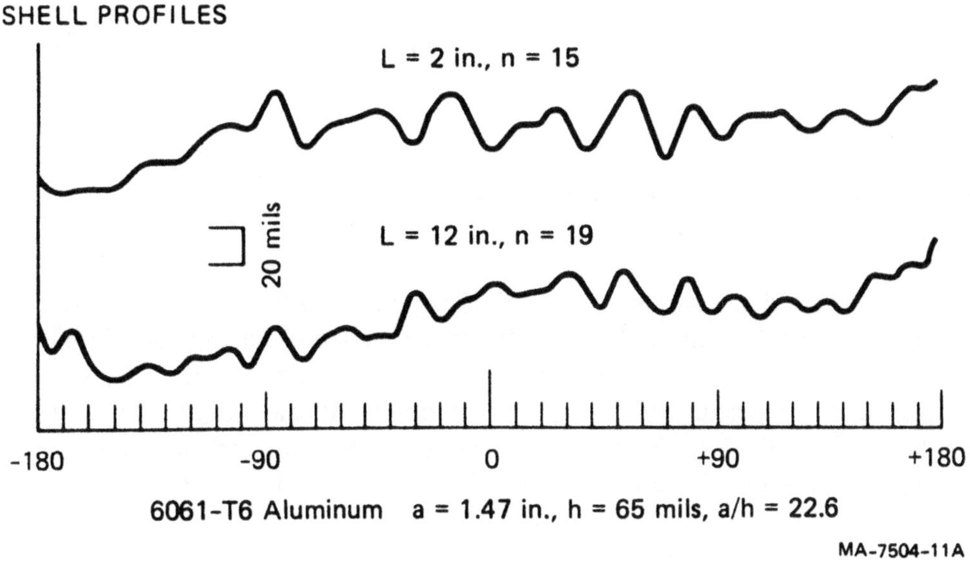


FIGURE 4.13 PROFILES OF SHORT AND LONG SHELLS (Shell 1 of Table 4.1)

Table 4.2

THEORETICAL AND EXPERIMENTAL MODE NUMBERS FOR CYLINDRICAL SHELLS OF 6061-T6 ALUMINUM

| | | | | Preferred Mode Number N | | | |
|---|---|---|---|---|---|---|---|
| Shell Number | Impulse ($N\text{-}s/m^2$) | Strain $\tau_f$ (%) | k(0) | $A_m(0)$ | $B_m(0)$ | Formulas | Experiment N |
| | | 9.5 | 0.48 | 13 | 11 | 13 | 15 |
| | | 9.2 | 0.42 | 14 | 12 | 14 | 15 |
| 1 | 6400 | 8.6 | 0.28 | 16 | 14 | 16 | 16 |
| | | 8.4 | 0.09 | 20 | 18 | 22 | 19 |
| | | 8.3 | 0.03 | 20 | 18 | 19 | 20 |
| | | 6.2 | 0.48 | 11 | 8 | 11 | 14 |
| 2 | 6600 | 6.0 | 0.42 | 12 | 9 | 12 | 15 |
| | | 5.6 | 0.28 | 14 | 11 | 14 | 15 |
| | | 5.2 | 0.48 | 10 | 7 | 10 | 11 |
| 3 | 6900 | 5.0 | 0.42 | 11 | 7 | 10 | 12 |
| | | 4.7 | 0.28 | 13 | 9 | 12 | 12 |
| | | 4.5 | 0.03 | 16 | 12 | 15 | 16 |

Average impulse.

the third way, the column headed "Formulas" refers to prediction of N by (4.1.104) or (4.1.108). The comparison of the predicted and experimental mode numbers in Table 4.2 shows general agreement.

## 4.2 VISCOPLASTIC FLOW BUCKLING WITH DIRECTIONAL MOMENTS

Figure 4.14 demonstrates that cylindrical shells made of strain-rate sensitive material such as fully annealed 1015 steel also experience dynamic plastic buckling when they are subjected to large enough uniform radially inward impulses. The buckling is reproducible and exhibits a predominant wavelength. A theory to describe this buckling is postulated here based on the resistive moments having a viscoplastic and a directional component. For simplicity of analysis, a linear viscoplastic constitutive relation is adopted. The preferred mode numbers compare favorably with mode numbers obtained from experiments on fully annealed 1015 steel cylinders with a length-to-diameter ratio of two (L/D = 2) and radius-to-thickness ratios of 10.5, 16.7, 22.8, and 35.6.

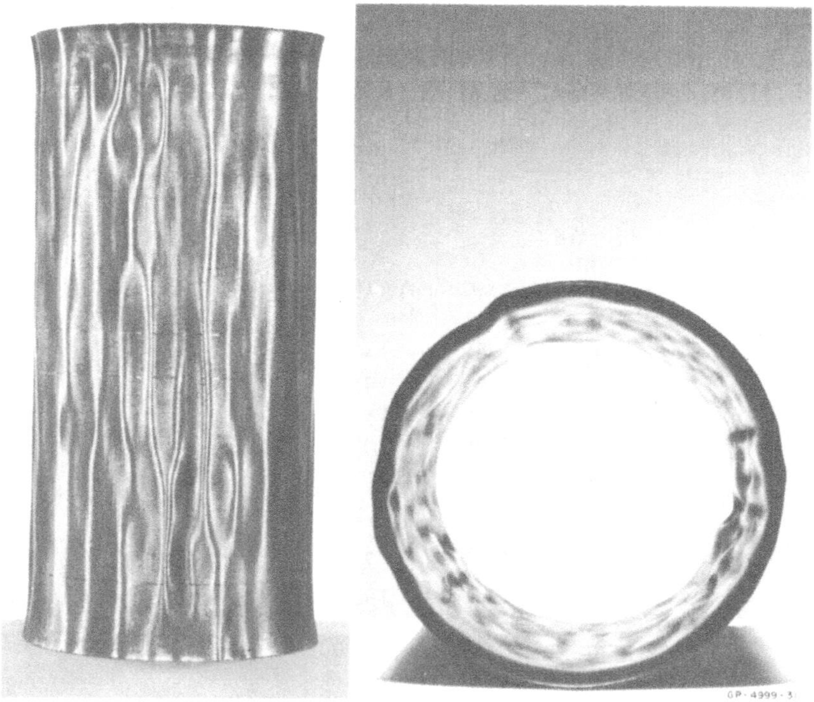

FIGURE 4.14 TYPICAL BUCKLED SHAPE OF A 1015 STEEL CYLINDRICAL SHELL SUBJECTED TO AN INWARD RADIAL IMPULSE

### 4.2.1 Viscoplastic Moments

Figure 4.15 illustrates how a resistive moment arises during flow buckling because of the viscoplastic material property. It is assumed that the compressive circumferential strain is increasing at every fiber of the shell. At two points A and B, on the outside and inside of a cross section where an outward portion of a wave exists, the strain rate is greater at B than at A. On the linearized stress, strain-rate curve the corresponding states are shown by the points A′ and B′. The higher strain rate at B given by the point B′ on the strain-rate line is associated with a higher compressive stress than at A. Thus a moment exists on the cross section, which is resisting the curvature increase. As deformation proceeds, the points A′ and B′ move to the left, as indicated in Figure 4.15, which is consistent with the shell coming to rest. The points separate because of curvature increase, so the moment increases with deformation.

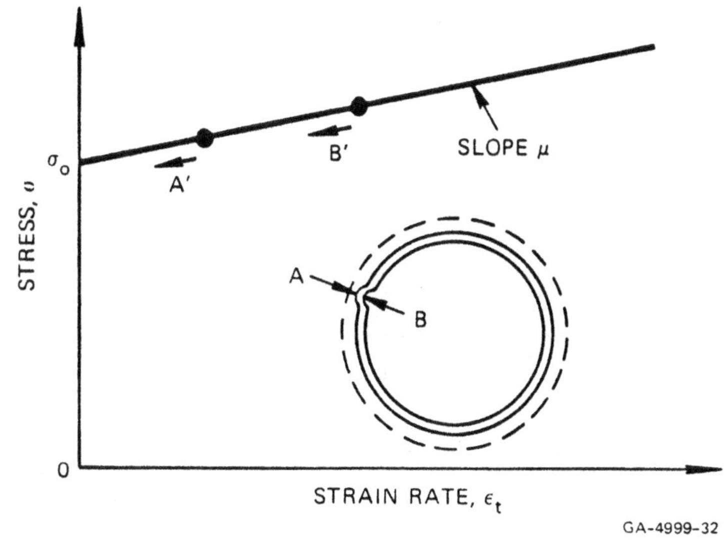


FIGURE 4.15 FIBER STRESSES ACCORDING TO THE LINEARIZED STRESS, STRAIN-RATE LAW

### 4.2.2 Theory of Viscoplastic Cylindrical Shells

The first portion of the theoretical treatment of Section 4.1.1 providing Equations (4.2.1) through (4.2.9) introduces incompressibility, generalized strain rate and stress, deviatoric stresses, and the flow law. Constitutive equations consistent with the flow law are taken in the form[3]

$$\gamma\dot{\epsilon}_x = \sigma_0 F \frac{\partial F}{\partial \sigma_x} \qquad \gamma\dot{\epsilon}_\theta = \sigma_0 F \frac{\partial F}{\partial \sigma_\theta} \qquad \gamma\dot{\epsilon}_z = \sigma_z F \frac{\partial F}{\partial \sigma_z} \qquad (4.2.1)$$

where $F = \sigma/\sigma_0 - 1 > 0$, $\sigma_0$ being the generalized stress at infinitesimally small strain rates and $\gamma$ the viscosity constant of the material. Equations (4.2.1) represent a special case of the constitutive equations of Perzyna,[3] and they provide the following relationship between the generalized strain rate (4.1.2) and the generalized stress (4.1.3)

$$\gamma\dot{\epsilon} = \sigma/\sigma_0 - 1 \tag{4.2.2}$$

The viscosity constant $\gamma$ is the slope of the straight line approximating the results of uniaxial stress experiments (Figure 4.16).

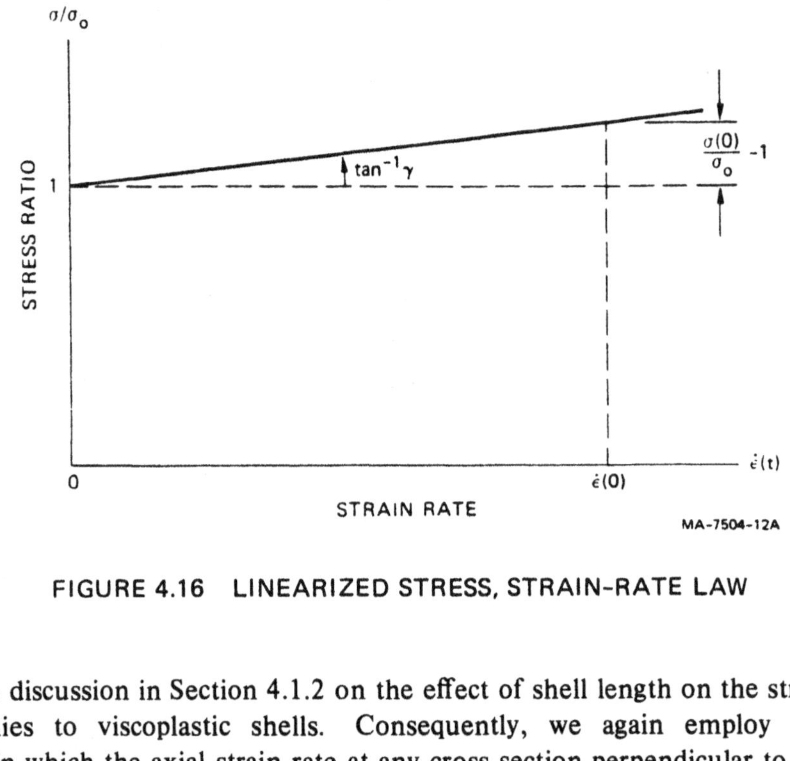


FIGURE 4.16 LINEARIZED STRESS, STRAIN-RATE LAW

The discussion in Section 4.1.2 on the effect of shell length on the strain rates also applies to viscoplastic shells. Consequently, we again employ Equation (4.1.11) in which the axial strain rate at any cross section perpendicular to the shell axis is proportional to the midsurface circumferential strain of the unperturbed motion. The proportionality parameter $k(x)$ lies in the range $0 < k < \frac{1}{2}$ covering very long to very short shells.

### 4.2.3 The Unperturbed Motion

If the shell cross section remains perfectly circular during its inward motion, the circumferential and axial strain rates are given by (4.1.12) and (4.1.13). Also, the generalized strain rate is given by (4.1.15). We substitute these strain rates into

(4.1.9) and (4.2.2) to give the stresses

$$\sigma_x = -(1 - 2k)\sigma^0/K_2 + (z/a)(3k^2/K_1K_2 + 2\gamma\dot{w}_0/3a)\sigma_0 \quad (4.2.3)$$

$$\sigma_\theta = -(2 - k)\sigma^0/K_2 + (z/a)(3k^2/K_1K_2 + 4\gamma\dot{w}_0/3a)\sigma_0 \quad (4.2.4)$$

$$\sigma = \sigma\left[1 + 2K_2(1 - K_3z/a)\gamma\dot{w}_0/3a\right] \quad (4.2.5)$$

$$\sigma^0 = \sigma_0(1 + 2K_2\gamma\dot{w}_0/3a) \quad (4.2.6)$$

By comparing (4.2.5) and (4.2.6), we see that $\sigma^0$ is the current generalized stress at the midsurface.

Equation (4.1.20) is again the equation of motion in which the circumferential membrane force, by integration of (4.2.4) through the shell thickness h is again (4.1.21), where now $\sigma^0$ is defined by (4.2.6). Hence, the equation of motion may be written as

$$\ddot{w}_0 + 2(2 - k)\sigma_0\gamma\dot{w}_0/3a^2\rho = -(2 - k)\sigma_0/K_2a\rho \quad (4.2.7)$$

The solution of (4.2.7) with initial conditions

$$w_0(0) = 0 \qquad \dot{w}_0(0) = V_0 \quad (4.2.8)$$

is

$$w_0 = \left(V_0 + \frac{3a}{2\gamma K_2}\right)\left[1 - e^{-2(2-k)\sigma_0\gamma t/3a^2\rho}\right]\frac{3a^2\rho}{2(2 - k)\sigma_0\gamma} - \frac{3at}{2\gamma K_2} \quad (4.2.9)$$

Solution (4.2.9) is useful if a numerical approach is taken to solve the buckling problem. Our approach is to continue the analysis as far as possible with the objective of obtaining useful design formulas. Our approach requires approximations. One approximation that allows analysis to proceed is based on the observation that for many practical materials the generalized stress does not vary greatly over a wide range of strain rates. Thus we take the unperturbed generalized stress $\sigma$ to be the constant stress $\bar{\sigma}$ averaged over the strain-rate range of the problem, that is, $\bar{\sigma} = [\sigma(0) - \sigma_0]/2$. This assumption reduces the equation of motion to

$$\ddot{w}_0 = -(2 - k)\bar{\sigma}/K_2a\rho \quad (4.2.10)$$

which, with initial conditions (4.2.8) has the solution

$$w_0/a = \tau(2 - \tau/\tau_f) \tag{4.2.11}$$

where

$$\tau = V_0 t/2a \qquad \tau_f = \rho V_0^2 K_2/2(2 - k)\bar{\sigma} \tag{4.2.12}$$

The unperturbed motion is similar to that described in Section 4.2.3 for strain-hardening materials except for the obvious difference in the manner of selecting the average generalized stress. Again, $\tau_f$ is the maximum midsurface circumferential strain to which the shell is driven by the impulsive load.

### 4.2.4 Perturbed Motion

Following the initial treatment in Section 4.1.5 provides formulas (4.1.13), (4.1.38), and (4.1.39) for the axial, circumferential, and generalized strain rates. Substitution of the generalized strain rate in the constitutive Equation (4.2.2) gives the generalized stress

$$\sigma = \sigma_0\left[1 + 2K_2\left\{\dot{w}_0 - K_3(z/a)(\dot{w}_0 + \dot{w}'')\right\}\gamma/3a\right] \tag{4.2.13}$$

Substitution of the strain rates and the generalized stress (4.2.13) in the stress formulas (4.1.9) gives the stresses

$$\begin{aligned}\sigma_x = &-(1 - 2k)\sigma^0/K_2 \\ &+ (z/a)(3k^2/K_1K_2 + 2\gamma\dot{w}_0/3a](1 + \dot{w}''/\dot{w}_0)\sigma_0\end{aligned} \tag{4.2.14}$$

$$\begin{aligned}\sigma_\theta = &-(2 - k)\sigma^0/K_2 \\ &+ (z/a)(3k^2/K_1K_2 + 4\gamma\dot{w}_0/3a](1 + \dot{w}''/\dot{w}_0)\sigma_0\end{aligned} \tag{4.2.15}$$

The membrane forces and bending moments resulting from stress distributions (4.2.14) and (4.2.15) through the shell thickness h are

$$\begin{aligned}N_x &= -h(1 - 2k)\bar{\sigma}/K_2 \\ N_\theta &= -h(2 - k)\bar{\sigma}/K_2\end{aligned} \tag{4.2.16}$$

$$M_x = (3k^2/K_1K_2 + 2\gamma\dot{w}_0/3a)(1 + \dot{w}''/\dot{w}_0)\sigma_0 h^3/12a$$

$$M_\theta = (3k^2/K_1K_2 + 4\gamma\dot{w}_0/3a)(1 + \dot{w}''/\dot{w}_0)\sigma_0 h^3/12a \qquad (4.2.17)$$

where we have again replaced the generalized stress at the midsurface with a constant average value.

The circumferential bending moment (4.2.17) that resists buckling may be considered as comprising two components, one a directional moment $M_d$ and the other a viscoplastic moment, $M_s$. The directional moment is described in Section 4.1.6. From (4.2.17) we obtain the ratio

$$M_s/M_d = (4\gamma K_1K_2/9k^2)\dot{w}_0/a \qquad (4.2.18)$$

which we use in Section 4.2.11 to investigate the relative importance of directional and viscoplastic moments for resisting buckling.

### 4.2.5 Governing Equation

In Section 4.1.7 we derived the equation of motion (4.1.52). We now eliminate from this equation the circumferential membrane force and bending moment by using the results (4.2.16) and (4.2.17). We also note that $N_\theta(1 + w_0/a) = \rho h\ddot{w}_0$. Thus we obtain

$$\rho h\ddot{w} + \bar{\sigma}h(2 - k)(w''/a)/K_2 a$$
$$+ (3k^2/K_1K_2\dot{w}_0 + 4\gamma/3a)\dot{w}''''\sigma_0 h^3/12a^3 = 0 \qquad (4.2.19)$$

which, in terms of the dimensionless variables u, $\xi$, $\tau_f$, and $\alpha$ introduced in Section 4.1.7 by (4.1.54) and (4.1.56), is

$$\ddot{u} - \alpha^2(3k^2/K_1\xi + 4\gamma V_0/3a)\dot{u}''''/(2 - k) + 2\tau_f u'' = 0 \qquad (4.2.20)$$

In the conversion from (4.2.19) to (4.2.20) we have assumed for simplicity that the coefficient $\sigma_0/\bar{\sigma}$ of the second term in (4.2.2) is approximately unity. By comparing (4.1.55) and (4.2.20), we see that, in addition to the inertial and buckling thrust terms, the terms having $\xi^{-1}$ are the same because they both stem from the directional part of the restoring moment.

### 4.2.6 Modal Solution

Following the procedure of Section 4.1.8 by substituting the displacement representation (4.1.57) in the governing Equation (4.2.20) leads to

$$\ddot{u}_n - (P_n + Q_n/\xi)\dot{u}_n - R_n^2 u_n = 0 \tag{4.2.21}$$

where

$$P_n = 4\alpha^2\gamma V_0 K_2 n^4/3(2-k)a \tag{4.2.22}$$

$$Q_n = 3\alpha^2 k^2 n^4/(2-k)K_1 \tag{4.2.23}$$

$$R_n^2 = 2\tau_f n^2 \tag{4.2.24}$$

The solution of (4.2.21) that satisfies the initially perturbed displacements and velocities (4.1.58) and (4.1.63) is

$$u_n(\xi) = A_n(\xi)a_n + B_n(\xi)b_n \tag{4.2.25}$$

in which the amplification functions are

$$\begin{aligned} A_n(\xi) &= \Gamma(1+a_n-b_n)(\omega_n\xi)^{1-b_n}e^{-(\lambda_n+\omega_n)+\lambda_n\xi} \\ &\Big[-\{(\lambda_n+1-b_n)U_n+\omega_n U_n'\}M_n(\xi) \\ &+\{(\lambda_n+1-b_n)M_n+\omega_n M_n'\}U_n(\xi)\Big]/\Gamma(2-b_n) \end{aligned} \tag{4.2.26}$$

$$\begin{aligned} B_n(\xi) &= \Gamma(1+a_n-b_n)(\omega_n\xi)^{1-b_n}e^{-(\lambda_n+\omega_n)+\lambda_n\xi} \\ &\Big[-U_n M_n(\xi)+M_n U_n(\xi)\Big]2\tau_f/\Gamma(2-b_n) \end{aligned} \tag{4.2.27}$$

In (4.2.26) and (4.2.27), the following shorthand notation is used to represent Kummer functions[2] (confluent hypergeometric functions) of the first and second kind

$$\begin{aligned} M_n(\xi) &= M(1+a_n-b_n, 2-b_n, \omega_n\xi) \\ U_n(\xi) &= U(1+a_n-b_n, 2-b_n, \omega_n\xi) \\ M_n &= M(1+a_n-b_n, 2-b_n, \omega_n) \\ U_n &= U(1+a_n-b_n, 2-b_n, \omega_n) \end{aligned} \tag{4.2.28}$$

In Equation (4.2.26) the primes over M and U denote differentiation with respect to $\omega_n\xi$; the result is then evaluated at $\xi = 1$. The remaining symbols are

$$a_n = -Q_n(1 - P_n/\omega_n)/2 \qquad b_n = -Q_n$$

$$\omega_n = (P_n^2 + 4R_n^2)^{1/2} \qquad \lambda_n = -(\omega_n - P_n)/2 \qquad (4.2.29)$$

When the strain rate contribution to the restoring moment is small relative to the directional contribution, the viscosity constant may be set equal to zero. Then according to (4.2.22) we have $P_n = 0$, which leads to $A_n = -Q_n/2$, $b_n = -Q_n$, $\omega_n = 2R_n$, and $\lambda_n = -R_n$. These special values allow the amplification functions to be written in terms of modified Bessel functions, resulting in Equations (4.1.66) and (4.1.67), with $S_n = 0$ $(\beta = 0)$, describing amplification against resistive directional moments only.

When the directional contribution to the restoring moment is small relative to the strain rate contribution, we may set $k = 0$ (long shell) so that, by (4.2.23), $Q_n = 0$. Thus $A_n = 0$ and $b_n = 0$, which give the amplification functions the special forms

$$A_n(\xi) = \left[(\omega_n + P_n)e^{(\omega_n - P_n)(1-\xi)/2} + (\omega_n - P_n)e^{-(\omega_n + P_n)(1-\xi)/2}\right]/2\omega_n \qquad (4.2.30)$$

$$B_n(\xi) = \left[e^{(\omega_n - P_n)(1-\xi)/2} - e^{-(\omega_n + P_n)(1-\xi)/2}\right]2\tau_f/\omega_n \qquad (4.2.31)$$

The above amplification functions are generally more easily obtained by direct solution of differential Equation (4.2.21) with $Q_n = 0$.

### 4.2.7 Amplification Functions

We again illustrate the nature of the amplification functions (4.2.26) and (4.2.27) by an example from a set of experiments on impulsively loaded cylindrical shells of fully annealed 1015 steel. The data for this example, shell 3a are given below

| | | | |
|---|---|---|---|
| shell radius | $a$ | = | 3.785 cm (1.49 inches) |
| shell thickness | $h$ | = | 0.226 cm (0.089 inch) |
| shell length | $L$ | = | 15.24 cm (6.0 inches) |
| flow stress | $\sigma_0$ | = | 600 MPa (87,020 psi) |
| viscosity | $\mu$ | = | $53.5 \times 10^6$ s$^{-1}$ |
| density | $\rho$ | = | 7.8 g/cm$^3$ (0.28 lb/in$^3$) |
| impulse | $I$ | = | 2038 N-s/cm$^2$ (0.296 psi-s) |

Equation (4.1.35) gives k(0) = 0.17 and (4.2.12) with $V_0 = I/\rho h$ = 116 m/s (4550 in./s) gives $\tau_f$ = 0.078, so the duration of motion is $t_f$ = 51 $\mu$s.

Instead of using Equations (4.2.26) and (4.2.27) for $A_n(\xi)$ and $B_n(\xi)$, we numerically integrated Equation (4.2.21) to obtain the amplification curves in Figures 4.17 and 4.18. Again these curves show how a strong preference develops for amplifying a narrow band of harmonics. The preferred modes are centered on n = 16 for $A_n$ and n = 14 for $B_n$. The variation of the preferred modes with $\xi$ are shown by the dashed lines through the peaks of the spectral curves. They show

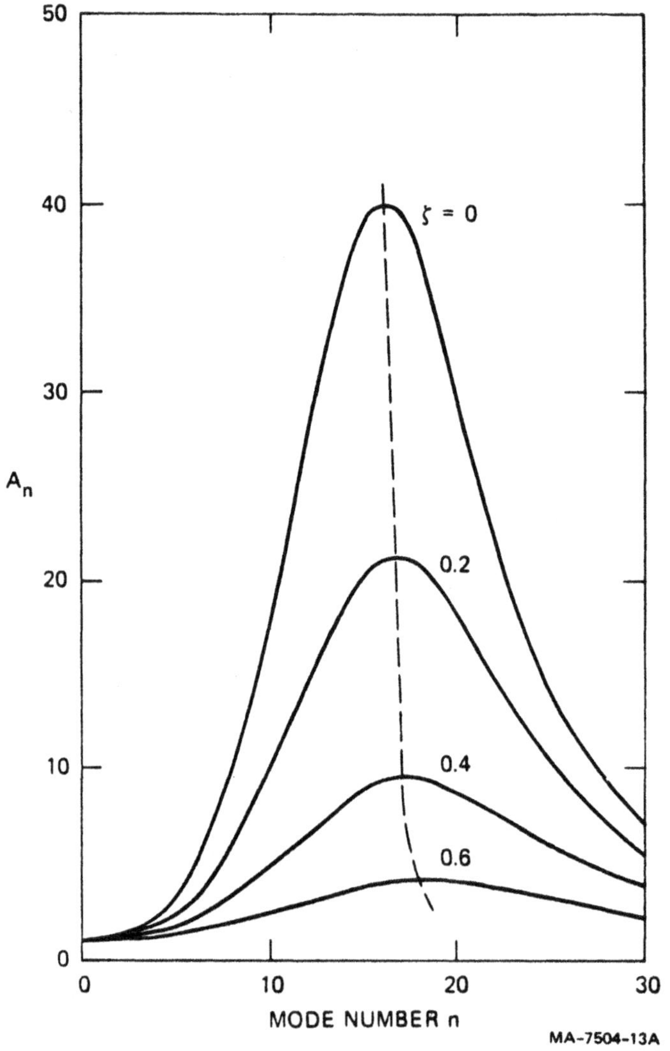


FIGURE 4.17 AMPLIFICATION FUNCTION FROM INITIAL DISPLACEMENTS (Shell 3a in Table 4.3)

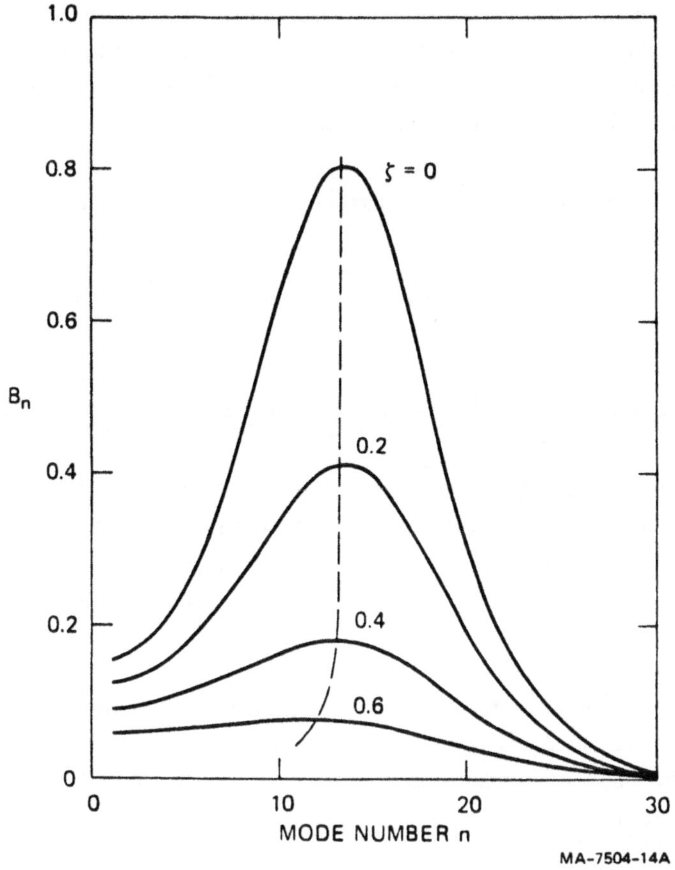


FIGURE 4.18 AMPLIFICATION FUNCTION FROM INITIAL VELOCITIES (Shell 3a in Table 4.3)

that the preferred mode numbers do not change substantially after $\xi = 0.4$ ($\tau = 0.6\tau_f$). Thus the choice appears to be made halfway through the duration of motion and held until motion ceases.

The amplification curves are typical representatives of the curves for all the steel shells of the experiments. The results are presented in Section 4.2.12, where the preferred mode numbers are compared with those predicted by our theory.

### 4.2.8 Approximate Solutions for Terminal Motion

Although it is simpler to generate the amplification curves in Figures 4.17 and 4.18 by numerical integration rather than by formulas (4.2.26) and (4.2.27) for $A_n(\xi)$ and $B_n(\xi)$ the formulas provide invaluable information. To determine the preferred mode number and the amplification caused by the impulsive loading, only the terminal amplification term is required. If in formulas (4.2.26) and (4.2.27) we let $\xi \rightarrow 0$ and employ the limiting values of the confluent hypergeometric functions, after removing derivatives by recurrence relations, we obtain

$$A_n(0) = e^{-(\lambda_n + \omega_n)}\Big[M(1 + a_n - b_n, 1 - b_n, \omega_n)$$

$$+ M(1 + a_n - b_n, 2 - b_n, \omega_n)\lambda_n/(1 - b_n)\Big] \tag{4.2.32}$$

$$B_n(0) = 2\tau_f e^{-(\lambda_n + \omega_n)} M(1 + a_n - b_n, 2 - b_n, \omega_n)/(1 - b_n) \tag{4.2.33}$$

From the relations (4.2.29) we see that the parameters and arguments for the functions M are positive. Whenever n is large enough, the functions become approximately $M(Q_n, Q_n, P_n) = e^{P_n}$, and $\lambda_n \approx 0$, $\omega_n \approx P_n$, and $(1 - b_n) \approx Q_n$, so $A_n(0) \rightarrow 1$ and $B_n(0) \rightarrow 0$ as $n \rightarrow \infty$, which shows that amplification of high modes is negligible.

When the viscoplastic moments are negligible the terminal amplification functions (4.2.32) and (4.2.33) become (4.1.70) and (4.1.71), formulas that were derived for the case of hardening moments being neglected in comparison with directional moments.

When the directional moments are negligible, the terminal amplification functions (4.2.32) and (4.2.33) reduce to

$$A_n(0) = \Big[(\omega_n + P_n)e^{(\omega_n - P_n)/2} + (\omega_n - P_n)e^{(\omega_n + P_n)/2}\Big]/2\omega_n \tag{4.2.34}$$

$$B_n(0) = 2\tau_f\Big[e^{(\omega_n - P_n)/2} - e^{-(\omega_n + P_n)/2}\Big]/\omega_n \tag{4.2.35}$$

in agreement with formulas (4.2.30) and (4.2.31) after setting $\xi = 0$. Formulas (4.2.34) and (4.2.35) may be simplified by noting that $\omega_n$ and $P_n$ are always positive, that $\omega_n > P_n$, and that in general the magnitudes thus make the first term dominant. Hence

$$A_n(0) = \Big[(\omega_n + P_n)/2\omega_n\Big]e^{(\omega_n - P_n)/2} \tag{4.2.36}$$

$$B_n(0) = (2\tau_f/\omega_n)e^{(\omega_n - P_n)/2} \tag{4.2.37}$$

### 4.2.9 Preferred Modes and Threshold Impulses

We do not derive here the formulas for the preferred modes and threshold impulses when viscoplastic and directional moments have comparable contributions to the resistive moment. However, they may be obtained numerically for a specified shell from formulas (4.2.32) or (4.2.33). Instead, we consider the two extremes of predominant viscoplastic or directional moment. In fact, we need treat only the case of a predominant viscoplastic moment because the case of a predominant directional moment is treated in Section 4.1.12 where we derived formula (4.1.92) for the preferred mode and formula (4.1.98) for the impulse-amplification function relationship; the threshold impulse depends on our choice of $A_n$, which in Section 4.1.14 was suggested as $A_n = 100$.

To obtain comparable preferred mode and threshold impulse formulas when the viscoplastic moment is dominant, we start with the terminal amplification function formulas (4.2.36) and (4.2.37). The preferred mode and threshold impulse can again be obtained numerically for each specified shell, but a large class of cases exists for which $(P_n/2R_n)$ is sufficiently less than unity to allow us to find the simple approximate relation

$$R_N = 2P_N \tag{4.2.38}$$

that maximizes $A_n(0)$ and $B_n(0)$ with respect to the mode number n. According to (4.2.38) these maxima are

$$A_N = (5/8)e^{3R_n/4} \qquad B_N = (\tau_f/R_n)e^{3R_N/4} \tag{4.2.39}$$

After substituting expressions (4.2.22) and (4.2.24) for $P_n$ and $R_n$ in formula (4.2.38), we obtain the preferred mode formula

$$N^3 = (3a/8\alpha^2\gamma)[(2 - k)\rho/K_2\overline{\sigma}]^{1/2} \tag{4.2.40}$$

In the derivation of (4.2.40) we eliminate $\tau_f$ by using (4.2.12). Within the restrictions imposed by our approximations, the result (4.2.40) shows that the preferred mode number is independent of the applied impulse, except through the choice of $\gamma$.

We again base our threshold impulse on the amplification of the initial displacement imperfections. When $R_n$ is eliminated from $A_n$, given by (4.2.39), and $\tau_f$ is eliminated from the result by using (4.2.12), we obtain

$$I = (4/3)[2(2 - k)/9K_2]^{1/3}(\overline{\sigma}^2h^2\rho\gamma)^{1/3}(h/a)\ln(8A_N/5) \tag{4.2.41}$$

Figure 4.19 shows the impulse-amplification curves obtained from formula (4.2.41) for the four shells listed in Table 4.3. If again we choose an amplification of $A_n = 100$ to determine the critical impulse, we obtain 7400, 14800, 23800, and 48600 dyne-s/cm$^2$ for shells 1, 2, 3, and 4, respectively.

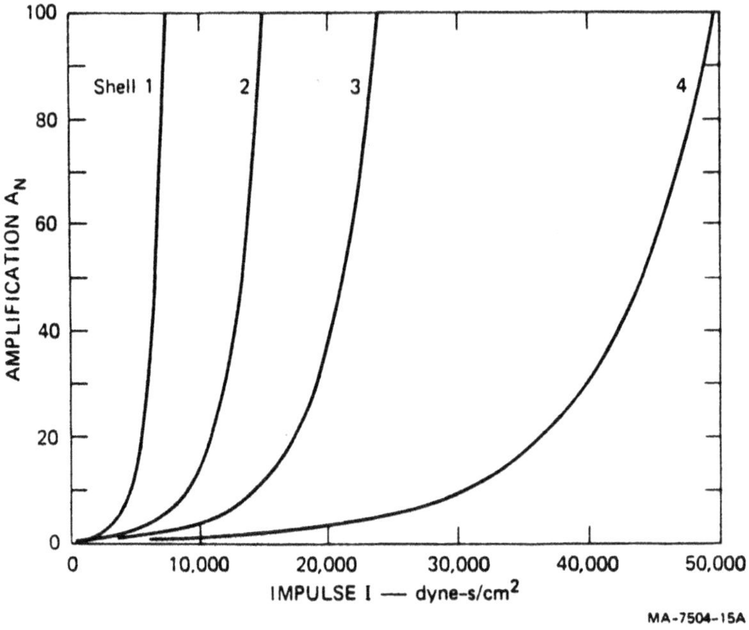


FIGURE 4.19 IMPULSE-AMPLIFICATION CURVES
(Shells of Table 4.3)

### 4.2.10 Displacement and Velocity Imperfections

It is of interest to determine the relative importance of displacement and velocity imperfections with regard to their influence on the final displacement amplitude. If we visualize the final shape as consisting only of the preferred mode

$$u(\theta, 0) = (A_N a_N + B_N b_N) \sin N\theta \tag{4.2.42}$$

where, for viscoplastic resistive moments only, the amplifications $A_N$ and $B_N$ are given by (4.2.39). Formulas (4.2.39) and (4.2.42) give

$$\max.u = \left[(5/8)a_N + \{(\tau_f/2)^{1/2}/N\}b_N\right]e^{3R_N/4} \tag{4.2.43}$$

For equal contributions to the final buckled amplitude, (4.2.43) shows that

$$b_N/a_N = (5/8)(2/\tau_f)^{1/2}N \tag{4.2.44}$$

In the case of the steel shells 1a and 4f of Table 4.3 the imperfection ratios in (4.2.44) are 91 and 56, respectively, indicating that for equal contributions to the final amplitude the velocity imperfection has to be a much larger percentage of the initial uniform velocity than the displacement imperfection is of the initial uniform shell radius.

### 4.2.11 Viscoplastic and Directional Moments

The relative contributions to the restoring moment of the viscoplastic and directional moments $M_s$ and $M_d$ can be assessed with the help of formula (4.2.18) for the ratio $M_s/M_d$. Because $\dot{w}_0 = V_0(1 - t/t_f)$, we have

$$M_s/M_d = (4\gamma V_0 K_1 K_2/9ak^2)(1 - t/t_f) \tag{4.2.45}$$

where $k = k(0)$ is given by (4.1.35) with $\Omega = \sqrt{3}/2$. Thus the moment ratio is a linear function of time and directional moments always dominate at the end of motion.

For the steel shells of Table 4.3, of Section 4.2.12, the products of the viscosity constant $\gamma$ and the initial strain rate $V_0/a$ lie in the narrow range $0.153 < \gamma V_0/a < 0.163$, so for illustrative purposes an average value of 0.158 is chosen in (4.2.45). Figure 4.20 shows the moment ratio (4.2.45) as a linear function of the time ratio $t/t_f$ for various values of the parameter k. By examining the positions of these lines relative to the line $M_s/M_d = 1$, we see that for the steel shells of Table 4.3, the viscoplastic moments dominate for $0 < k \gtrsim 0.3$ ($L/D \gtrsim 1.27$) and that for $0.4 < k < 0.5$ the directional moments dominate. The experimental shells have $L/D = 2$, which corresponds to $k(0) = 0.17$; thus the restoring moments were primarily viscoplastic.

### 4.2.12 Comparison of Theory and Experiment

Table 4.3 contains the main experimental data for 16 fully annealed 1015 steel shells, 6 inches long.

Figure 4.21 is a stress-strain diagram obtained from a static tensile test on one of several specimens cut from the cylinders. It shows negligible strain hardening below about 3% strain. According to the experimental results of Manjoine,[4] the material is extremely sensitive to rate of strain; both the yield stress and the length of the horizontal or ideally plastic portion of the stress-strain curve increase with strain rate.

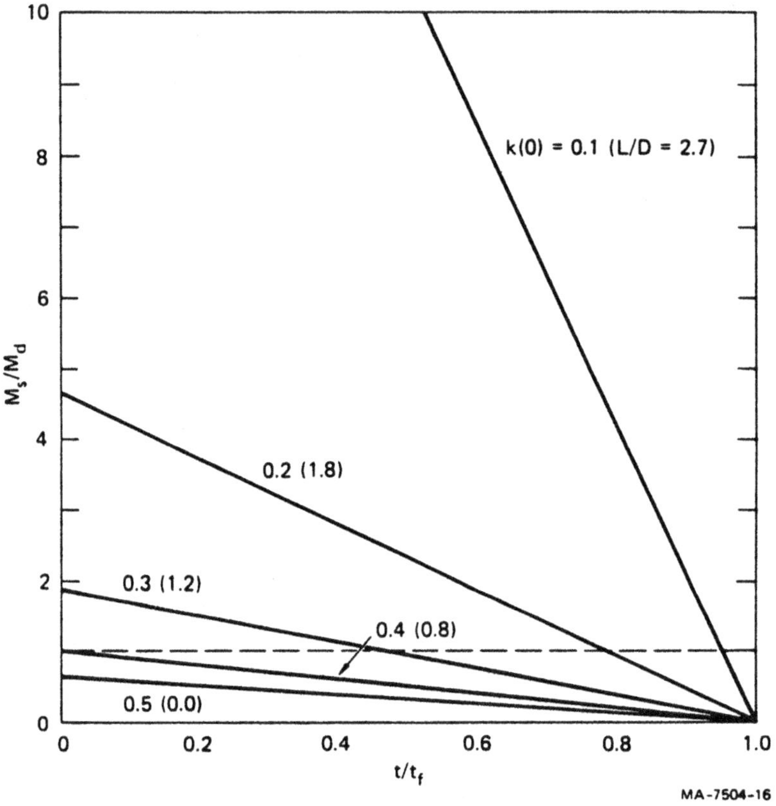


FIGURE 4.20 TIME VARIATION OF STRAIN RATE/DIRECTIONAL MOMENT RATIO FOR FIVE LENGTH/DIAMETER RATIOS, EQUATION (4.2.45)

Table 4.3

EXPERIMENTAL DATA FOR CYLINDRICAL SHELLS OF 1015 STEEL

| Shell Number | Radius a (cm) | Thickness h (cm) | a/h | Impulse I $(N\text{-}s/m^2)$ | Initial Velocity $V_o$ (m/s) | Final strain $\epsilon_\theta(t_f)$ (%) | Mode N |
|---|---|---|---|---|---|---|---|
| 1a | 3.795 | 0.107 | 35.6 | 708.9 | 85.2 | 9.9 | 26 |
| 1a | 3.795 | 0.107 | 35.6 | 607.6 | 73.0 | 6.0 | 26 |
| 2a | 3.815 | 0.168 | 22.8 | 1382.3 | 105.7 | 9.5 | 19 |
| 2b | 3.815 | 0.168 | 22.8 | 744.3 | 56.9 | 4.5 | 19 |
| 2c | 3.815 | 0.168 | 22.8 | 744.3 | 56.9 | 4.5 | 19 |
| 2d | 3.815 | 0.168 | 22.8 | 638.0 | 48.8 | 3.1 | 22 |
| 2e | 3.815 | 0.168 | 22.8 | 638.0 | 48.8 | 2.7 | 22 |
| 3a | 3.780 | 0.213 | 16.7 | 2038.2 | 115.6 | 9.3 | 13 |
| 3b | 3.780 | 0.214 | 16.7 | 1514.2 | 85.9 | 5.0 | 14 |
| 3c | 3.780 | 0.214 | 16.7 | 1456.0 | 82.6 | 5.5 | 14 |
| 4a | 3.835 | 0.366 | 10.5 | 3319.6 | 116.4 | 8.7 | 9 |
| 4b | 3.835 | 0.366 | 10.5 | 2096.4 | 73.5 | 4.7 | 9 |
| 4c | 3.835 | 0.366 | 10.5 | 2038.4 | 71.5 | 4.3 | 9 |
| 4d | 3.835 | 0.366 | 10.5 | 1747.1 | 61.2 | 3.5 | 9 |
| 4e | 3.835 | 0.366 | 10.5 | 1513.8 | 53.1 | 3.5 | 9 |
| 4f | 3.835 | 0.366 | 10.5 | 1513.8 | 53.1 | 2.4 | 9 |

Shell data: 1015 steel, fully annealed, $\rho = 7.8\ g/cm^3$, length 6 inches, k(0) = 0.17.

Table 4.3 (concluded)

EXPERIMENTAL DATA FOR CYLINDRICAL SHELLS OF 1015 STEEL

| Shell Number | Radius a (inch) | Thickness h (inch) | a/h | Impulse I $(dyne\text{-}s/cm^2)$ | Initial Velocity $V_o$ inch/s | Final strain $\epsilon_\theta(t_f)$ (%) | Mode N |
|---|---|---|---|---|---|---|---|
| 1a | 1.494 | 0.042 | 35.6 | 7,089 | 3,354 | 9.9 | 26 |
| 1b | 1.494 | 0.042 | 35.6 | 6,076 | 2,875 | 6.0 | 26 |
| 2a | 1.502 | 0.066 | 22.8 | 13,823 | 4,162 | 9.5 | 19 |
| 2b | 1.502 | 0.066 | 22.8 | 7,443 | 2,241 | 4.5 | 19 |
| 2c | 1.502 | 0.066 | 22.8 | 7,443 | 2,241 | 4.5 | 19 |
| 2d | 1.502 | 0.066 | 22.8 | 6,380 | 1,921 | 3.1 | 22 |
| 2e | 1.502 | 0.066 | 22.8 | 6,380 | 1,921 | 2.7 | 22 |
| 3a | 1.488 | 0.089 | 16.7 | 20,382 | 4,551 | 9.3 | 13 |
| 3b | 1.488 | 0.089 | 16.7 | 15,142 | 3,381 | 5.0 | 14 |
| 3c | 1.488 | 0.089 | 16.7 | 14,560 | 3,251 | 5.5 | 14 |
| 4a | 1.510 | 0.144 | 10.5 | 33,196 | 4,581 | 8.7 | 9 |
| 4b | 1.510 | 0.144 | 10.5 | 20,964 | 2,893 | 4.7 | 9 |
| 4c | 1.510 | 0.144 | 10.5 | 20,384 | 2,813 | 4.3 | 9 |
| 4d | 1.510 | 0.144 | 10.5 | 17,471 | 2,411 | 3.5 | 9 |
| 4e | 1.510 | 0.144 | 10.5 | 15,138 | 2,089 | 3.5 | 9 |
| 4f | 1.510 | 0.144 | 10.5 | 15,138 | 2,089 | 2.4 | 9 |

Shell data: 1015 steel, fully annealed, $\rho = 7.8\ g/cm^3$, length 6 inches, k(0) = 0.17.

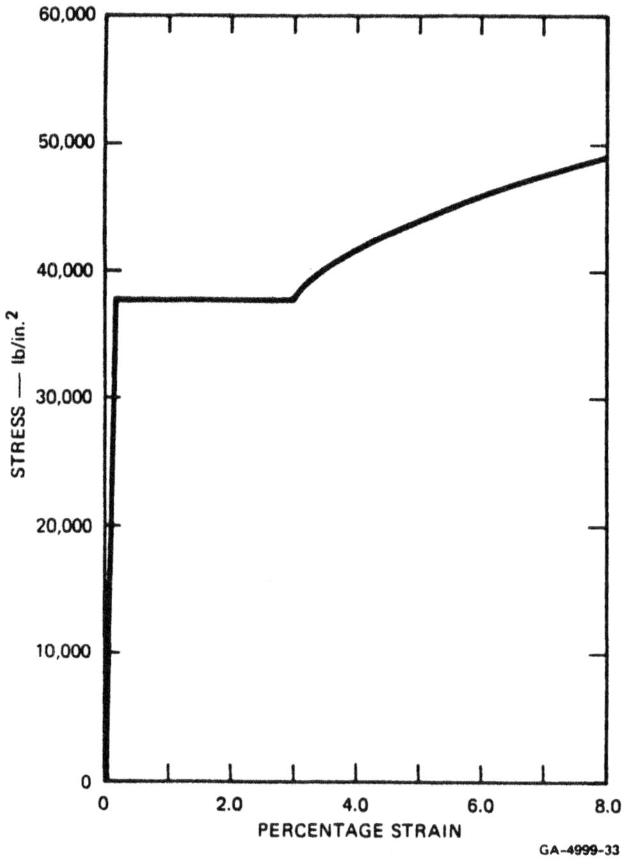


FIGURE 4.21 STRESS-STRAIN RELATIONSHIP FOR FULLY ANNEALED 1015 STEEL

Based on these results, an empirical relation between yield stress and strain rate has been suggested by Symonds and Bodner[5] in the form

$$\sigma/\sigma_y = 1 + (\dot{\epsilon}/D)^{1/p} \tag{4.2.46}$$

where $\sigma_y$ is the static yield stress, and p and D are empirical constants. For mild steel with $\sigma_y$ = 30,000 psi (207 MPa), these constants are p = 5 and D = 40.4 $s^{-1}$. The stress strain-rate relationship (4.2.46) is shown in Figure 4.22. In the theory, the linear strain-rate law (4.2.2) is used. We have chosen this line to be the tangent to the curve (4.2.46) at the point with its abscissa equal to the initial strain rate. The slope of this line is

$$\mu = \gamma\sigma_0 = \sigma_y/pD^{1/p}\dot{\epsilon}_0^{1-1/p} \tag{4.2.46a}$$

where $\dot{\epsilon}_0 = \dot{\epsilon}(0) = V_0/a = I/\rho ha$. The intersection of the line with the stress axis is

$$\sigma_0 = \left[1 + (\dot{\epsilon}_0/D)^{1/p}(1 - 1/p)\right]\sigma_y \qquad (4.2.47)$$

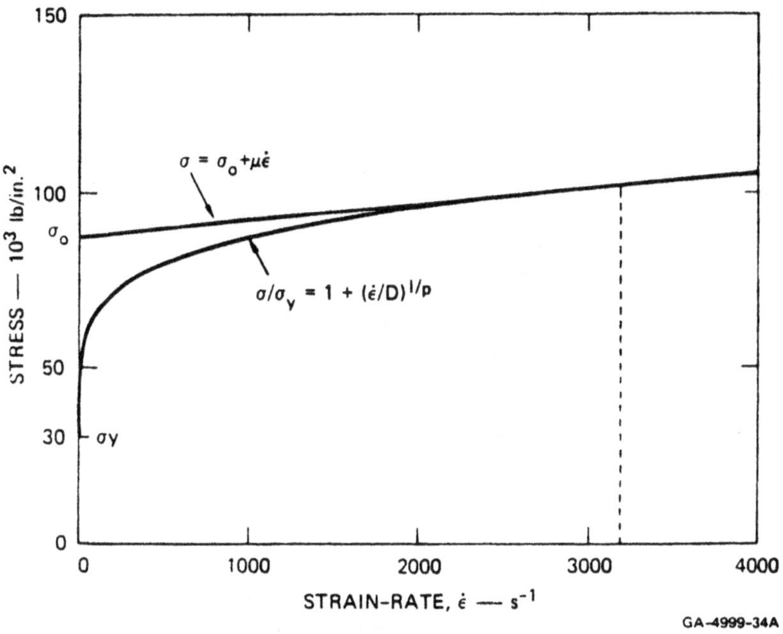


FIGURE 4.22 STRESS, STRAIN-RATE RELATIONSHIP FOR FULLY ANNEALED 1015 STEEL

The experimental technique is that described in Section 3.2.7, and the data are listed in Table 4.3. Theoretical results are listed in Table 4.4, and compared with experimental mode numbers; the predictions are given by formula (4.2.40). Generally, the theoretical and experimental mode numbers agree. We recall that formula (4.2.40) assumes predominance of the viscoplastic contribution to the restoring moment and that $P_N/2R_N$ is sufficiently less than unity to allow us to obtain (4.2.40) without much error. All the shells have $L/D = 2$ giving $k = 0.17$, and this low value was seen in Section 4.2.10 (Figure 4.20) to justify the assumption of viscoplastic predominance. As for the ratio $P_N/2R_N$, each case must be verified; for the shells in Table 4.3, the assumption is satisfactory.

Figures 4.23 and 4.24 show the developed final shapes of shells 3a and 4a. The shapes represent two examples of the experimental data from which the number of waves are obtained. Figure 4.25 shows the result of a harmonic analysis

Table 4.4

THEORETICAL AND EXPERIMENTAL MODE NUMBERS
FOR CYLINDRICAL SHELLS OF 1015 STEEL

| Shell Number | Impulse ($N\text{-}s/m^2$) | Strain-Rate Law Parameters<br>$\dot{\epsilon}(o)$ ($s^{-1}$) | $\sigma(o)$ (MPa) | $\sigma_o$ (MPa) | $\mu$ ($N\text{-}s/m^2$) | $\gamma$ ($\mu s$) | Final Strain $\tau_f$ (%) | Theoretical Mode* N | Experimental Mode N |
|---|---|---|---|---|---|---|---|---|---|
| 1a | 708.9 | 2245 | 669 | 576 | 41,157 | 71.4 | 4.9 | 23 | 26 |
| 1b | 607.6 | 1924 | 655 | 565 | 46,535 | 82.3 | 3.7 | 22 | 26 |
| 2a | 1382.3 | 2771 | 689 | 592 | 35,366 | 59.7 | 7.4 | 18 | 19 |
| 2b | 744.3 | 1492 | 632 | 547 | 57,082 | 104.3 | 2.3 | 15 | 19 |
| 2c | 744.3 | 1492 | 632 | 547 | 57,082 | 104.3 | 2.3 | 15 | 22 |
| 2d | 638.0 | 1279 | 620 | 537 | 64,528 | 120.2 | 1.7 | 14 | 22 |
| 2e | 638.0 | 1279 | 620 | 537 | 64,528 | 120.2 | 1.7 | 14 | 22 |
| 3a | 2038.2 | 3058 | 698 | 600 | 32,126 | 53.6 | 8.7 | 15 | 13 |
| 3b | 1514.2 | 2272 | 670 | 577 | 40,744 | 70.6 | 5.0 | 14 | 14 |
| 3c | 1456.0 | 2185 | 666 | 574 | 42,053 | 73.2 | 4.6 | 14 | 14 |
| 4a | 3319.6 | 3034 | 697 | 599 | 32,333 | 53.9 | 8.8 | 11 | 9 |
| 4b | 2096.4 | 1916 | 654 | 565 | 46,741 | 82.8 | 3.7 | 10 | 9 |
| 4c | 2038.4 | 1863 | 652 | 563 | 47,775 | 84.9 | 3.6 | 10 | 9 |
| 4d | 1747.1 | 1597 | 638 | 552 | 54,049 | 97.9 | 2.7 | 9 | 9 |
| 4e | 1513.8 | 1384 | 626 | 542 | 60,598 | 111.8 | 2.0 | 9 | 9 |
| 4f | 1513.8 | 1384 | 626 | 542 | 60,598 | 111.8 | 2.0 | 9 | 0 |

* Equation (4.2.40).

Table 4.4 (concluded)

THEORETICAL AND EXPERIMENTAL MODE NUMBERS
FOR CYLINDRICAL SHELLS OF 1015 STEEL

| Shell Number | Impulse ($dyne\text{-}s/cm^2$) | Strain-Rate Law Parameters<br>$\dot{\epsilon}(o)$ ($s^{-1}$) | $\sigma(o)$ (psi) | $\sigma_o$ (psi) | $\mu$ ($lb\text{-}s/in^2$) | $\gamma$ ($\mu s$) | Final Strain $\tau_f$ (%) | Theoretical Mode* N | Experimental Mode N |
|---|---|---|---|---|---|---|---|---|---|
| 1a | 7,089 | 2,245 | 97,000 | 83,602 | 5.97 | 71.4 | 4.9 | 23 | 26 |
| 1b | 6,076 | 1,924 | 94,968 | 81,974 | 6.75 | 82.3 | 3.7 | 22 | 26 |
| 2a | 13,823 | 2,771 | 99,883 | 85,906 | 5.13 | 59.7 | 7.4 | 18 | 19 |
| 2b | 7,443 | 1,492 | 91,745 | 79,396 | 8.28 | 104.3 | 2.3 | 15 | 19 |
| 2c | 7,443 | 1,492 | 91,745 | 79,396 | 8.28 | 104.3 | 2.3 | 15 | 22 |
| 2d | 6,380 | 1,279 | 89,871 | 77,897 | 9.36 | 120.2 | 1.7 | 14 | 22 |
| 2e | 6,380 | 1,279 | 89,871 | 77,897 | 9.36 | 120.2 | 1.7 | 14 | 22 |
| 3a | 20,382 | 3,058 | 101,276 | 87,021 | 4.66 | 53.6 | 8.7 | 15 | 13 |
| 3b | 15,142 | 2,272 | 97,163 | 83,730 | 5.91 | 70.6 | 5.0 | 14 | 14 |
| 3c | 14,560 | 2,185 | 96,639 | 83,311 | 6.10 | 73.2 | 4.6 | 14 | 14 |
| 4a | 33,196 | 3,034 | 101,161 | 86,929 | 4.69 | 53.9 | 8.8 | 11 | 9 |
| 4b | 20,964 | 1,916 | 94,911 | 81,929 | 6.78 | 82.8 | 3.7 | 10 | 9 |
| 4c | 20,384 | 1,863 | 94,548 | 81,638 | 6.93 | 84.9 | 3.6 | 10 | 9 |
| 4d | 15,138 | 1,384 | 92,588 | 80,070 | 7.84 | 97.9 | 2.7 | 9 | 9 |
| 4e | 15,138 | 1,384 | 90,819 | 78,655 | 8.79 | 111.8 | 2.0 | 9 | 9 |
| 4f | 15,138 | 1,384 | 90,819 | 78,655 | 8.79 | 111.8 | 2.0 | 9 | 0 |

* Equation (4.2.40).

### 4.2.1 Viscoplastic Moments

Figure 4.15 illustrates how a resistive moment arises during flow buckling because of the viscoplastic material property. It is assumed that the compressive circumferential strain is increasing at every fiber of the shell. At two points A and B, on the outside and inside of a cross section where an outward portion of a wave exists, the strain rate is greater at B than at A. On the linearized stress, strain-rate curve the corresponding states are shown by the points A′ and B′. The higher strain rate at B given by the point B′ on the strain-rate line is associated with a higher compressive stress than at A. Thus a moment exists on the cross section, which is resisting the curvature increase. As deformation proceeds, the points A′ and B′ move to the left, as indicated in Figure 4.15, which is consistent with the shell coming to rest. The points separate because of curvature increase, so the moment increases with deformation.

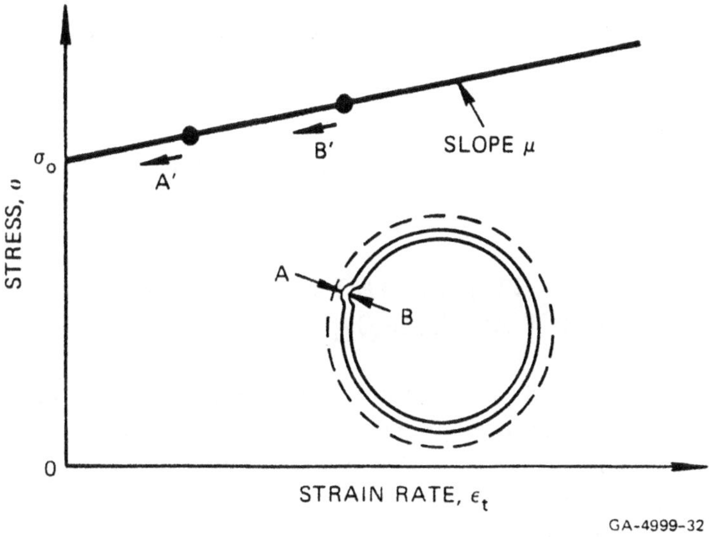


FIGURE 4.15 FIBER STRESSES ACCORDING TO THE LINEARIZED STRESS, STRAIN-RATE LAW

### 4.2.2 Theory of Viscoplastic Cylindrical Shells

The first portion of the theoretical treatment of Section 4.1.1 providing Equations (4.2.1) through (4.2.9) introduces incompressibility, generalized strain rate and stress, deviatoric stresses, and the flow law. Constitutive equations consistent with the flow law are taken in the form[3]

$$\gamma\dot{\epsilon}_x = \sigma_0 F \frac{\partial F}{\partial \sigma_x} \qquad \gamma\dot{\epsilon}_\theta = \sigma_0 F \frac{\partial F}{\partial \sigma_\theta} \qquad \gamma\dot{\epsilon}_z = \sigma_z F \frac{\partial F}{\partial \sigma_z} \tag{4.2.1}$$

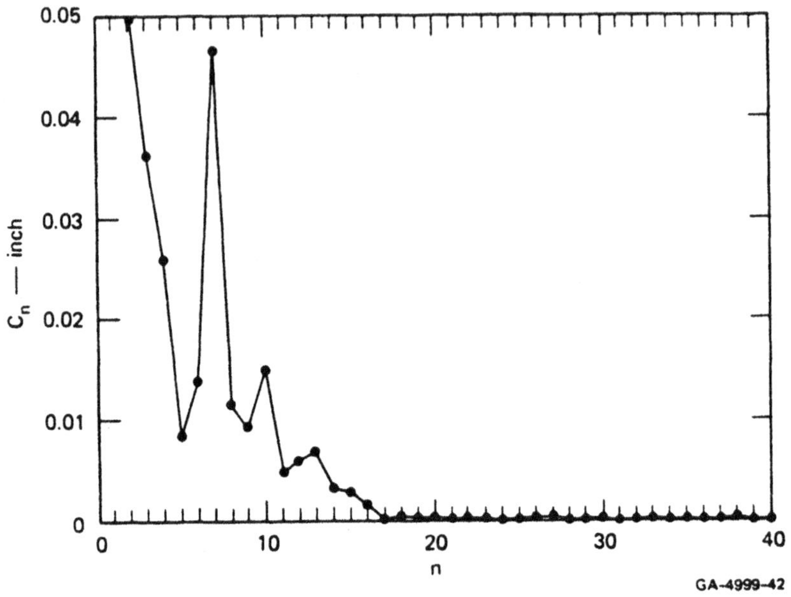


FIGURE 4.25 RESULT OF HARMONIC ANALYSIS ON SHELL 4

## 4.3 CRITICAL VELOCITY FOR COLLAPSE OF CYLINDRICAL SHELLS WITHOUT BUCKLING

High speed collapse of shells of circular cross section occurs in many practical devices. Such devices are used in explosive closure for obtaining gas-tight pipe seals, in magnetic field constriction for producing intense transient magnetic fields, in shock tubes driven by rapidly collapsing a cylindrical reservoir to produce high-pressure and high-velocity gas flows, and in armor-piercing and oil well perforation by metallic jets produced by rapidly collapsing conical shells. In these devices, performance is optimized whenever plastic buckling is absent during the collapse motion.

The new feature in the treatment of dynamic buckling is that the inward unperturbed displacement is large, which results in an appreciable thickening of the shell. Because of this thickening the thrust and the resistive motion of the perturbed state increase, the dependence on the thickness being, respectively, linear and cubic. This dependence on thickness means that the stability improves as motion proceeds so that it is possible for departures from the circular form to be held within acceptable values. Also, growth of departures requires time, which again suggests that they may be held within acceptable values, in this case by sufficiently high collapse velocities.

A critical collapse velocity will be defined for a specific amplification factor (say, 100) and a specified shell (both geometry and properties specified). Collapse velocities above critical will result in acceptable departures from circularity; those below critical will not.

In the theoretical treatment to follow, we ignore the directional contribution to the restoring moment or we consider the shells long enough to allow us to neglect this contribution. Also, we take the collapse velocity to be constant; in other words, we assume that only a small part of the initial kinetic energy is absorbed as plastic work. We shall treat the two special cases of a linear strain-hardening law and a linear strain-rate law.

Let the initial and current average radii of the cylindrical shell be $r_o$ and $r(t)$, and let the initial and current wall thickness be $h_o$ and $h(t)$. Because we are considering departures from the circular form that are small compared with the radius, these radii and thicknesses are related approximately by the plastic incompressibility condition.

$$h(t) = h_o r_o / r(t) \tag{4.3.1}$$

Let the initial departure of shell cross section from the circular form be $w_i(\theta)$ measured radially inward, and let additional radially inward deflections that arise during collapse be $w(\theta,t)$. The total inward departure from a circle of radius $r(t)$ is therefore $w + w_i$. Then, the current curvature is

$$\kappa = 1/r + (w + w'' + w_i + w_i'')/r^2 \tag{4.3.2}$$

where primes denote differentiation with respect to the angular coordinate $\theta$.

The equation of motion for an element of the collapsing shell is readily found with the aid of Figure 4.26. Neglecting rotary inertia we have

$$Q = \frac{\partial M}{\partial \lambda} \qquad \frac{\partial Q}{\partial \lambda} + N\frac{\partial \phi}{\partial \lambda} - p = \rho h \ddot{(r - w)} \tag{4.3.3}$$

where dots denote differentiation with respect to time. In (4.3.3), $d\lambda$ is the arc length corresponding to $d\theta$ and subtending an angle $d\phi$ at the instantaneous center of curvature so

$$\kappa = \frac{\partial \phi}{\partial \lambda} \tag{4.3.4}$$

Because the displacements $w$ and $w_i$ are small, we make the approximation $d\lambda \sim r d\theta$.

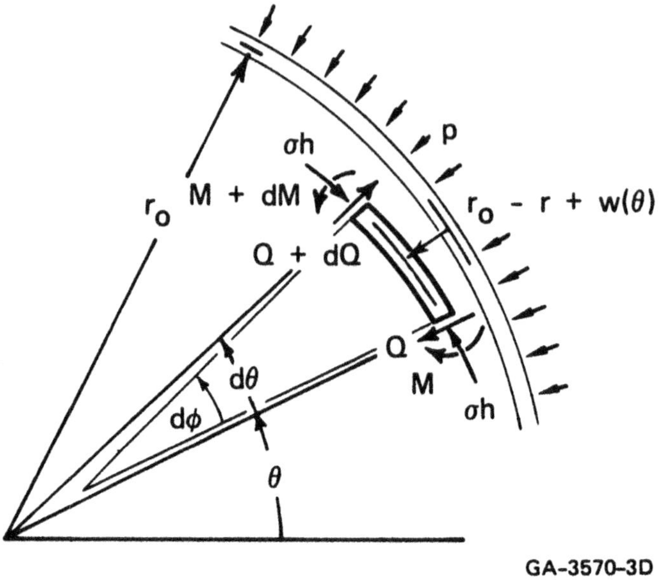


FIGURE 4.26 NOTATION OF RING ELEMENT IN MOTION

We now combine (4.3.2), (4.3.3), and (4.3.4) by eliminating $\phi$, $\kappa$, and Q to obtain

$$M''/r^2 + N[1/r + (w + w'' + w_i + w_i'')/r^2] = \rho h(r - w)^{\cdot\cdot} + p \qquad (4.3.5)$$

Again, because the displacements w and $w_i$ are small the thrust N is the same as the thrust in the equation governing unperturbed motion

$$N/r = \rho h\ddot{r} + p \qquad (4.3.6)$$

Equation (4.3.6) allows us to simplify (4.3.5) to give

$$M''/r^2 + N(w + w'' + w_i + w_i'')/r^2 = -\rho h\ddot{w} \qquad (4.3.7)$$

It remains for us to establish expressions for the bending moment and thrust to convert Equation (4.3.7) into a partial differential equation governing the perturbed displacement, w. These expressions are derived below for linear strain hardening alone and for a linear stress-strain rate law.

### 4.3.1 Strain-Hardening Moments Only

If no strain-rate reversal occurs over a cross section of the shell, the moment-curvature relationship is

$$M = E_h I(\kappa - \kappa_i) \tag{4.3.8}$$

where $E_h$ is the hardening modulus, $I = h^3/12$, and $\kappa_i$ is the initial curvature. The current curvature $\kappa$ is given by (4.3.2). Let the compressive stress at the midsurface of the shell be $\bar{\sigma}$. Then the thrust is simply

$$N = \bar{\sigma} h \tag{4.3.9}$$

Considerable simplification is introduced if we restrict our theory to materials with stress-strain curves that permit the hardening portion to be approximated by a straight line with a slope that does not cause the flow stress to rise substantially above the yield stress in the strain range of interest. Then in (4.3.8), $E_h$ is a constant and in (4.3.9) we may take $\bar{\sigma}$ to be a constant, chosen as an average stress over the strain range. In this way, we linearize the problem.

Substitution of M and N given by (4.3.8) and (4.3.9) into (4.3.7) and the use of the curvature expression (4.3.2) and the incompressibility relation (4.3.1) lead to the equation

$$E_h h_o^2 r_o^2 (w + w'')''/12r^6 + \bar{\sigma}(w + w'' + w_i + w_i'')/r^2 + \rho\ddot{w} = 0 \tag{4.3.10}$$

If we introduce the dimensionless quantities

$$\begin{aligned} u &= w/r_o \quad & u_i &= w_i/r_o \quad & \xi &= r/r_o \quad & \tau &= \alpha c_h t/r_o \\ c_h^2 &= E_h/\rho \quad & \alpha^2 &= h_o^2/12r_o^2 \quad & \beta &= E_h/\bar{\sigma} \end{aligned} \tag{4.3.11}$$

Equation (4.3.10) becomes

$$\xi^6 \ddot{u} + u'''' + (1 + s^2\xi^4)u'' + s^2\xi^4 u = -s^2\xi^4(u_i + u_i'') \tag{4.3.12}$$

where

$$\xi = r/r_o = 1 - Vt/r_o = 1 - V\tau/\alpha c_h = 1 - q\tau \tag{4.3.13}$$

$$q = V/c_h \tag{4.3.14}$$

$$s^2 = 1/\alpha^2\beta = 12(r_o/h_o)^2(\bar{\sigma}/E_h) \tag{4.3.15}$$

The buckling displacement and initial departure from circularity may be represented by

$$u(\theta,\tau) = \sum_{n=2}^{\infty} [F_n(\tau) \cos n\theta + G_n(\tau) \sin n\theta] \qquad (4.3.16)$$

$$u_i(\theta) = \sum_{n=2}^{\infty} [A_n \cos n\theta + B_n \sin n\theta] \qquad (4.3.17)$$

the first mode being omitted because it does not contribute to buckling. Substituting (4.3.16) and (4.3.17) into (4.3.12) yields

$$\xi^6 \ddot{f}_n + (n^2 - 1)(n^2 - s^2\xi^4) f_n = s^2\xi^4(n^2 - 1) \qquad (4.3.18)$$

where

$$f_n = F_n(\tau)/A_n \qquad (4.3.19)$$

is the amplification function. As long as the coefficient of $f_n$ in (4.3.18) remains positive, the amplification function will not experience substantial growth; that is, collapse will remain stable whenever

$$n > s\xi^2 = \xi^2/\alpha\sqrt{\beta} \qquad (4.3.20)$$

The minimum value of $\xi$ occurs when the inner radius is zero and hence total collapse has occurred. From geometrical considerations, we then have $\min\xi = \sqrt{3}\,\alpha$, so according to (4.3.20) modes that are stable at the termination of collapse have mode numbers that satisfy the inequality

$$n > 3\alpha/\sqrt{\beta} = (h_o/r_o)(\bar{\sigma}/E_h)/2 \qquad (4.3.21)$$

This result shows that in practical cases almost all modes are stable at the end of collapse.

For negative values of the coefficient of $f_n$ in (4.3.18), the amplification function can grow substantially. Initially, when $\xi = 1$, modes with numbers satisfying

$$n < s = 1/\alpha\sqrt{\beta} \qquad (4.3.22)$$

are unstable. However, $\xi = r/r_o$ decreases from unity to $\sqrt{3}\,\alpha$, which is generally a small value; thus (4.3.21) or (4.3.22) implies that many modes may be initially unstable, but almost all modes are eventually stable.

For each mode, it is of interest to determine $f_n$ when $n = s\xi^2$, that is, at the value of $\xi$ when the types of growth change from being exponential in character to being trigonometrical. These values of $f_n$ obtained by numerical integration of (4.3.12) are given in Figure 4.27, for the choice $s = 30$, as a function of q. This parameter, according to (4.3.14), represents the collapse velocity V. For each n, $f_n$ decreases with increasing q because the growth duration decreases with increasing collapse velocity. At each q, the value of n for which $f_n$ is maximum is $n = 10$. This value is achieved when $\xi = \sqrt{n/s}$, which for $s = 30$ and $n = 10$, results in $\xi = r/r_o = 0.58$.

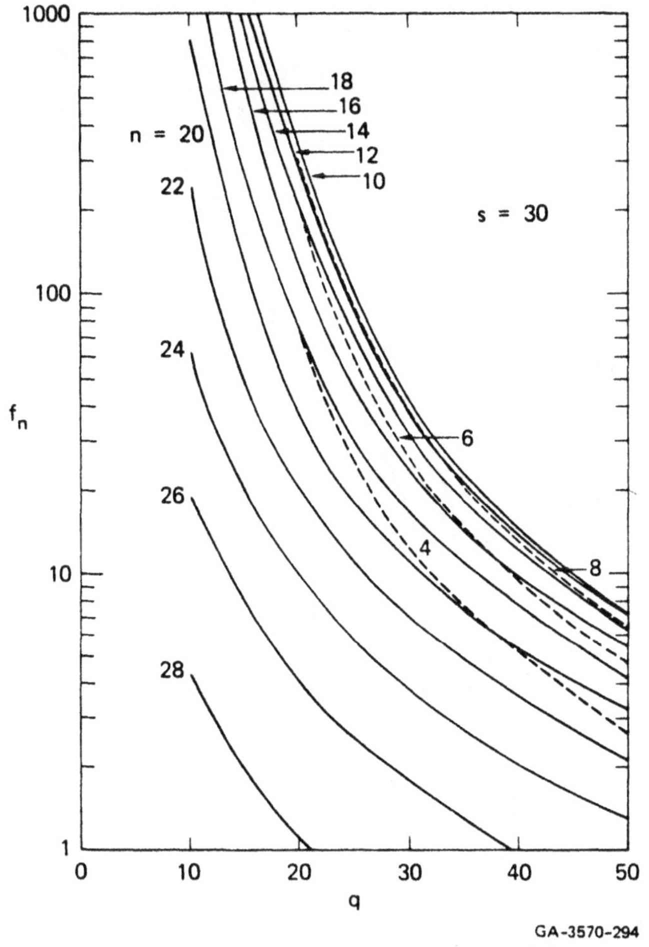


FIGURE 4.27 FINAL MAGNITUDE OF $f_n$ FOR s = 30
Critical mode is n = 10.

Curves such as those in Figure 4.27 were generated for $15 < s < 200$. From these curves, the values of q for $f_n = 10$ and $f_n = 100$ were determined and are shown in Figure 4.28. For a particular value of s, the critical velocity represented by q is greater for $f_n = 10$ than for $f_n = 100$ because of the shorter durations required to restrict the amplification. The critical mode numbers increase with s, but are nearly the same for both curves; an increase in s can be the result of decreasing the shell thickness or reducing the hardening modulus, as can be seen in (4.3.15), so higher mode numbers should be expected with increasing s.

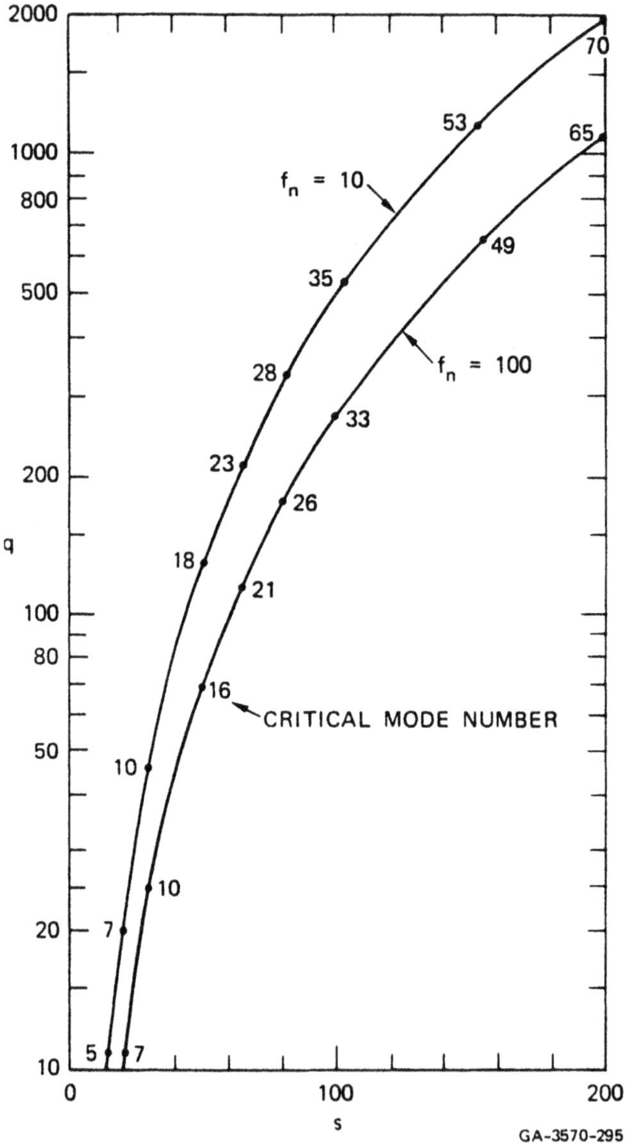


FIGURE 4.28 CRITICAL COLLAPSE VELOCITY CURVES FOR $f_n$ = 10 AND 100

To determine the initial collapse velocity of a specific cylindrical shell, we first calculate the value of s by substituting $r_o$, $h_o$, $\overline{\sigma}$, and $E_h$ in formula (4.3.15). For example, if the radius-to-thickness ratio $r_o/h_o = 30$ and the hardening modulus, average flow stress ratio is $E_h/\overline{\sigma} \simeq 4$, we obtain $s = 52$. If the critical collapse velocity criterion is $f_n = 10$, the curve in Figure 4.28 gives $q = 130$. If the plastic wave speed in our example is $c_h = (E_h/\rho)^{1/2} = 5 \times 10^4$ cm/s, we can now obtain V from Equation (4.3.14) as $V = 6.2 \times 10^4$ cm/s = 0.6 mm/μs = 1800 ft/sec.

Direct experimental evidence is lacking to check the results of the preceding analysis. However, collapse velocities for copper shaped-charge liners are about 1 mm/μs (3300 ft/sec), which is in qualitative agreement with the foregoing result.

### 4.3.2 Strain Rate Moments Only

We assume here that the constitutive equation can be adequately represented by the linear stress/strain-rate law

$$\sigma = \sigma_o + \mu\dot{\epsilon} \tag{4.3.23}$$

where $\sigma_o$ is the flow stress extrapolated linearly to static strain rates, and $\mu$ is the viscosity parameter. At a material point in an elemental sector of the cylindrical shell located a distance z measured radially inward from the center or midsurface area of the sector, the circumferential strain rate is approximately

$$\dot{\epsilon}(\theta, z, t) = \dot{r}(1 + 2z/r)/r - z\dot{\kappa}_b \tag{4.3.24}$$

where $\theta$ is the angular coordinate, t is time, and $\kappa_b$ is the curvature rate of the elemental sector caused by departure from circularity. Time differentiation of the current curvature (4.3.2) gives

$$\dot{\kappa} = -\dot{r}/r^2 - (w + w'' + w_i + w_i'')2\dot{r}/r^3 + (w + w'')^{\dot{}}/r^2 \tag{4.3.25a}$$

$$= -\dot{r}/r^2 + \dot{\kappa}_b$$

so

$$\dot{\kappa}_b = -(w + w'' + w_i + w_i'')2\dot{r}/r^3 + (w + w'')^{\dot{}}/r^2 \tag{4.3.26}$$

The circumferential bending moment on a shell element of unit width is

$$M = -\int_{-h/2}^{h/2} \sigma(z)\, dz = \mu I(-2\dot{r}/r^2 + \dot{\kappa}_b)$$

where we have used (4.3.23) and (4.3.24) for the stress and strain rate, and (4.3.25b) to introduce the curvature rate $\dot{\kappa}$. Also $I = h^3/12$ is again the second moment of area of the cross section.

Substituting M from (4.3.27), using (4.3.25a) for $\dot{\kappa}$ and $N = \sigma^\circ h$ in the equation of motion (4.3.7), and setting $\dot{r} = -V$ lead to

$$\mu h \left[ (w + w'')^{\cdot} + (w + w'' + w_i + w_i'')\,2V/r \right]''/12r^4$$

$$+ \sigma^\circ (w + w'' + w_i + w_i'')/r^2 = -\rho \ddot{w} \qquad (4.3.28)$$

In (4.3.28), $\sigma^\circ$ is the midsurface compressive stress where the compressive strain rate is $V/r$, so the law (4.2.23) gives

$$\sigma^\circ = \sigma_o + \mu V/r \qquad (4.3.29)$$

We now introduce the dimensionless quantities

$$u = w/r_o \qquad u_i = w_i/r_o \qquad \xi = r/r_o \qquad \tau = \mu_o t/\rho r_o^2$$

$$\alpha^2 = h_o^2/12 r_o^2 \qquad \nu = \mu/\mu_o \qquad \beta = \rho V r_o/\mu_o \qquad \gamma = \rho r_o^2 \sigma_o/\mu_o^2 \qquad (4.3.30)$$

where $\sigma_o$ is a reference viscosity parameter. After substituting (4.3.29) in (4.3.28) and converting to the dimensionless quantities (4.3.30), we obtain

$$\xi^6 \ddot{u} + \nu\alpha^2 \left[ (u + u'')^{\cdot} + (2\beta/\xi)(u + u'' + u_i + u_i'') \right]''$$

$$+ \xi^4 \gamma (1 + \nu\beta/\gamma\xi)(u + u'' + u_i + u_i'') = 0 \qquad (4.3.31)$$

where

$$\dot{u} = \partial u/\partial \tau \qquad u' = \partial u/\partial \theta$$

In (4.3.31), we can replace $\xi = r/r_o = 1 - Vt = 1 - \beta\tau$ with the variable $\epsilon$ through the relation $\xi = e^{-\epsilon}$. When this is done, (4.3.31) becomes

$$\ddot{u} + \dot{u} + (\alpha\nu e^5 \xi/\beta) \left[ (u + u'')^{\cdot} + 2(u + u'' + u_i + u_i'') \right]''$$

$$+ (\gamma/\beta^2)(1 + \nu\beta e^\epsilon/\gamma)(u + u'' + u_i + u_i'') = 0 \qquad (4.3.32)$$

where

$$\dot{u} = \partial u/\partial \epsilon \qquad u' = \partial u/\partial \theta$$

The variable $\epsilon = \ln(1/\xi) = \ln(r_o/r)$ is the average true circumferential strain.

We now solve (4.3.32) for the initial conditions

$$u(\theta,0) = 0 \qquad \dot{u}(\theta,0) = V_i(\theta) \tag{4.3.33}$$

the initial displacement imperfections being represented by $u_i(\theta)$ in the equation itself.

Let us represent the initial displacements and velocities by

$$u_i(\theta) = \sum_{n=1}^{\infty} (A_n \cos n\theta + B_n \sin n\theta) \tag{4.3.34}$$

$$u(\theta,o) = \sum_{n=1}^{\infty} (C_n \cos n\theta + D_n \sin n\theta) \tag{4.3.35}$$

In physical terms, $A_n$ and $B_n$ are fractions of the initial radius $r_o$, and $C_n$ and $D_n$ are fractions of the initial velocity V.

Let us represent the buckling displacement similarly by

$$u(\theta,o) = \sum_{n=1}^{\infty} \left[F_n(\epsilon) \cos n\theta + G_n(\epsilon) \sin n\theta\right] \tag{4.3.36}$$

in which the coefficients $F_n$ and $G_n$ are to be determined. Substituting the series representations (4.3.34) and (4.3.36) into the governing equation (4.3.32) and setting to zero the coefficients of cos n $\theta$ and sin n $\theta$ lead to

$$\ddot{F}_n + a_n(\epsilon)\dot{F}_n + b_n(\epsilon)F_n = -b_n(\epsilon)A_n \tag{4.3.37}$$

$$\ddot{G}_n + a_n(\epsilon)\dot{G}_n + b_n(\epsilon)G_n = -b_n(\epsilon)B_n \tag{4.3.38}$$

$$a_n(\epsilon) = 1 + (\alpha^2\nu/\beta)e^{5\epsilon}\, n^2(n^2 - 1) \tag{4.3.39}$$

$$b_n(\epsilon) = \left[2(\alpha^2\nu/\beta)e^{5\epsilon}\, n^2 - (\gamma/\beta^2)\,(1 + \nu\beta e^{\epsilon}/\gamma)\right](n^2 - 1) \tag{4.3.40}$$

The initial conditions for $F_n$ and $G_n$ from (4.3.33) and (4.3.35) are

$$F_n(o) = 0 \qquad \dot{F}_n(o) = C_n \qquad G_n(o) = 0 \qquad \dot{G}_n(o) = D_n \tag{4.3.41}$$

By introducing the function $H_n$ defined by $H_n(\epsilon) = F_n(\epsilon) + A_n$ into Equation (4.3.37) we obtain the set of homogeneous equations

$$\ddot{H}_n + a_n(\epsilon)\dot{H}_n + b_n(\epsilon)H_n = 0 \tag{4.3.42}$$

where $a_n(\epsilon)$ and $b_n(\epsilon)$ are given by (4.3.39) and (4.3.40). A fundamental pair of solutions $H_n^{(1)}$ and $H_n^{(2)}$ to each equation of the set (4.3.42) exists if the initial conditions are

$$H_n^{(1)}(o) = 1 \qquad \dot{H}_n^{(1)}(o) = 0 \tag{4.3.43}$$

$$H_n^{(2)}(o) = 0 \qquad \dot{H}_n^{(2)}(o) = 1 \tag{4.3.44}$$

Consequently, the solution of (4.3.42) with initial conditions

$$H_n(o) = A_n \qquad \dot{H}_n(o) = C_n \tag{4.3.45}$$

is

$$H_n(\epsilon) = A_nH_n^{(1)}(\epsilon) + C_nH_n^{(2)}(\epsilon) \tag{4.3.46}$$

If we start with $H_n(\epsilon) = G_n(\epsilon) + B_n$, the foregoing procedure leads to

$$H_n(\epsilon) = B_nH_n^{(1)}(\epsilon) + D_nH_n^{(2)}(\epsilon) \tag{4.3.47}$$

Because the current departure from circularity is the sum of the initial displacement (4.3.34) and the buckling displacement (4.3.36), the required solution is

$$\begin{aligned} u_i(\theta) + u(\theta,\epsilon) &= \sum_{n=1}^{\infty} \Big[H_n^{(1)}(\epsilon)\{A_n \cos n\theta + B_n \sin n\theta\} \\ &\quad + H_n^{(2)}(\epsilon)\{C_n \cos n\theta + D_n \sin n\theta\}\Big] \end{aligned} \tag{4.3.48}$$

The advantage of introducing function $H_n$ is that the fundamental set of equation (4.3.42) with initial conditions (4.3.43) and (4.3.44) need be solved only once for all initial displacements and velocities. Note that solution (4.3.48) brings out clearly the amplification effects of the functions $H_n^{(1)}(\epsilon)$ and $H_n^{(2)}(\epsilon)$; the former amplifies the initial shape and the latter amplifies the initial velocity needed to produce the current departure from a circular shape.

The character of the amplification functions can be illustrated by solving numerically the fundamental set of Equations (4.3.42) with initial conditions (4.3.43) and (4.3.44). For this purpose the dimensional and material properties chosen are as follows;

| | | |
|---|---|---|
| shell radius | $r_o$ | $=$ 2.517 inches (6.393 cm) |
| shell thickness | $h_o$ | $=$ 0.141 inch (0.358 cm) |
| copper density | $\rho$ | $=$ 0.321 lb/in$^3$ (8.9 g/cm$^3$) |
| flow stress | $\sigma_o$ | $=$ 45,000 psi (310 MPa) |
| viscosity | $\mu_o$ | $=$ 3.625 psi-s (1724 N-s/m$^2$) |
| collapse velocity | $V$ | $=$ 3,281 ft/s (1000 m/s) |

Corresponding to these physical quantities are the dimensionless parameters defined in (4.3.30).

$$\alpha = 0.0162 \qquad \beta = 22.76 \qquad \gamma = 17.46 \qquad \nu = 0.01 \text{ (and 1.0)}$$

The values of $\mu_o$ and $\nu = 1$ correspond to OFHC copper properties for a strain rate of $10^3$ s. Because the compressive circumferential strain rate is initially $V/r_o = 15.6 \times 10^3$ sec$^{-1}$, $\nu$ is less than unity. Unfortunately, data are not available in the literature for estimating $\nu$, so a suitably small value of $\nu = 0.01$ was chosen to obtain the influence on the amplification functions.

For mathematical convenience, the amplification functions have been expressed in terms of the average compressive circumferential strain, $\epsilon$. To assist physical interpretation, it is recalled that the average radius r and the time t corresponding to the strain $\epsilon$ are determined by $r = \xi r_\rho$, $\xi = e^{-\epsilon}$, $t = (r_o - r)/V = (1 - \xi)r_o/V$. Collapse is complete when the inner radius of the shell is zero and $\xi = (h_o/2r_o)^{1/2}$, from simple geometric considerations. Thus, for the foregoing data, min $\xi = 0.168$, max $\epsilon = 1.78$, and max $t = 53\ \mu$s.

For brevity, the remaining discussions is confined to the amplification function $H_n^{(1)}(\epsilon)$; the overall behavior of $H_n^{(2)}(\epsilon)$ is similar.

Figure 4.29 shows the amplification function $H_n^{(1)}(\epsilon)$ plotted against strain or dimensionless time for several mode numbers, all for $\nu = 0.01$. The maximum value reached is $H_n^{(1)} = 55$ when the strain is $\epsilon = 0.85$ in mode $n = 48$. Figure 4.30 shows $H_n^{(1)}$ plotted against mode number with strain as a parameter. At a given time (or $\epsilon$ value), certain modes are being preferred in the buckling process; these modes are grouped about a central preferred mode where the peak amplification occurs. The preferred mode numbers decrease with time. For example, the central preferred mode number decreases from 75 to 42 as the strain increases from 0.4 to 1.2, the greatest amplification being 55 when $\epsilon = 0.8$ ($\xi = r/r_o = 0.45$) in mode 52.

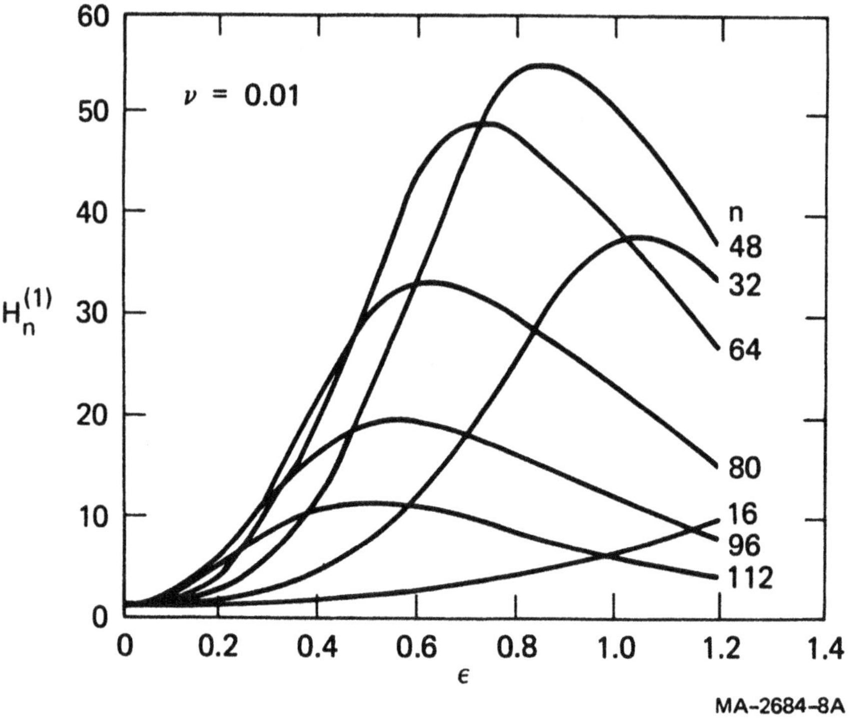


FIGURE 4.29 TIME VARIATION OF INITIAL SHAPE AMPLIFICATION FUNCTION ($\nu$ = 0.01)

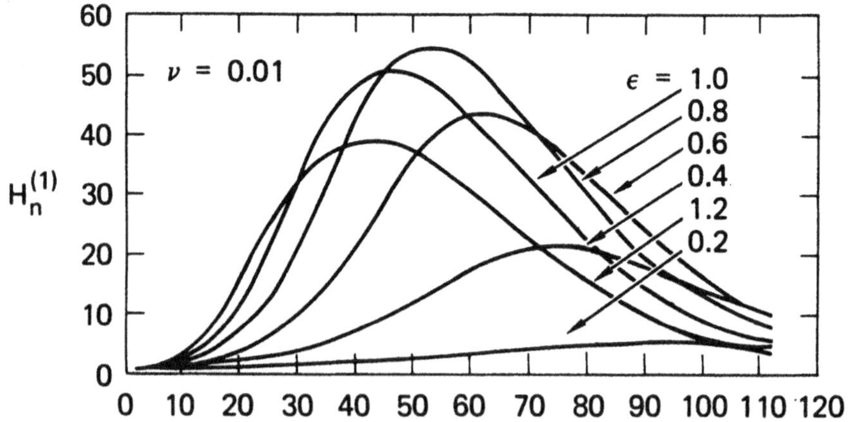


FIGURE 4.30 MODE VARIATION OF INITIAL SHAPE AMPLIFICATION FUNCTION ($\nu$ = 0.01)

Figures 4.31 and 4.32 show how the curves in Figure 4.29 and 4.30 change when the viscosity parameter is increased by a factor of 100 to become $\nu = 1.0$. Qualitative similarities exist, but the amplification factors are drastically reduced because of the increased bending resistance to buckling. For Figure 4.31 the maximum value of $H_n^{(1)}$ is now 3.3 in mode 12, when the strain is 0.8 (c.f., $H_n^{(1)} = 55$, $n = 48$, $\epsilon = 0.85$ for $\nu = 0.01$). From Figure 4.32, the central preferred mode number decreases from 24 to 7 as $\epsilon$ increases from 0.2 to 1.2.

The maximum value of the amplification function in the sense just described is the value chosen to define the critical collapse velocity. This maximum is designated $H_N^{(1)}$, N being the preferred mode determined by the location of the peak of the envelope of the curves shown in Figure 4.30 for $\nu = 0.01$.

Figure 4.33 shows three curves that relate the maximum amplification factor $H_N^{(1)}$ to the constant imploding velocity V represented by the dimensionless parameter $\beta = \rho V r_o/\mu_o$, each curve being for a specific value of the parameter $\alpha = h_o/\sqrt{12}\ r_o$, or radius-to-thickness ratio $r_o/h_o$ as labelled. The remaining parameters have the values $\nu = 0.01$ and $\gamma = 17.46$. The curves may be considered as an illustration of the effect of shell thickness on the relationship between $H_N^{(1)}$ and $\beta$ or V. The value of N attached to each curve is the preferred mode number for buckling. Although no obvious theoretical reason exists for the mode number being the same along each curve, calculations showed that this was the case. The central curve corresponds to the previously illustrated example, in that $r_o/h_o = 17.85$ gives $\alpha = 0.0162$ and a point on the curve is $\beta = 22.76$ and $H_N^{(1)} = 55$.

The curves show how the amplification factor decreases as the velocity increases. They also show how instability increases with radius-to-thickness ratio. For example, when $\beta = 22.76$, the amplification factor increases from 11 to 470 as $r_o/h_o$ increases from 10 to 30.

Figure 4.34 shows the relationship between the amplification factor and the collapse velocity when the viscosity parameter has been increased by an order of magnitude to $\nu = 0.1$. The stability has essentially doubled throughout in that the critical velocity associated with a specific amplification has halved.

Figure 4.35 shows three curves that relate the radius-to-thickness ratio of a cylindrical shell to the collapse velocity, each curve being for a specific amplification factor and all curves being for $\nu = 0.01$ and $\gamma = 17.46$. The curves were obtained from cross plots from Figure 4.33 at amplification factors of 10, 100, and 1000. Figure 4.36 shows a similar set of curves obtained from Figure 4.34 for a viscosity parameter value of $\nu = 0.1$.

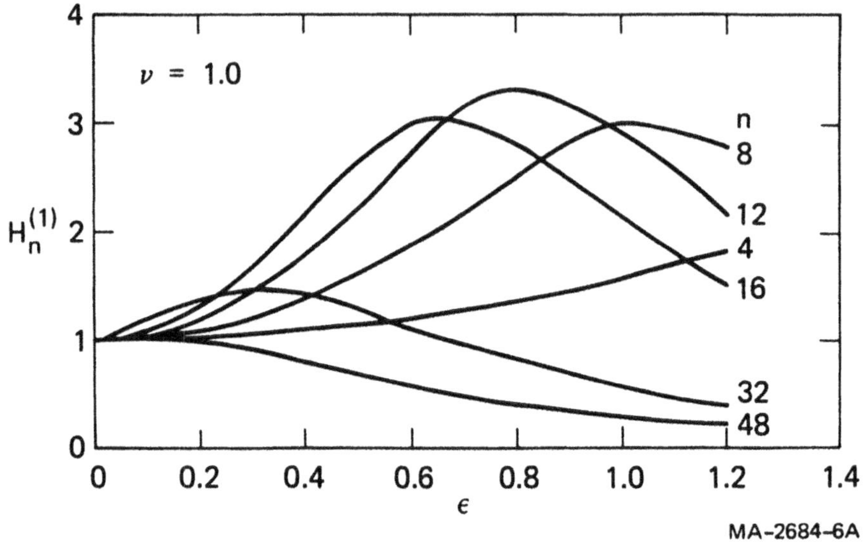


FIGURE 4.31 TIME VARIATION OF INITIAL SHAPE AMPLIFICATION FUNCTION (ν = 1.0)

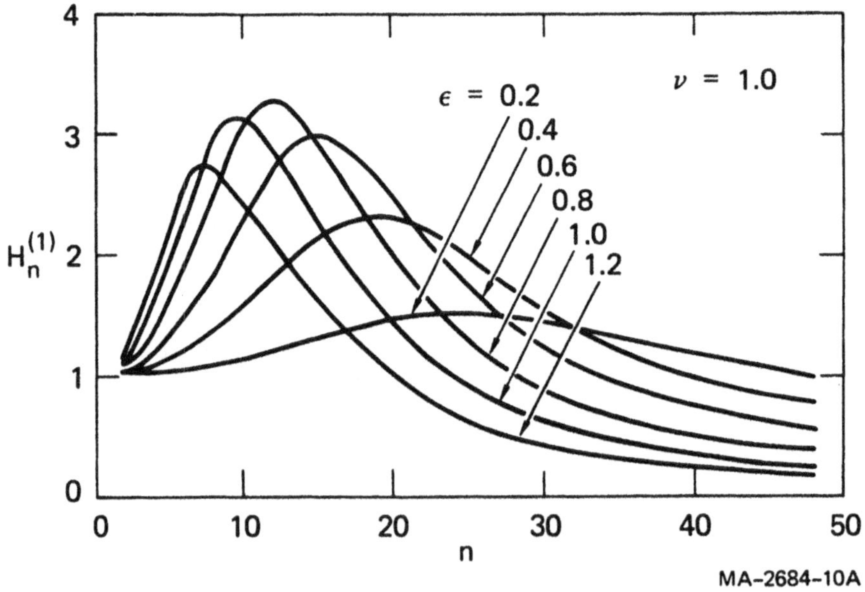


FIGURE 4.32 MODE VARIATION OF INITIAL SHAPE AMPLIFICATION FUNCTION (ν = 1.0)

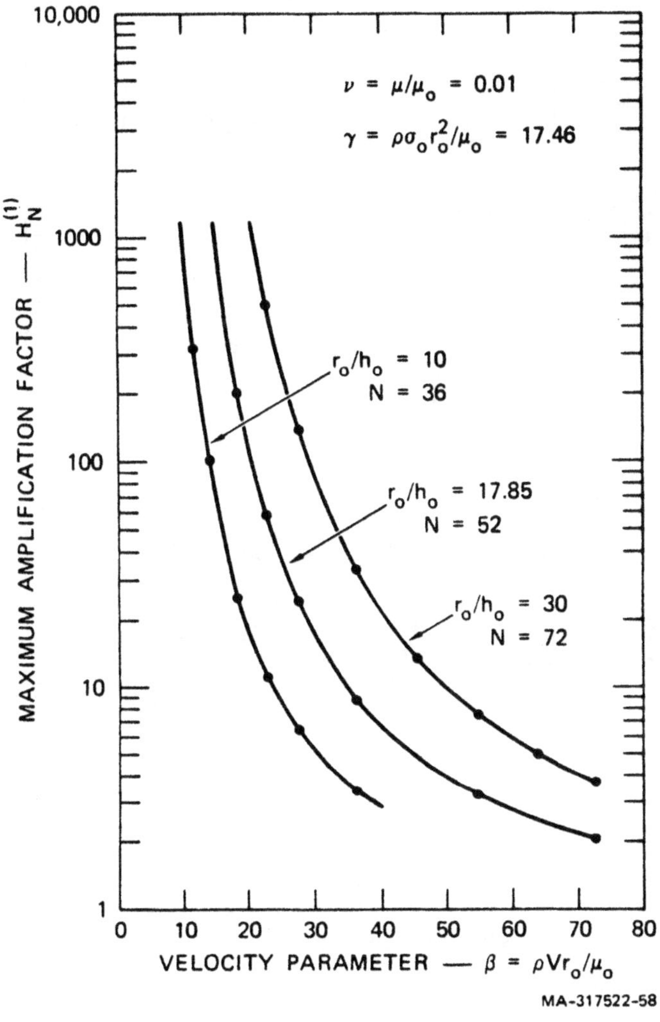


FIGURE 4.33 RELATIONSHIP BETWEEN MAXIMUM AMPLIFICATION FACTOR AND VELOCITY ($\nu$ = 0.01)

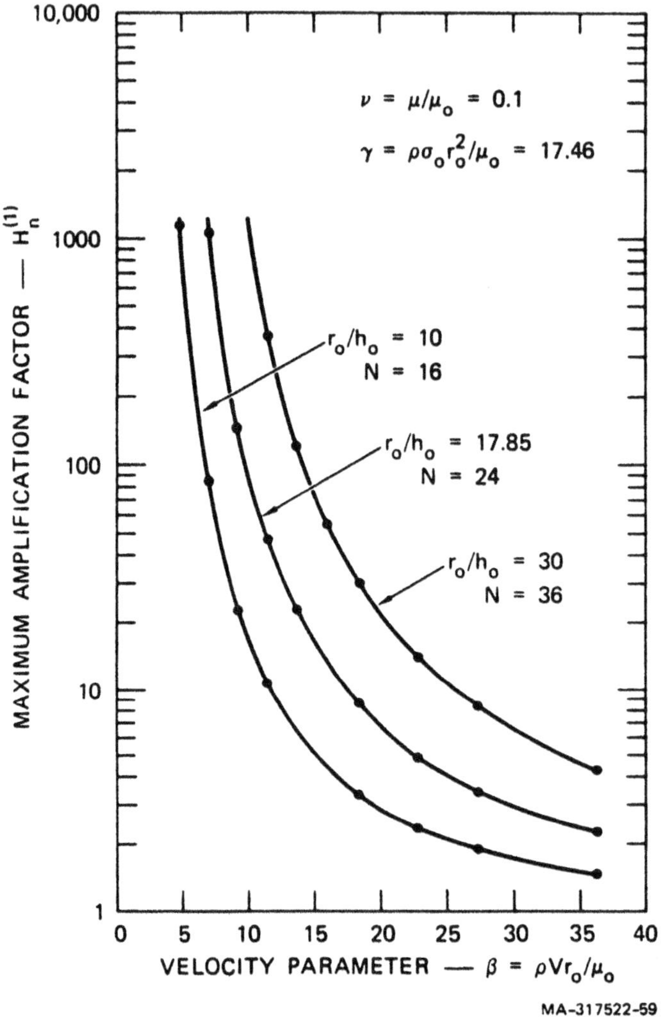


FIGURE 4.34 RELATIONSHIP BETWEEN MAXIMUM AMPLIFICATION FACTOR AND VELOCITY PARAMETER ($\nu = 0.1$)

Curves such as those constructed for Figure 4.35 and 4.36 for OFHC copper may be used for stability design of devises that involve high-speed collapse of shells of circular cross section. For example, a trial design with the values of the numerical example just discussed, having a radius-to-thickness ratio of $r_o/h_o = 17.85$ ($\alpha = 0.0162$) and manufactured with enough precision to allow an amplification of $H_n^{(1)} = 1000$, has a critical velocity parameter of about $\beta = 15$, according to Figure 4.35. With a copper density of $\rho = 8.9$ g/cm$^3$, an initial radius of $r_o = 6.393$ cm (2.517 inches), and a reference viscosity of $\mu_o = 0.25 \times 10^6$ dyne-s/cm$^2$, the value $\beta = 15$ gives a critical collapse velocity of 0.67 mm/$\mu$s. Higher collapse velocities should therefore result in stable collapse.

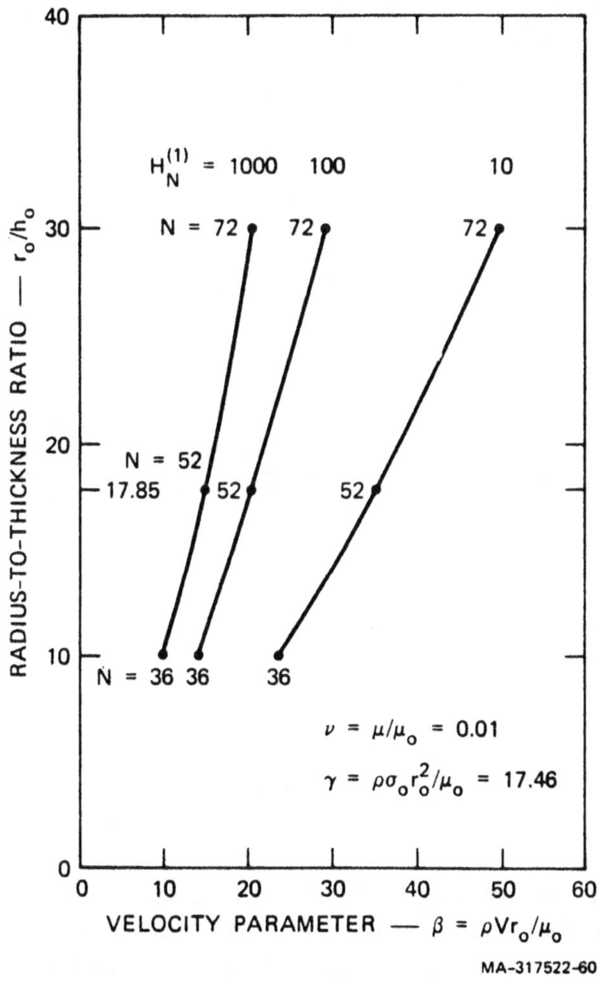


FIGURE 4.35 CRITICAL VELOCITY CURVES ($\nu = 0.01$)

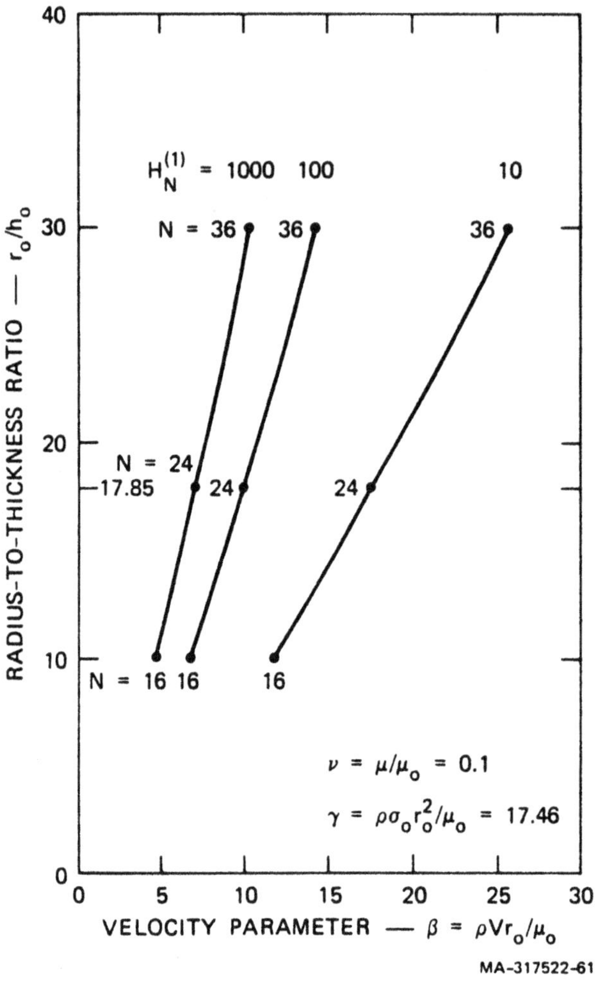


FIGURE 4.36 CRITICAL VELOCITY CURVES ($\nu$ = 0.1)

## REFERENCES


1. J.N. Goodier, "Dynamic Buckling of Rectangular Plates and Sustained Plastic Compressive Flow," in *Engineering Plasticity,* Cambridge University Press, 1968, Proceedings of an International Conference of Plasticity held in Cambridge, England, March 1968.

2. M. Abramovitz and L.A. Stegun, *Handbook of Mathematical Functions,* Applied Mathematics Series 55, National Bureau of Standards, August 1966.

3. P. Perzyna, "The Constitutive Equations for Rate Sensitive Plastic Materials," Quarterly of Applied Mechanics, **20**, pp. 321-332, January 1963.

4. M.J. Manjoine, "Influence of Rate of Strain and Temperature on Yield Stresses in Mild Steel," J. Appl. Mech., **11**, pp. 211-218, 1944.

5. S.R. Bodner and P.S. Symonds, "Experimental and Theoretical Investigations of the Plastic Deformation of Cantilever Beams Subjected to Impulse Loading," J. Appl. Mech., **29**, pp. 719-728, December 1962.

# 5. DYNAMIC BUCKLING OF CYLINDRICAL SHELLS UNDER AXIAL IMPACT

A great deal of work has been done on buckling of cylindrical shells under dynamic axial loads. The motivation for much of this work stems from the design of missiles to sustain the sudden application of thrust from rocket engine ignition. The thrust-time history is idealized as a step or a linear ramp increase to a constant operating thrust. The concern is that the dynamic buckling load may be smaller than the buckling load for a slowly applied thrust. Other applications occur in transportation accidents, where the loading is from impact and often exceeds the static buckling load.

In this chapter, as in the rest of the monograph, we are concerned mainly with buckling in which the axial load is higher than the static buckling load. We also include the classical theory for static buckling (perfect shells) and the essentials of Budiansky and Hutchinson's[1] theory for dynamic buckling under step loads, which can occur at loads lower than the static buckling load of even imperfect shells. But emphasis is on impact buckling at high loads. Final buckles in shells under these loads are similar in form to the familiar diamond pattern of static buckling, but the wavelengths are much shorter and their aspect ratio is much larger (they are flatter, with a circumferential wavelength several times the axial wavelength). Thicker shells buckle in symmetric modes during plastic axial flow.

We consider impact at both elastic (Section 5.1) and plastic (Section 5.2) axial stress levels. In each case, we determine the modes of buckling and the time at the onset of large deformation. These results can be used for design to avoid threshold buckling and also to provide the initial conditions for analysis of plastic collapse buckling (Section 5.3) in modes determined by the threshold buckling theory. The experiments discussed in Sections 5.1 and 5.2 were designed to limit the load duration so that terminal observations could be made of threshold buckling not obscured by the complex folding and stretching of collapse. The buckling experiments discussed in Section 5.3 were allowed to continue through collapse. A simple kinematic theory is used to calculate average collapse forces, which are much lower than the impact forces that initiate buckling.

## 5.1 DYNAMIC BUCKLING OF CYLINDRICAL SHELLS UNDER ELASTIC AXIAL IMPACT

### 5.1.1 Analytical Formulation

In experiments described in Section 1.2, Figure 1.3, and discussed in more detail in Section 5.1.5, cylindrical shells are impacted at one end by a massive ring, which sends a step axial stress wave up the shell. The shell buckles before any signal is received from the opposite end (free in these experiments), so in the theory the shell is taken arbitrarily long. Also, as in the elastic buckling of bars in Section 2.2, the thrust is assumed uniform throughout the length of the shell, thus neglecting any effects of the moving axial stress front. As will be shown, this is a reasonable assumption for the shell for the same reason as for the bar: the buckle wavelengths are very short compared with the distance the stress wave travels while the buckles form. Thus, in the analytical formulation we consider a long, thin cylindrical shell subjected to a suddenly applied, constant compressive force uniform throughout the shell.[2]

For studying the early displacements, a small-deflection linear theory suffices. The experiments show that the wavelengths of buckling established by this early motion carry over to large deflection buckling and even to the permanent plastic buckle pattern. Also, the dominant buckle wave numbers in the circumferential direction are $n > 4$, so the Donnell equations derived in Section 3.5.2 for buckling under radial loads apply here also, with the applied in-plane force being $\bar{N}_x$ rather than $\bar{N}_\theta$. With time denoted by t, axial force $\bar{N}_x$ per unit length taken positive in compression, and coordinates taken as in Figure 3.37, the appropriate Donnell equations for axial loading are obtained from Equations (3.5.37) and (3.5.40) as follows:

$$D\nabla^4 w + \bar{N}_x \frac{\partial^2}{\partial x^2}(w + w_i) + \rho h \frac{\partial^2 w}{\partial t^2} - \frac{1}{a}\frac{\partial^2 F}{\partial x^2} = 0 \qquad (5.1.1)$$

$$\nabla^4 F = -\frac{Eh}{a}\frac{\partial^2 w}{\partial x^2} \qquad (5.1.2)$$

where

$$\nabla^4 = \left(\frac{\partial^2}{\partial x^2} + \frac{\partial^2}{a^2 \partial \theta^2}\right)^2$$

In these equations, x is axial coordinate, $\theta$ is circumferential coordinate, w is positive inward displacement measured from an unstressed initial imperfection displacement $w_i$, $\rho$ is material density, E is Young's modulus, h is wall thickness, a is radius, and F is Airy's stress function for in-plane force resultants produced by the buckling deformation [see Equations (3.5.36)].

It is convenient to use dimensionless variables

$$\xi = x\left(\frac{\bar{N}_x}{D}\right)^{1/2}, \quad \eta = a\theta\left(\frac{\bar{N}_x}{D}\right)^{1/2}, \quad \tau = \frac{\bar{N}_x}{(\rho h D)^{1/2}}\cdot t \qquad (5.1.3)$$

Then Equations (5.1.1) and (5.1.2) become

$$\nabla^4 w + \frac{\partial^2}{\partial \xi^2}(w + w_i) + \ddot{w} - \frac{1}{a\bar{N}_x}\frac{\partial^2 F}{\partial \xi^2} = 0 \tag{5.1.4}$$

$$\nabla^4 F = \frac{EhD}{a\bar{N}_x}\frac{\partial^2 w}{\partial \xi^2} \tag{5.1.5}$$

in which

$$\nabla^4 = \left(\frac{\partial^2}{\partial \xi^2} + \frac{\partial^2}{\partial \eta^2}\right)^2, \qquad \ddot{w} = \frac{\partial^2 w}{\partial \tau^2} \tag{5.1.6}$$

To simplify the mathematics, the ends of the shell are taken to be simply supported. The boundary conditions, from Equations (3.5.41), are therefore

$$w = \partial^2 w/\partial x^2 = 0 \quad \text{at} \quad x = 0, L \tag{5.1.7}$$

As already discussed, L is taken arbitrarily large. The condition at the impacted end of the shells in the experiments is more closely approximated by a clamped boundary (zero rotation), but the experiments show that this boundary condition does not affect the buckling significantly because buckling extends over many axial waves. The conditions of simple support are satisfied by

$$w(\xi, \eta, \tau) = \sum_{m=1}^{\infty}\sum_{n=1}^{\infty} W_{mn}(\tau) \sin\alpha_m\xi \sin\beta_n\eta \tag{5.1.8}$$

$$F(\xi, \eta, \tau) = \sum_{m=1}^{\infty}\sum_{n=1}^{\infty} F_{mn}(\tau) \sin\alpha_m\xi \sin\beta_n\eta \tag{5.1.9}$$

in which

$$\alpha_m = \frac{m\pi}{L}\left(\frac{D}{\bar{N}_x}\right)^{1/2}, \qquad \beta_n = \frac{n}{a}\left(\frac{D}{\bar{N}_x}\right)^{1/2} \tag{5.1.10}$$

The initial displacement is also expanded into a Fourier series:

$$w_o(\xi, \eta) = \sum_{m=1}^{\infty}\sum_{n=1}^{\infty} a_{mn} \sin\alpha_m\xi \sin\beta_n\eta \tag{5.1.11}$$

Equation (5.1.9) is now substituted into Equation (5.1.5), and the result is substituted, with Equations (5.1.8) and (5.1.11), into Equation (5.1.4) to give the following equation for the amplitudes $W_{mn}$ of the normal modes:

$$\ddot{W}_{mn} + \left[(\alpha_m^2 + \beta_n^2)^2 - \alpha_m^2 + \frac{EhD}{a^2\bar{N}_x^2}\,\frac{\alpha_m^4}{(\alpha_m^2 + \beta_n^2)^2}\right] W_{mn} = \alpha_m^2 a_{mn} \qquad (5.1.12)$$

## 5.1.2 Static Buckling

For static buckling, the inertia terms $\ddot{W}_{mn}$ are zero and, with finite $a_{mn}$, the modal amplitudes $W_{mn}$ grow without bound at eigenvalues for which the coefficient of $W_{mn}$ vanishes. Setting the coefficient equal to zero gives for the eigenvalues:

$$\frac{\bar{N}_x}{D} = Q + \frac{Eh}{a^2 D} \cdot \frac{1}{Q} \qquad (5.1.13)$$

in which

$$Q = \left(\frac{m^2\pi^2}{L^2} + \frac{n^2}{a^2}\right)^2 \left(\frac{m\pi}{L}\right)^{-2} = \pi^2\left(\frac{m^2}{L^2} + \frac{n^2}{\pi^2 a^2}\right)^2 \left(\frac{m}{L}\right)^{-2} \qquad (5.1.14)$$

The static buckling load is the lowest of these eigenvalues. The quantities $L/m$ and $\pi a/n$ are the half-wavelengths of the buckles in the axial and circumferential directions, respectively. Thus, the value of Q that minimizes $\bar{N}_x$ remains nearly fixed as the shell length is increased; the wavelengths $L/m$ and $\pi a/n$ stay nearly the same and m increases to compensate for the increase in L. For long shells m is therefore large and $L/m$ and hence Q can be treated as continuous variables. To find the minimum buckling load, we therefore set $d(\bar{N}_x/D)/dQ = 0$, which yields

$$Q = \left(\frac{Eh}{a^2 D}\right)^{1/2} \qquad (5.1.15)$$

and hence

$$\bar{N}_{xst} = \frac{2(EhD)^{1/2}}{a} = \frac{Eh^2}{a}\,\frac{1}{\sqrt{3(1 - \nu^2)}} \qquad (5.1.16)$$

This is the classical static buckling load for thin cylindrical shells. Equations (5.1.14) and (5.1.15) show that this same eigenvalue results for a large set of critical wavelength combinations specified by Equation (5.1.15). For symmetric buckling (accordian fashion, n = 0), the resulting axial half-wavelength is

$$\ell_o = \frac{L}{m_{cr}} = \frac{\pi\sqrt{ah}}{[12(1-\nu^2)]^{1/4}} = 1.728\sqrt{ah} \tag{5.1.17}$$

in which $\nu = 0.3$ in the numerical result. The general relationship (5.1.15) between axial and circumferential wavelengths now becomes simply

$$\frac{\ell_\theta}{\ell_x} = \left(\frac{\ell_x}{\ell_o} - 1\right)^{-1/2} \tag{5.1.18}$$

in which $\ell_x = L/m$ and $\ell_\theta = \pi a/n$.

Equation (5.1.18) shows that the axial half-wavelength of nonsymmetric buckling is always greater than the symmetric half-wavelength $\ell_o$. As $\ell_x/\ell_o \rightarrow 1$, the aspect ratio $\ell_\theta/\ell_x$ approaches infinity. For an aspect ratio of unity, the axial wavelength is twice the symmetric axial wavelength. To obtain aspect ratios significantly less than unity, the axial wavelength must be very large. For example, when the circumferential wavelength is half the axial wavelength, the axial wavelength is five times its symmetric value.

These wavelength expressions also show one reason why critical loads for thin cylindrical shells are very sensitive to initial imperfections. For example, with $a/h = 500$, Equation (5.1.17) gives a symmetric half-wavelength $\ell_o = 0.077$ a. For a square buckle pattern (unity aspect ratio), the axial and circumferential half-wavelengths are $\ell_x = \ell_\theta = 2\ell_o = 0.155$ a. The corresponding arc length is $\Delta\theta = 8.9°$. Thus, even a very localized imperfection encompasses a buckle and could affect the critical load. Furthermore, Equation (5.1.16) shows that the critical load is inversely proportional to a. A small inward dent would greatly increase a and hence decrease the critical load for the section of shell near the dent.

A large-deflection analysis by von Kárman, Dunn, and Tsien[3] shows that, once buckling has started, the critical load drops even further. Thus, the local buckling quickly spreads throughout the shell, and equilibrium eventually bifurcates to a large deflection post-buckled shape. Typically, this shape is a diamond pattern with an aspect ratio near unity and a half-wavelength several times the value $2\ell_o$ of the small-deflection theory. Another reason for the sensitivity to imperfections is the large-deformation nonlinear behavior of the shell even in a fixed buckle pattern. This is discussed further in Section 5.1.7.

Almroth, Holmes, and Brush[4] performed experiments on carefully prepared nickel shells that demonstrate this sequence of events. They found that, when these shells were loaded to about 75% of their experimental buckling loads, a small perturbation introduced by a mouth-blown puff of air or a slight touch of a finger would cause the shell to buckle suddenly. They took high-speed motion pictures (8000 frames/s) of such buckling, triggered by an electromagnet to allow timing of the photography. An example is shown in Figure 5.1. Buckling starts in a local area and spreads over the shell in a wedge-shaped front. This initial buckling is in a short-wavelength pattern. Later, the motion becomes very complex, and the shell eventually comes to rest in a large-deflection pattern of longer wavelengths, about twice the wavelength of the initial buckling.

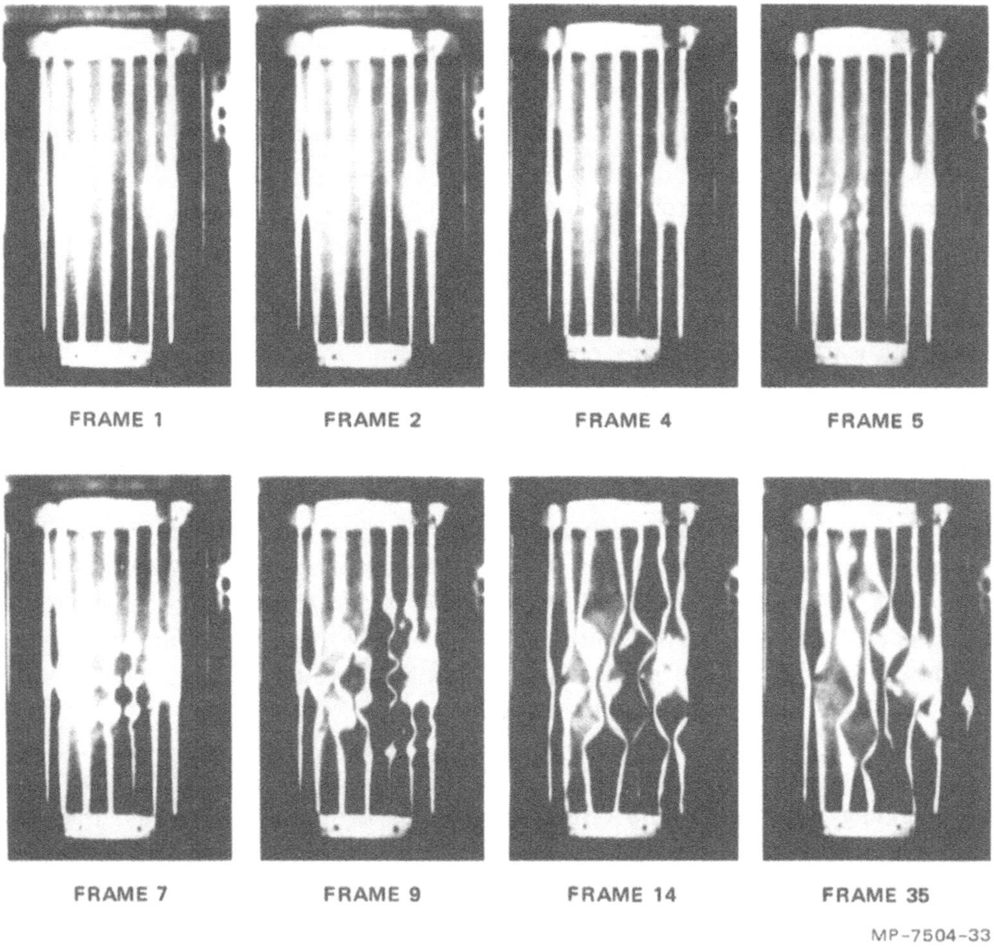


FIGURE 5.1 BUCKLE INITIATION AND PROGRESSION FOR "STATIC" BUCKLING

These shells were 6 inches (152 mm) in diameter with a 0.0036-inch-thick (91 $\mu$m) wall. The theoretical small-deflection half-wavelength for unity aspect ratio buckling is therefore $2\ell_o = 2(1.728)(3 \times 0.0036)^{1/2} = 0.36$ inch (9.1 mm).

The observed half-wavelengths of the initial buckling ranged from 0.3 to 0.4 inch (7.6 to 10 mm). By the time the buckles were large enough to be visible, they were approaching a diamond pattern rather than the rectangular pattern of Equation (5.1.8). Nevertheless, the agreement with the theoretical buckle size is significant.

The circumferential arc length of a buckle half-wavelength in the initial pattern is 0.36 inch/3 inch = 0.12 radians. The rise of the shell above a chord spanning this length is 3 inch × [1 − cos(0.06)] = 0.0054 inch (0.137 mm). Similarly, the rise above a full-wavelength chord is 0.0216 inch (0.55 mm). It is apparent that local imperfections or perturbations much smaller than the wall thickness could change the radius of curvature enough to substantially reduce the local critical buckling load.

In the Almroth, Holmes, and Brush tests, critical buckling loads for nine shells ranged from 45% to 73% of the classical buckling load. Shells with lower loads were judged before the test to have small but visible imperfections. In tests reported in the literature on actual prototype shells, not made with the exacting care in these tests, critical loads typically range from about 20% to 40% of the classical load. The load sustained in the post-buckled state is about 10% of the classical buckling load, as in the Almroth, Holmes, and Brush experiments.

### 5.1.3 Amplification Functions for Dynamic Buckling

We return now to Equation (5.1.12) for dynamic motion. In the limit as $L \to \infty$, $\alpha_m$ becomes a continuous variable. Because the wavelengths of buckling are small compared with the shell circumference, it is convenient to also treat $\beta_n$ as a continuous variable. In the dynamic analysis, $W_{mn}(\tau)$ is then denoted by $W(\alpha,\beta,\tau)$ and the subscripts on $\alpha$ and $\beta$ are dropped. We observe further that, from the static analysis, the shell parameter coefficient in Equation (5.1.2) can be expressed by

$$\frac{EhD}{a^2\bar{N}_x^2} = \frac{1}{4}\left(\frac{\sigma_{cl}}{\sigma}\right)^2 \tag{5.1.19}$$

in which $\sigma$ is the unperturbed axial stress and $\sigma_{cl}$ is the axial static buckling stress from Equation (5.1.16), which gives

$$\sigma_{cl} = \frac{Eh}{a}\,\frac{1}{\sqrt{3(1-\nu^2)}} \tag{5.1.20}$$

Equation (5.1.12) for dynamic motion is now written

$$\ddot{g} + k(\alpha, \beta) g = \alpha^2 \tag{5.1.21}$$

in which

$$k(\alpha, \beta) = (\alpha^2 + \beta^2)^2 - \alpha^2 + \frac{1}{4}\left(\frac{\sigma_{cl}}{\sigma}\right)^2 \frac{\alpha^4}{(\alpha^2 + \beta^2)^2} \tag{5.1.22}$$

and

$$g(\alpha, \beta, \tau) = \frac{W(\alpha, \beta, \tau)}{a(\alpha, \beta)} \tag{5.1.23}$$

is the amplification function, which gives the growth of initial imperfections $a(\alpha,\beta)$ under a given axial stress $\sigma = \bar{N}_x/h$.

By proceeding just as for the elastic bar in Section 2.2, we find that the solution to Equation (5.1.21) subject to initial conditions $g(\alpha,\beta,0) = \dot{g}(\alpha,\beta,0) = 0$ is

$$g(\alpha, \beta, \tau) = \frac{\alpha^2}{k(\alpha, \beta)} \left[1 - \begin{matrix}\cosh p\tau \\ \cos p\tau\end{matrix}\right] \tag{5.1.24}$$

in which

$$p = \left|k(\alpha, \beta)\right|^{1/2}$$

The hyperbolic form is taken for $k(\alpha,\beta) < 0$ and the trigonometric form is taken for $k(\alpha,\beta) > 0$. The equation $k(\alpha,\beta) = 0$ gives the static solution.

Figure 5.2 gives example plots of $g(\alpha,\beta,\tau)$ for $\tau = 4$ and $\tau = 8$. The dependence of these amplitudes on the axial wave number $\alpha$ is similar to that in the buckling bar, exhibiting a pronounced hump of "preferred" modes near $\alpha = 1/\sqrt{2}$. The circumferential wave number of the most amplified mode is $\beta = 0$, i.e., a symmetric mode, but there is an appreciable bandwidth of amplified modes in both the axial and circumferential directions.

Comparing the curves for $\tau = 4$ and $\tau = 8$, we see that as the motion proceeds, the bandwidth in both directions decreases, tending in the limit to produce a fixed axial wavelength corresponding to wave number $\alpha \approx 1/\sqrt{2}$, but tending toward longer wavelengths in the circumferential direction, ultimately approaching a simple symmetric pattern corresponding to $\beta = 0$ (i.e., $n = 0$). If the imperfections are large enough, however, nonlinear effects would dominate before the symmetric mode is attained, producing final buckles with a finite wavelength in both directions. Buckling of both types was observed in the experiments described in the following text.

From the definition of $k(\alpha,\beta)$ in Equation (5.1.22), we see that the normalized amplification curves $g(\alpha,\beta,\tau)$ depend only on the ratio $\sigma/\sigma_{cl}$ of the applied axial stress to the classical static buckling stress. Larger values give faster growth and narrower bandwidths, but the general shape of the amplification curves remains about the same for any $\sigma/\sigma_{cl} > 1$. In the next section, it will be shown that the expected values of the buckle wavelengths are very nearly proportional to $(D/\bar{N}_x)^{1/2} = r/\epsilon^{1/2}$, where $r = h/[12(1 - \nu^2)]^{1/2}$, and $\epsilon = \sigma/E$ is the strain from the axial thrust.

Figure 5.3 gives plots of the maximum value of g (at the peak of the hump in Figure 5.2) against time for several values of $\sigma/\sigma_{cl}$. As $a/h \rightarrow \infty$, we see from

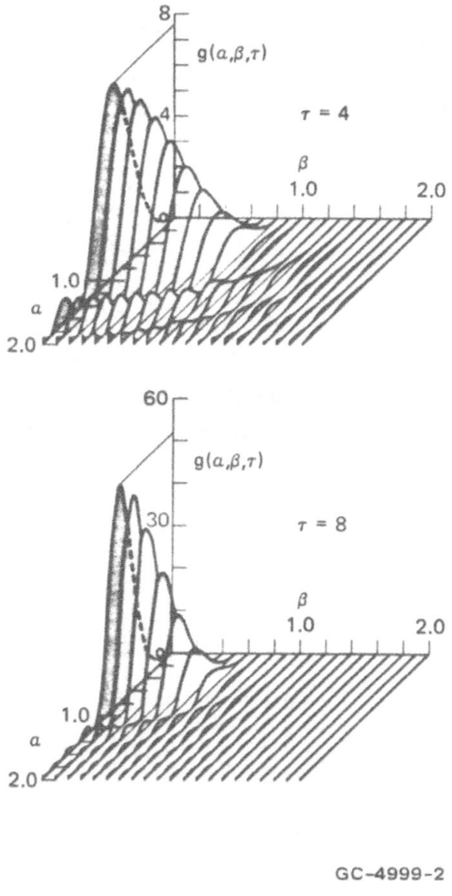


GURE 5.2 AMPLIFICATION FUNCTION g VERSUS AXIAL AND CIRCUMFERENTIAL WAVE NUMBERS α AND β FOR $\sigma/\sigma_{cl} = 2.5$

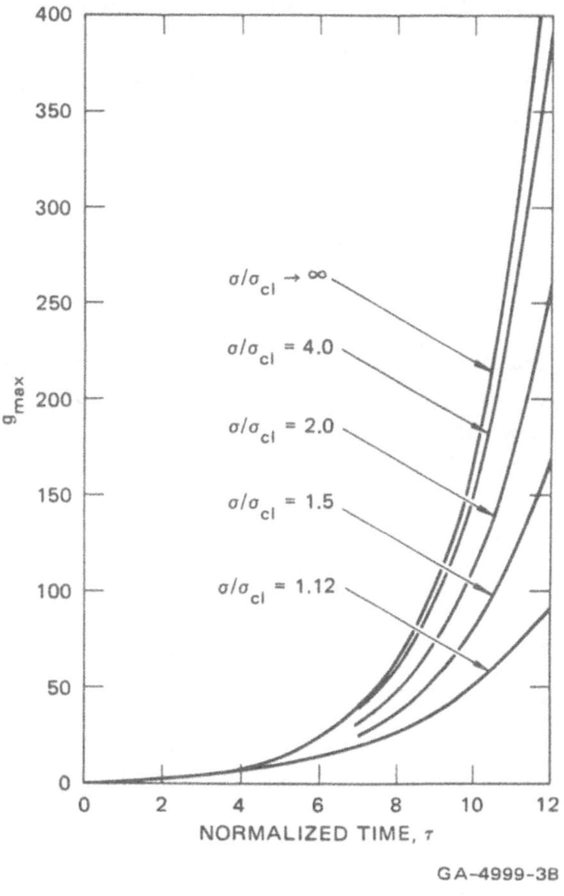


FIGURE 5.3 GROWTH OF MAXIMUM AMPLIFICATION WITH TIME

Equation (5.1.20) that $\sigma_{cl} \to 0$, so the curve for $\sigma/\sigma_{cl} \to \infty$ is that for a buckling plate (or rectangular bar if the factor $1 - \nu^2$ is omitted). For $\sigma/\sigma_{cl} > 2$, the time required for large amplification is only slightly greater than for the plate or bar.

### 5.1.4 Buckling from Random Imperfections

To compare theory with experiment, it is necessary to assume some form of imperfections in order to specify $a(\alpha,\beta)$ and hence to compute $W(\alpha,\beta,\tau)$ from Equation (5.1.23). The experiments on impact buckling of bars described in Section 2.2 demonstrated that a good description of the observed buckles was obtained by assuming imperfections in the form of white noise. These allow the buckle wavelengths to be dictated by the magnitude of the thrust, giving a random scatter of wavelengths with a mean and standard deviation both inversely proportional to the square root of the thrust, in agreement with experimental observations. It is expected that such an assumption will also be reasonable in the present problem. Thus we assume that $a(\alpha,\beta)$ are random normal with zero mean and constant variance over all $\alpha$ and $\beta$ in the amplified band of interest. With this assumption, the power spectral density of the modal amplitudes $W(\alpha,\beta,\tau)$ is proportional to $g^2(\alpha,\beta,\tau)$.

Having the power spectral density, we can compute the statistics of the buckled shape. For the bar, complete statistics of the wavelengths were computed using a Monte Carlo technique. The simpler problem of determining the mean wavelengths can be solved analytically and will be undertaken here for the shell. From Rice[5], the expected number of zeros in an interval $(s_1, s_2)$ of a random function $f = F(s)$ is

$$\int_{s_1}^{s_2} ds \int_{-\infty}^{\infty} |q| P(f, q; s) dq \tag{5.1.25}$$

where $P(f,q; s)$ is the probability density function for the variables

$$f = F(s)$$

$$q = \frac{dF}{ds}$$

For our problem, we have then that the expected number of zeros in the $\xi$-direction in the interval $(0,\ell)$ for a specified value of $\eta$ and $\tau$ is

$$R = \int_0^{\ell} d\xi \int_{-\infty}^{\infty} |z| P(w, z; \xi, \eta, \tau) dz \tag{5.1.26}$$

where $\ell = L(\bar{N}_x/D)^{1/2}$, and $P(w,z;\ \xi,\eta,\tau)$ is the probability density function of the variables

$$w = w(\xi,\eta,\tau)$$

$$z = \frac{\partial w}{\partial \xi}$$

and $\xi$, $\eta$, $\tau$ are carried along as parameters. Since it has been assumed that the initial imperfections have a Gaussian probability distribution (or, alternately, if we apply the central limit theorem for more general imperfection statistics), then the final buckled form will also have a Gaussian distribution. Thus, the distribution function has the form

$$P(w,z;\ \xi,\eta,\tau) = \frac{1}{2\pi(\mu_{11}\mu_{22} - \mu_{12}^2)^{1/2}} \times \exp\left[\frac{-\mu_{11}w^2 + 2\mu_{12}wz - \mu_{22}z^2}{2(\mu_{11}\mu_{22} - \mu_{12}^2)}\right] \qquad (5.1.27)$$

where

$$\mu_{11} = \langle w^2 \rangle$$
$$\mu_{22} = \langle z^2 \rangle$$
$$\mu_{12} = \langle wz \rangle$$

and $\langle\ \rangle$ indicates ensemble average.

The results are simplified if it is assumed that the imperfections and final buckled form are stationary (in space) so that in place of Equation (5.1.8) we can write

$$w(\xi,\eta,\tau) = \sum_{m=1}^{\infty} \sum_{n=1}^{\infty} W_{mn}(\tau)\ \sin(\alpha_m\xi - \phi_m)\ \sin(\beta_n\eta - \theta_n) \qquad (5.1.28)$$

where the $\phi_m$ and $\theta_n$ are each uniformly distributed over the interval $(0,2\pi)$. With this assumption, the statistics of the buckled form are independent of $\xi$ and $\eta$, which we would expect to be justifiable for waves at a sufficient distance from the ends of the cylinder. In fact, the results of a Monte Carlo computation including end effects for the bar, which is similar in form to the cylinder, indicate satisfactory agreement with the stationary process assumption even for the first wave from the simple support.

Using Equation (5.1.28), we find that

$$\mu_{11} = \langle w^2 \rangle = \int_0^\infty \int_0^\infty W^2(\alpha, \beta, \tau) d\alpha \, d\beta$$

$$= \sigma^2 \int_0^\infty \int_0^\infty g^2(\alpha, \beta, \tau) d\alpha \, d\beta \tag{5.1.29a}$$

$$\mu_{22} = \langle z^2 \rangle = \int_0^\infty \int_0^\infty \alpha^2 W^2(\alpha, \beta, \tau) d\alpha \, d\beta$$

$$= \sigma^2 \int_0^\infty \int_0^\infty \alpha^2 g^2(\alpha, \beta, \tau) d\alpha \, d\beta \tag{5.1.29b}$$

$$\mu_{12} = \langle wz \rangle = 0 \tag{5.1.29c}$$

where

$$\sigma^2 = \langle a^2(\alpha, \beta) \rangle = \text{constant}$$

and the sums have been replaced by integrals.

Making use of $\mu_{12} = 0$ from Equation (5.1.29c) in Equation (5.1.27) and substituting the resulting expression into Equation (5.1.26) gives the expected number of zeros in the length $\ell$:

$$R = \frac{\ell}{\pi} \left( \frac{\mu_{22}}{\mu_{11}} \right)^{1/2} \tag{5.1.30}$$

The mean wavelength, as measured between alternate zero crossings, is simply

$$\lambda_\xi = \frac{2\ell}{R} = 2\pi \left( \frac{\mu_{11}}{\mu_{22}} \right)^{1/2} \tag{5.1.31}$$

or, upon using Equations (5.1.29),

$$\lambda_\xi = 2\pi \left[ \frac{\int_0^\infty \int_0^\infty g^2(\alpha, \beta, \tau) d\alpha \, d\beta}{\int_0^\infty \int_0^\infty \alpha^2 g^2(\alpha, \beta, \tau) d\alpha \, d\beta} \right]^{1/2} \tag{5.1.32}$$

The mean wavelength in the circumferential direction can similarly be shown to be

$$\lambda_\eta = 2\pi \left[ \frac{\int_0^\infty \int_0^\infty g^2(\alpha, \beta, \tau) d\alpha \, d\beta}{\int_0^\infty \int_0^\infty \beta^2 g^2(\alpha, \beta, \tau) d\alpha \, d\beta} \right]^{1/2} \tag{5.1.33}$$

Equations (5.1.32) and (5.1.33) were integrated numerically over the area $0 \leqslant \alpha \leqslant 2$, $0 \leqslant \beta \leqslant 2$ of significant amplification, and the results are presented in Figure 5.4. Figure 5.4(a) shows the variation of $\lambda_\xi$ and $\lambda_\eta$ with $\tau$. Bands have been drawn that contain the curves for $\sigma/\sigma_{cl}$ in the range from 1.12 to 4.0. The narrowness of these bands indicates that the most significant parameters affecting wavelength are those given in the normalization in Equation (5.1.3).

Also from Figure 5.4(a), we see that the mean axial wavelength increases only slightly with $\tau$ for $\tau > 4$ and quickly approaches a "preferred" wavelength. For ratios of $\sigma/\sigma_{cl}$ nearing unity, the preferred wavelength becomes the classical static wavelength $\lambda_\xi = 2\sqrt{2}\pi$. The mean circumferential wavelength, however, increases with $\tau$ without approaching an asymptotic value. Thus, as discussed in the preceding section, the mean circumferential wavelengths actually observed in large-deflection buckling will probably depend on the magnitude of the initial imperfections, which determines the duration for which this small-deflection theory is applicable. Smaller imperfections can grow for a longer time, and from Figure 5.4(a), we would expect to see longer circumferential wavelengths.

Figure 5.4(b) gives a plot of the aspect ratio $\lambda_\eta/\lambda_\xi$. If nonlinear effects begin to dominate at, say, $\tau = 7$, and subsequent buckling proceeds with a fixed pattern (one would expect a fixed pattern to be established eventually, as confirmed in the experiments), Figure 5.4.(b) indicates that the aspect ratio would be about 3.3. This is discussed again in reference to the experiments.

### 5.1.5 Impact Experiments

A small-deflection linear theory applied to the static buckling of cylindrical shells under end loads is notoriously inadequate for predicting experimental buckling loads except for very nearly perfect shells. Also, the shells ultimately buckle into a form very different from the buckling mode of the small-deflection theory. It is not obvious, then, that the simple linear theory given here should reasonably predict the large-deflection dynamic buckling of such a shell. It was thought, however, that the small-deflection theory would give promise for the dynamic problem because, once started in the linear theory pattern, the shell would continue to deform in this pattern and not have time to convert to another pattern. The experiments described here were run simultaneously with the theoretical investigation to determine whether such a theory should be pursued at all.

Early experimenters reported only the final shapes (diamond buckles) after very severe and prolonged buckling and showed high-speed (about 15,000 frames/s) motion pictures, which were nevertheless at a speed too slow to show the early buckling process. To record the early motion, the experimental arrangement shown in Figure 5.5 incorporated three unique features:

(1) The shells were free at the end opposite the impact so that the compressive impact stress would have a duration (at the impacted end), at most, equal to the transit time 2L/c of the longitudinal stress wave up and down the shell. This allowed terminal observation of early permanent buckling not obliterated by later folding.

(2) The impacted end was bonded with epoxy cement to a relatively massive inside ring and to a thin outside ring to provide a clamped boundary to the shell. Without these rings, severe crimping at the end rapidly lowered the thrust and made comparison with theory impossible.

(3) The massive end ring was accelerated explosively so that the time and simultaneity of impact could be controlled to within about 2 $\mu$s. This allowed the use of a Beckman-Whitley framing camera running at 240,000 frames/s, fast enough to see the details of early wave formation.

Test shells were made from 0.0027-inch-thick (69 $\mu$m) 5052-H19 aluminum sheet rolled to a 3-inch (76-mm) diameter with a lapped seam held with cloth tape. The steel ring at the lower end of each shell served as the "impacting" mass and weighed 12 times the weight of the shell so that its change in velocity during the impact, and hence the change in impact stress, was small. The ring was accelerated by placing it on the heavy steel anvil bar and detonating a sheet explosive charge on the opposite end of the anvil. The explosive sent a step-fronted shock into the anvil. This shock entered the ring and bounced it off in much the same way as end pellets are bounced from a Hopkinson bar. The pressure gradient behind the shock was shallow enough that the reverberating stresses in the ring were small, so to a good approximation the ring was stress-free when it "impacted" the cylinder. To ensure good transmission of the shock from the anvil to the ring, the contact surfaces were lapped.

The ring velocity from various thicknesses of sheet explosive was determined in a separate series of experiments. Since the ring is very massive compared with

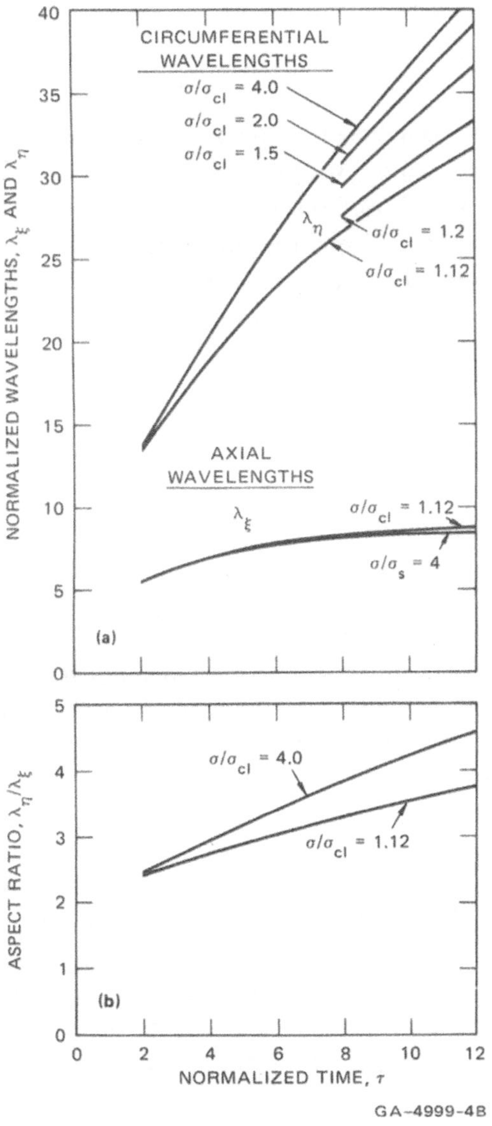


FIGURE 5.4 AXIAL AND CIRCUMFERENTIAL MEAN WAVELENGTHS VERSUS TIME

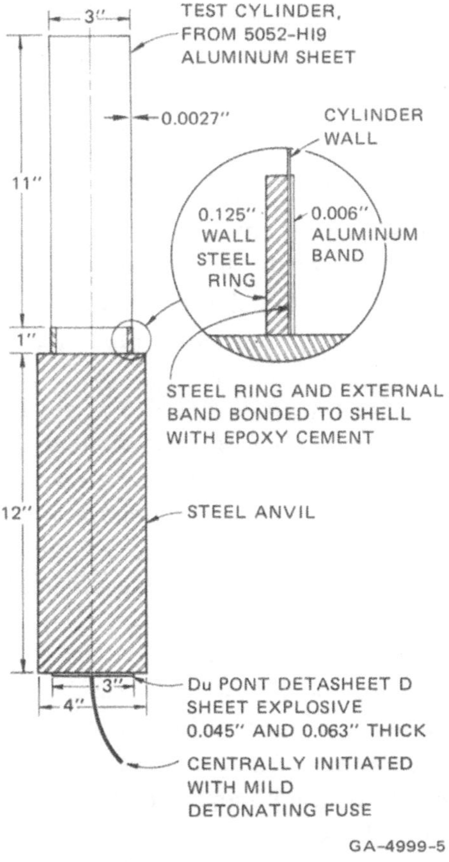


FIGURE 5.5 EXPERIMENTAL ARRANGEMENT (1 inch = 25.4 mm)

the shell, the impact stress in the shell is assumed to be $\rho cv$, where $\rho c$ is the acoustic impedance of aluminum shell and v is the velocity of the steel ring. The axial strain corresponding to this stress is

$$\epsilon = \frac{\rho cv}{E} = \frac{v}{c}$$

where c is the axial wave speed in the shell.

Figure 1.3 shows a sequence of framing camera pictures taken in an experiment in which v = 340 inch/s (8600 mm/s), which, with c = 200,000 inch/s (5 mm/$\mu$s), gives $\epsilon = 0.0017$. Substituting a/h = 1.50/0.0027 = 555 into Equation (5.1.20) yields $\epsilon_{cl} = 0.00109$ so that $\epsilon/\epsilon_{cl} = 1.56$. Thus the impact stress is 1.56 times the classical static buckling stress. Only alternate frames from the original record are shown, giving 8.33 $\mu$s between frames. Exposure time was about 1.4 $\mu$s per frame.

Normalized time $\tau$ can be computed from Equation (5.1.3), which yields $\tau = (c\epsilon/r)t = 0.38t$, with t in $\mu$s. On the original record, very tiny displacements could be discerned at t = 7 $\mu$s ($\tau = 2.6$). At t = 11.2 $\mu$s ($\tau = 4.2$) in Figure 1.3, small wrinkles near the bottom of the shell are clearly visible. By t = 27.8 $\mu$s ($\tau = 10.6$), these wrinkles are definitely taking on the diamond pattern, indicating that nonlinear effects are predominating.

Thus, from $\tau = 2.6$ to $\tau = 10.6$, the displacements grow from being just visible to amplitudes so large that nonlinear effects dominate. This agrees very well with the period of first rapid growth given by the theory. From Figure 5.3 at $\sigma/\sigma_{cl} = 1.5$, we see that the amplification grows from $g_{max} = 3$ at $\tau = 2.6$ to $g_{max} = 100$ at $\tau = 10.6$. It is probable, as discussed later, that nonlinear effects begin to dominate at an intermediate time of about $\tau = 7$.

At frame t = 27.8 $\mu$s, the beginning of buckling farther up the shell is evident. In the following frames, these buckles stay fixed in position and grow in amplitude. Other buckles appear at areas in between, at which initial imperfections were probably smaller. The tiny ripples just above the buckles at the lower end characteristically appeared in all the experiments. They are most visible at t = 44.5 $\mu$s.

To compare the wavelengths in Figure 1.3 with theory, the expected value of the axial wavelength is determined by substituting $\lambda_\xi = 8.9$ from Figure 5.4 into Equation (5.1.3), which yields

$$\lambda_x = \left(\frac{D}{\bar{N}_x}\right)^{1/2} \lambda_\xi = \frac{r}{\epsilon^{1/2}} \lambda_\xi = 8.9 \frac{r}{\epsilon^{1/2}} \tag{5.1.34}$$

With $\epsilon = 0.0017$ and $h = 0.0027$ inch (0.069 mm), this gives $\lambda_x = 0.18$ inch (4.6 mm). From Figure 1.3, the average length from 15 waves is $\lambda_x = 0.20$ inch (5.1 mm), only 10% greater than the theoretical mean.

The permanent buckles remaining in several shells after impact are shown in Figure 5.6. By comparison with Figure 1.3, it is evident that most of the buckling observed in Figure 1.3 is elastic. In general, most of the permanent buckles were confined to the area closest to the impacted end. The most striking feature of these buckles is their very small size and large aspect ratio as compared with static buckles. Figure 5.7 shows a statically buckled shell of the same material, wall thickness, and diameter as those in Figures 1.3 and 5.6. The aspect ratio of these buckles is near unity, and the wavelength is about 1.3 inch (33 mm), seven times the axial wavelength of the dynamic buckles.

The dynamic buckles in Figure 5.6 have a mean aspect ratio of 3.3 and axial wavelengths of 0.18 inch (4.6 mm). Several shells are shown to illustrate the range of observed aspect ratio. There is little doubt that the shape of the permanent buckles was strongly influenced by large elastic and plastic deformation, but it is significant that the large aspect ratio is suggested by the linear theory of the preceding sections.

Circumferential wavelengths from these and two other shells are summarized in Figure 5.8. Aspect ratio is plotted rather than wavelength to emphasize the difference between these buckled forms and static buckles. The values range from 2.2 to 7.1 with a mean of 3.3. These are much larger than the static ratio of about 1. It is difficult to compare these wavelengths with the linear theory because, as shown in Figure 5.4, the theoretical wavelengths continually increase with time. However, from Figure 5.4(b), the ratio of the theoretical average circumferential wavelength to the average axial wavelength is 3.3 at $\tau = 7$. Also, the photographs in Figure 1.3 show that this is about the time at which the buckle amplitudes first become distinct.

### 5.1.6 Formula for Threshold Buckling

The value $\tau = 7$ corresponds to a peak amplification of $g_{max} = 25$ for $\sigma/\sigma_{cl} = 1.5$, obtained from the curve in Figure 5.3. Thus, as in the buckling of shells under lateral pressure in Chapter 3, a critical condition for buckling can be taken as an amplification near this value. Figure 5.3 also shows that for $\sigma/\sigma_{cl}$ ranging from 1.12 to infinity, the value of $\tau$ for $g_{max} = 25$ ranges only from 6 to 8. Threshold buckling can therefore be calculated with the critical condition $\tau_{cr} = 7$, which, from

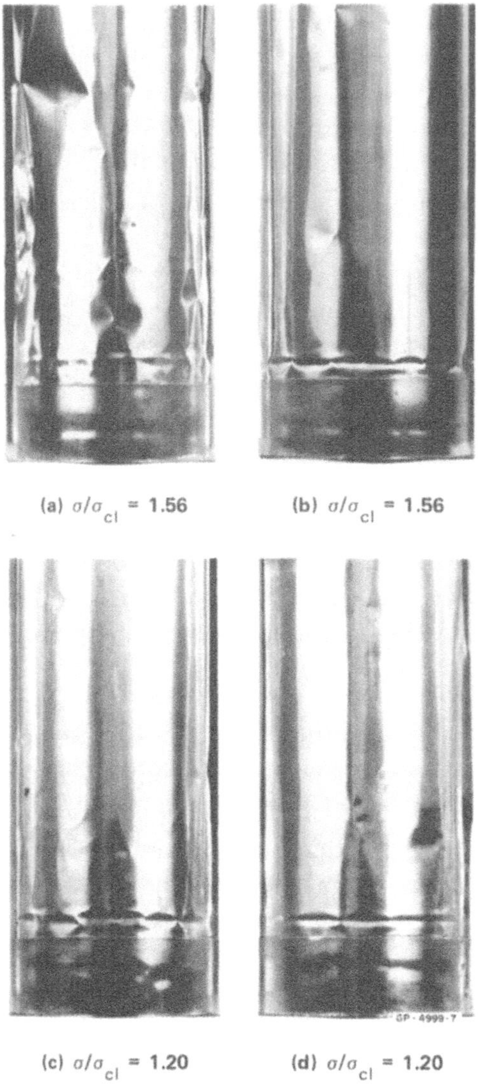


FIGURE 5.6 PERMANENT BUCKLES FROM AXIAL IMPACT (opposite end was free)

FIGURE 5.7 STATIC BUCKLING OF AN IDENTICAL SHELL

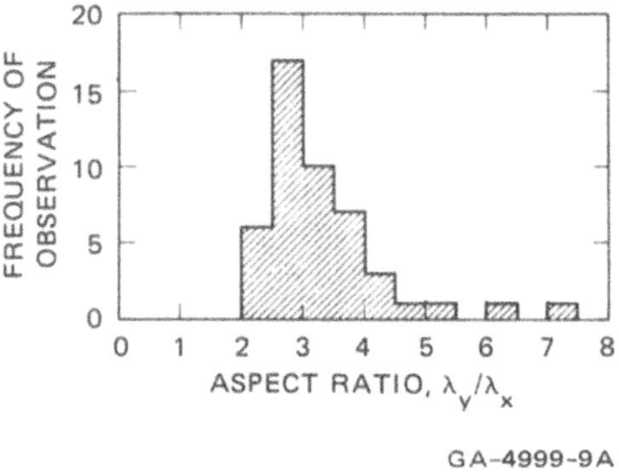


FIGURE 5.8 HISTOGRAM OF OBSERVED ASPECT RATIOS

Equation (5.1.3) gives the formula

$$\overline{N}_x t = 7(\rho h D)^{1/2}$$

or

$$\sigma_x t = \frac{7}{\sqrt{12}} \rho c h \approx 2\rho c h \tag{5.1.35}$$

This is the same formula as for a bar of thickness h except here $c^2 = E/\rho(1 - \nu^2)$ rather than $E/\rho$, as can be confirmed by comparison with Equation (2.2.28), with $\tau = 7$ and $r = h/\sqrt{12}$. The only significant difference is that, for the cylindrical shell, buckling will not occur unless $\sigma$ is greater than the static buckling stress. The curves in Figure 5.3 show that $\sigma$ does not have to be much greater than the static buckling stress, since the curves rapidly approach the curve $\sigma/\sigma_{cl} \rightarrow \infty$ for the bar as $\sigma$ increases.

Because of the extreme sensitivity of the cylindrical shell to initial imperfections, this stipulation on $\sigma$ should be applied to the *local* static buckling stress, not to the classical buckling stress of the perfect shell as given by Equation (5.1.20). The experiments of Almroth, Holmes, and Brush[4] described in Section 5.1.2 demonstrate why this is so. In experiments of the type in Figure 5.1, the axial stress ranged from about 35% to 60% of $\sigma_{cl}$ for the perfect shell, and yet the shells buckled dynamically in a manner similar to that shown in Figure 1.3. The difference is that the buckling in Figure 5.1 propagates outward from a local region rather than appearing nearly simultaneously throughout the shell as in Figure 1.3. As discussed in Section 5.1.2, this behavior is attributed to the local classical buckling stress in the Figure 5.1 shell being about half that of the perfect shell. The theory here can be applied if this reduction is caused by small imperfections increasing the local radius of curvature to twice that of the perfect shell.

In any event, Equation (5.1.35) is conservative because it is essentially the same result as for a long plate, for which $a \rightarrow \infty$. It provides a conservative estimate for loading in which $\sigma > \sigma_s$, where $\sigma_s$ is the static buckling stress of the imperfect shell. For smaller $\sigma$, it is not applicable and gives unduly conservative results. These values correspond to large t, so loading is essentially a step to a stress that is maintained indefinitely. Critical loads in this range are discussed in the next section.

### 5.1.7 Dynamic Buckling Under Step Loads

In addition to increasing the local radius of curvature, imperfections trigger buckling because of their contribution to the modal imperfection coefficients $a_{mn}$ in Equation (5.1.11). When these imperfections are coupled with the nonlinear behavior of the shell for large deflections, the result can be a decrease in the dynamic buckling load below the static buckling load of even the imperfect shell. This occurs when the load is suddenly applied and then maintained for a long time, as in a step load. In this section we present the essentials of Budiansky and Hutchinson's theory for such buckling.[1,6]

Their theory makes use of the simple one-degree-of-freedom model for nonlinear buckling used by von Kármán, Dunn, and Tsien.[3] This model is the three-hinge, rigid-rod column shown in Figure 5.9. The spring is taken as a cubic-softening spring having the force-displacement characteristic

$$F = KL(\xi - b\xi^3) \ , \quad b > 0 \tag{5.1.36}$$

in which $\xi = x/L$, with x measured from an unloaded initial position $\bar{x}$ as shown in the figure.

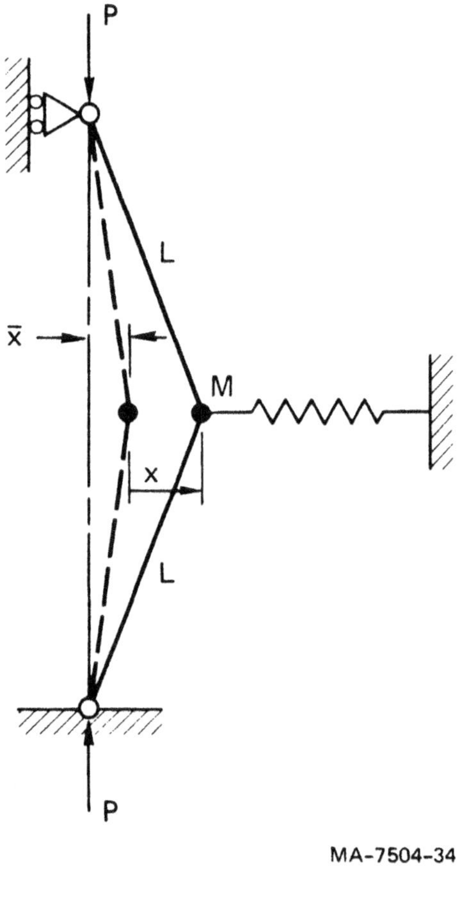


FIGURE 5.9 SINGLE-DEGREE-OF-FREEDOM RIGID-ROD COLUMN WITH IMPERFECTION AND NONLINEAR SPRING

For small rotations, static equilibrium relates P and the displacement $\xi$ by

$$(1 - P/P_C)\xi - b\xi^3 = (P/P_C)\bar{\xi} \tag{5.1.37}$$

in which $P_C = KL$. This terminology for KL follows from the limiting behavior of the perfect column ($\bar{\xi} = 0$), for which the load can increase to $P_C$ without deflection but cannot exceed $P_C$. The graph of Equation (5.1.37) for $\bar{\xi} = 0$ in Figure 5.10 shows that the equilibrium load for a finite deflection is less than $P_C$, and the load continues to decrease with increasing deflection $\xi$. Thus, motion is unstable and $P_C$ is the buckling load for the perfect column.

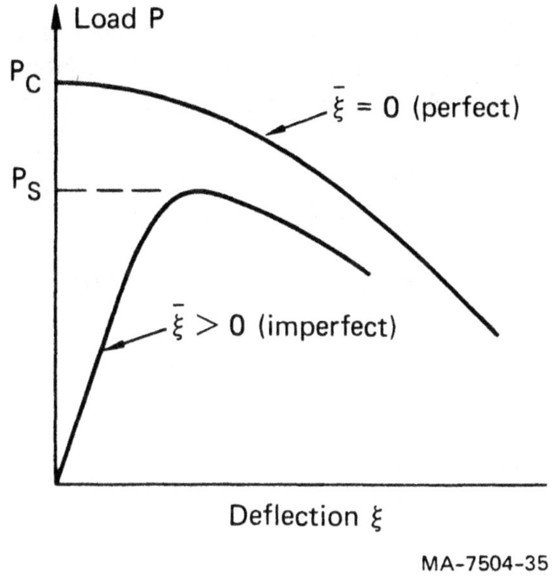


FIGURE 5.10 LOAD-DEFLECTION CURVES FOR RIGID-ROD COLUMN WITH CUBIC SOFTENING SPRING

The imperfect column ($\bar{\xi} \neq 0$) begins to deflect as soon as the load begins to increase and hence the column has a range of stable equilibrium. As the graph in Figure 5.10 shows, the deflection curve eventually has a horizontal tangent and further deflection is unstable. Furthermore, the buckling load, $P_S$, at this tangent point is less than that of the perfect column.

This behavior represents the essentials of static axial buckling of a cylindrical shell (and other imperfection-sensitive structures). The load $P_C$ represents the classical buckling load of the perfect shell, and $P_S$ represents the actual static buckling load of the imperfect shell. The details for the shell are much more complicated, as we have seen. For example, the rigid-rod column model does not account directly for the decrease in buckling load simply because an imperfection can increase the radius of curvature over a buckle wavelength. Also, as shown in Figure 5.1, the mode of final buckling for a step load is different from the mode in the initial motion, so a single-degree-of-freedom model is an oversimplification. Nevertheless, by making the spring and imperfection parameters b and $\bar{\xi}$ large enough (in fact, simply the parameter $b\bar{\xi}^2$, as we will see), we can make the buckling reduction factor $P_S/P_C$ low enough to account for the cumulative effect of several types of imperfection.

Budiansky and Hutchinson's idea was to see whether these difficulties in static buckling analysis can be circumvented for dynamic buckling analysis by taking $P_S/P_C$ as known. Their objective was to express the dynamic buckling load by an expression between $P_D/P_S$ and $P_S/P_C$, with no explicit reference to the shell imperfections. Then the study of imperfections can be confined to the static problem, which is simpler both theoretically and experimentally. It turns out that such expressions can indeed be found.

To find $P_S$, $z = \xi/\bar{\xi}$ is introduced into Equation (5.1.37) to obtain

$$(1 - P/P_C)\, z - (b\bar{\xi}^2)\, z^3 = P/P_C \tag{5.1.38}$$

from which it is evident that $P_S/P_C$ depends only on $b\bar{\xi}^2$. Setting $dP/dz = 0$ yields the desired equation for $P_S/P_C$:

$$(1 - P_S/P_C)^{3/2} = \frac{3\sqrt{3}}{2}\, b^{1/2}\bar{\xi}\,(P_S/P_C) \tag{5.1.39}$$

To find $P_D$, consider a time-dependent load $P(t)$ and introduce inertia into the system by a mass M at the central hinge. Then Equation (5.1.38) becomes

$$\ddot{z} + (1 - P/P_C)\, z - (\beta\bar{\xi}^2)\, z^3 = P/P_C \tag{5.1.40}$$

in which the dot indicates differentiation with respect to $t\sqrt{K/M}$. For step loading, $z = \dot{z} = 0$ at $t = 0$. Then, with the identity $\ddot{z} = \dot{z}d\dot{z}/dz$, the equation can be integrated once to obtain

$$\dot{z}^2 + (1 - P/P_C)\, z^2 - \frac{1}{2}(\beta\bar{\xi}^2)\, z^4 = 2(P/P_C)\, z \tag{5.1.41}$$

At loads below the dynamic buckling load, the steady-state motion is periodic and Equation (5.1.41) defines its limit cycle in phase space $z$, $\dot{z}$. The maximum value, $z_{max}$, of this limit cycle occurs when $\dot{z} = 0$, which gives

$$(1 - P/P_C)\, z_{max}^2 - \frac{1}{2}(\beta\bar{\xi}^2)\, z_{max}^4 = 2(P/P_C)\, z_{max} \tag{5.1.42}$$

The critical dynamic load, $P_D$, is defined as the load for which the amplitude (and period) of this limit cycle is infinite, so the motion diverges rather than approaching a limit cycle. This occurs under the condition $dP/dz_{max} = 0$ applied to Equation (5.1.42), with the result

$$(1 - P_D/P_C)^{3/2} = \frac{3\sqrt{6}}{2} b^{1/2}\bar{\xi}(P_D/P_C) \tag{5.1.43}$$

The key feature of this simple model is that the imperfection parameter $b\bar{\xi}^2$ can be eliminated between Equations (5.1.39) and (5.1.43), giving the desired relationship between the static and dynamic loads with no explicit dependence on the imperfections. The result is

$$P_D/P_S = \frac{\sqrt{2}}{2}\left(\frac{P_C - P_D}{P_C - P_S}\right)^{3/2} \tag{5.1.44}$$

This function is plotted in Figure 5.11 (solid curve) along with a similar result (dashed curve) for a quadratic-softening spring, which by an analogous derivation

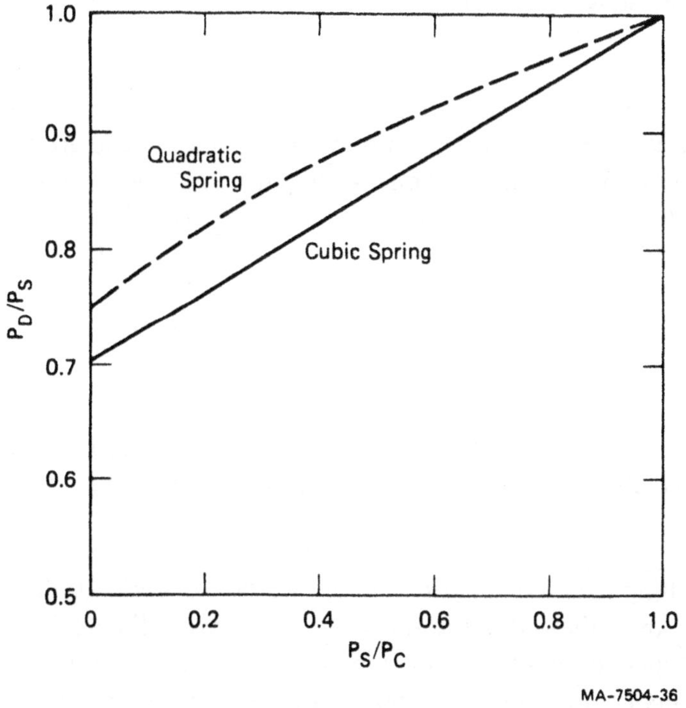


FIGURE 5.11 DYNAMIC BUCKLING LOADS FOR RIGID-ROD COLUMN MODELS

gives the expression

$$P_D/P_S = \frac{3}{4}\left(\frac{P_C - P_D}{P_C - P_S}\right)^2 \tag{5.1.45}$$

The two curves give essentially the same result, within the accuracy of an assumed single-degree-of-freedom representation of the actual complex buckling process. As the static buckling load becomes a smaller fraction of the buckling load of the perfect structure (as $P_S/P_C \rightarrow 0$), the dynamic load also becomes a smaller fraction of the imperfect structure static buckling load. However, no matter how small $P_S/P_C$ is, the further reduction because of dynamics is never below $\sqrt{2}/2 = 0.707$ for the cubic spring and $3/4 = 0.75$ for the quadratic spring.

Budiansky and Hutchinson derived an analogous expression for the cylindrical shell under sudden axial compression. They used the Donnell equations as in Section 5.1.1 here, but with the addition of second-order terms to introduce the nonlinear effects of large deformations. Motion was assumed to take place with two degrees of freedom, one being $\xi_1$ in the symmetric pattern with axial wavelength $2\ell_o$, and the other being $\xi_2$ in the square nonsymmetric pattern with axial and circumferential wavelengths each $4\ell_o$. These modes and wavelengths are as described in Section 5.1.2.

Curves of the type shown in Figure 5.11 were found by numerical integration of the coupled dynamic equations of motion for $\xi_1$ and $\xi_2$ and are repeated here in Figure 5.12. For initial imperfections with ratio $\bar{\xi}_1/\bar{\xi}_2 = 2$, there is very little reduction in dynamic load below the static buckling load. When imperfections are taken to be entirely in the nonsymmetric load, i.e., $\bar{\xi}_1/\bar{\xi}_2 = 0$, they find an explicit solution by the same method used to solve Equation (5.1.40), with the result

$$P_C/P_S = \frac{\sqrt{2}}{2}\left(\frac{P_C - P_D}{P_C - P_S}\right)^2 \tag{5.1.46}$$

This is the same result as for the quadratic-softening spring, in Equation (5.1.45), but with the coefficient $\sqrt{2}/2$ found for the cubic-softening spring in Equation (5.1.44).

The resulting curve in Figure 5.12 is nearly the same as the solid curve in Figure 5.11 for the cubic-softening spring. The curves in Figure 5.12 show that the dynamic load reduction is greater for imperfections in the nonsymmetric mode than for imperfections in the symmetric mode. For prudent design, the curve for nonsymmetric imperfections is therefore recommended. Furthermore, comparison of Figures 5.11 and 5.12 shows that the simple rigid-rod column model in Figure 5.9 with a cubic-softening spring gives a good representation of the calculated dynamic buckling in the shell.

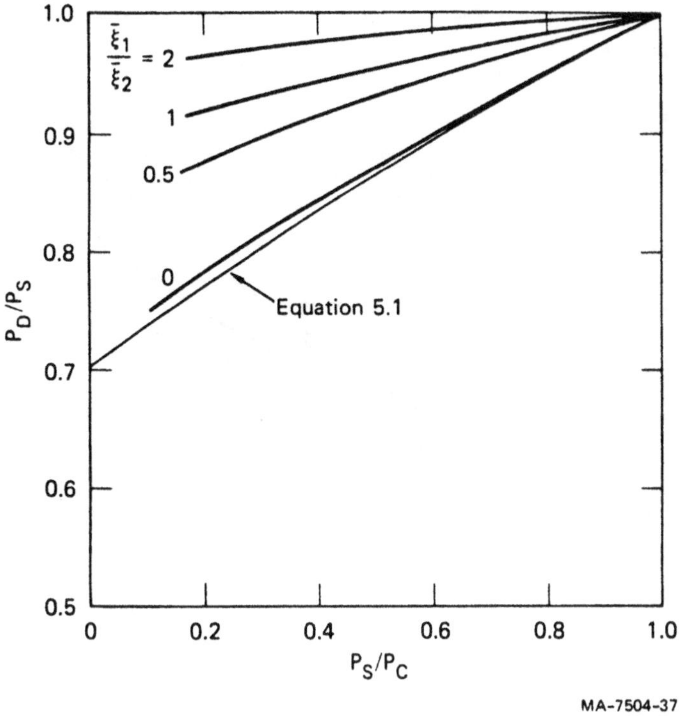


FIGURE 5.12 DYNAMIC BUCKLING LOADS FOR AXIAL STEP LOAD ON CYLINDRICAL SHELL

Budiansky and Hutchinson integrated the equations of motion for the rigid-rod column model for rectangular and triangular loading pulses of finite duration.[6] Their results for the rectangular pulse are summarized in Figure 5.13, in which $T_o = 2\pi/\omega$ is the free vibration period of the unloaded structure. For the rigid-rod column, $\omega = \sqrt{K/M}$, but the comparison above for the cylindrical shell suggested that Figure 5.13 could be applied more generally. The curves in Figure 5.13 are for the limiting case in which $P_S/P_C \rightarrow 0$ and give a generalization of the conservative result in Figure 5.10 at $P_S/P_C = 0$. Since typically $P_S/P_C \approx 0.25$, the curves refer to practical cases of interest.

For the cylindrical shell, the natural period of the square nonsymmetric buckling mode, from Equation (5.1.12) with $\overline{N}_x \rightarrow 0$ and $L/m = \pi a/n = 4\ell_o$ from Equation (5.1.17), is

$$T_o = 2\pi/\omega_2 = 2\sqrt{2} \cdot \pi a/c \qquad (5.1.47)$$

To obtain a simple formula for critical loads in which $T_o/T > 3$, observe that in this range the curves in Figure 5.13 are nearly straight lines that project from the origin. For the cubic model in this range, this line is $P_D/P_S = T_o/3T$. With $T_o$

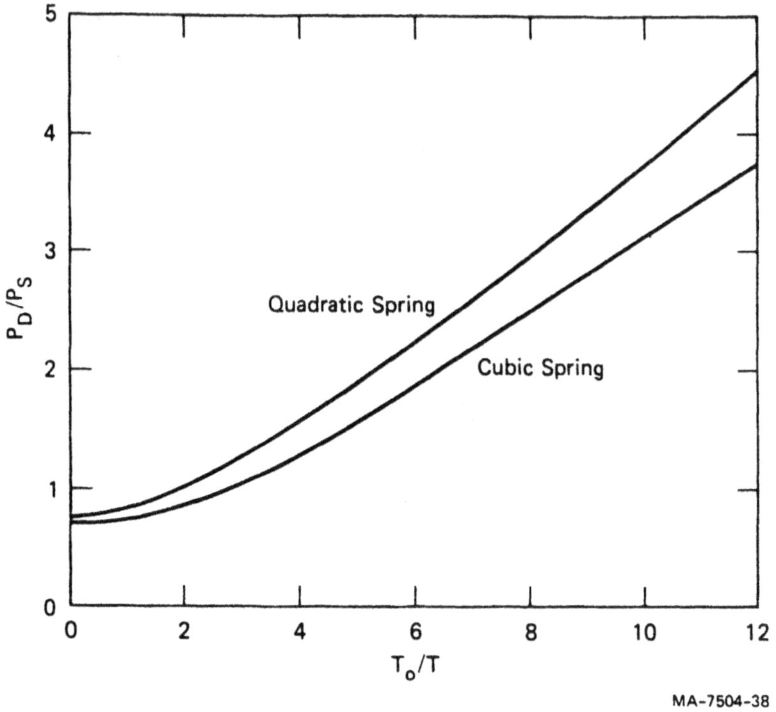


FIGURE 5.13 DYNAMIC BUCKLING LOADS FOR RIGID-ROD COLUMN UNDER RECTANGULAR PULSES (FOR $P_s/P_c \rightarrow 0$)

from Equation (5.1.47) and $\sigma_{cl} = Eh/a\sqrt{3(1 - \nu^2)}$, this gives for the critical dynamic axial stress

$$\sigma_D = \pi \left(\frac{2}{3}\right)^{3/2} \frac{\rho ch}{T} \cdot \frac{\sigma_S}{\sigma_C}$$

or

$$\sigma_D T = 1.710 \; \rho ch \; (\sigma_S/\sigma_C) \qquad (5.1.48)$$

If one uses the symmetric mode period $T_o = \sqrt{2}\pi a/c$, from Equation (5.1.12) with $n = 0$, $L/m = \ell_o$, the same formula results but with a more conservative coefficient, namely, $1.710/2 = 0.855$.

Equation (5.1.48) has the same form as $\sigma_D T = 2 \rho ch$, given as Equation (5.1.35) in Section 5.1.6 and found by using a critical amplification buckling criterion. However, Equation (5.1.48) contains the static load reduction factor $\sigma_S/\sigma_C$, which is typically near 0.25 in common shells. With this value, Equation

(5.1.48) becomes $\sigma_D T = 0.43\ \rho ch$. The corresponding equation for the symmetric mode becomes $\sigma_D T = 0.21\ \rho ch$. The response mode for this last equation is close to the critical mode found in Section 5.1.6; the aspect ratio of the most amplified mode is about $\ell_\theta/\ell_x = 3$ [see Figure 5.4(b) for $\tau = 7$] and $\ell_o/\sqrt{2} < \ell_S < \ell_o$, depending on $\sigma_D$.

Thus, for common shells, the critical load estimated by Equation (5.1.48) is a factor of 10 lower than the critical load given by Equation (5.1.35) derived with a critical amplification criterion. The value of $\tau$ corresponding to $\sigma_D T = 0.21\ \rho ch$ is $\tau = 0.7$, which from Figure 5.3 gives virtually no amplification. Furthermore, as pointed out in Section 5.1.6, the formula $\sigma_D T = 2\rho ch$ is also the critical load formula for a bar, or plate, which corresponds to $a \rightarrow \infty$. Thus, there is no reason to suspect that the critical amplification formula is unconservative because of any peculiarity of complex nonlinear shell response; the finite radius of the shell makes the shell stronger than the plate.

For pulse loads, it therefore seems more reasonable to use the critical amplification criterion and not express the dynamic buckling load in terms of the imperfect shell static buckling load. Thus, in place of Figure 5.13, we suggest use of the plot in Figure 5.14. For large $P_D/P_S$, the plot is the straight line

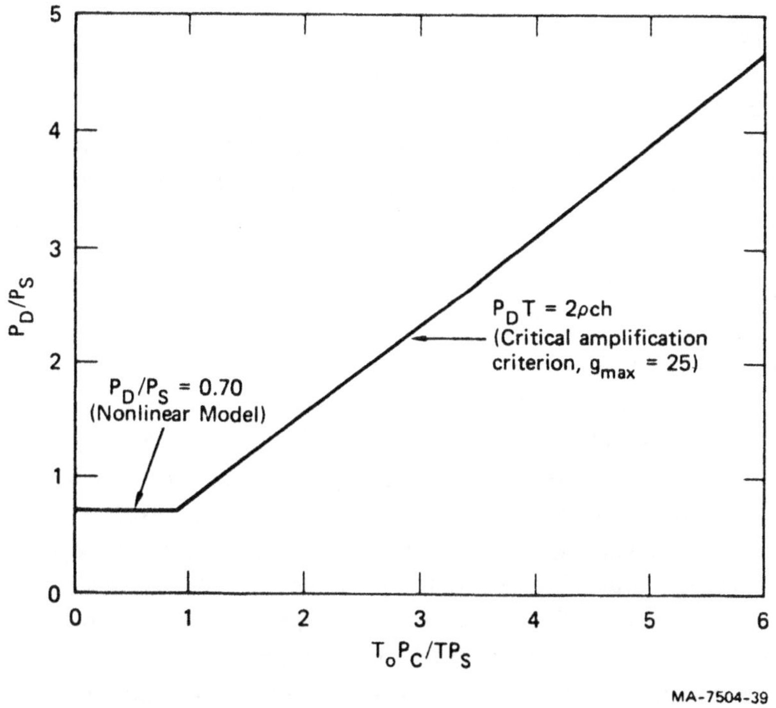


FIGURE 5.14 DYNAMIC BUCKLING LOADS FOR CYLINDRICAL SHELL UNDER RECTANGULAR PULSE LOADS OF DURATION T (FREE VIBRATION PERIOD OF SYMMETRIC BUCKLING LOAD IS $T_o$)

$$\frac{P_D}{P_S} = \frac{\sqrt{6}}{\pi}\left(\frac{T_o}{T}\right)\left(\frac{P_C}{P_S}\right) \qquad (5.1.49)$$

which is Equation (5.1.35) with $T_o$ taken as the free vibration period $\sqrt{2}\,\pi a/c$ of the symmetric buckling mode. This formula is taken to hold for $P_D$ greater than the reduced buckling load $P_D/P_S = \sqrt{2}/2$ given by Equation (5.1.46) for $P_S/P_C \rightarrow 0$. For longer pulse durations (smaller $T_o/T$), this straight line is terminated and the critical load is taken as the conservative value $P_D/P_S = \sqrt{2}/2$ from the step-load theory. For common shells in which $P_S/P_C \approx 0.25$, the corner of the resulting plot in Figure 5.12 occurs at a pulse duration $T \approx 4T_o$. For a one-meter-diameter shell, $T \approx 4\sqrt{2}\;\pi(1\text{ m})/5000\text{ m/s} = 0.00355$ s.

## 5.2 AXIAL PLASTIC FLOW BUCKLING OF CYLINDRICAL SHELLS[7]

### 5.2.1 Introduction

Cylindrical shells with low radius-to-thickness ratios buckle axisymmetrically when subjected to axial impact at velocities sufficient to cause moderate plastic strains. Figure 5.15 shows this flow buckling in the final axisymmetric shapes of tubes of 6061-T6 aluminum alloy, which before impact (left-hand tube) were 4 inches long (102 mm), 1 inch outside diameter (25 mm), and 1/10 inch (2.5 mm) thick (radius/thickness $\approx$ 5). Each of the three plastically deformed tubes was impacted at 342 ft/s (104 m/s); the shortening in each case is 0.55 inch, giving an average longitudinal strain of about 14%. The wavelengths are fairly uniform along the length of each tube, and they are reproducible.

In the development of the theory, it is assumed that during deformation the shell is contained between a flat rigid target and a heavy mass that approaches the target as depicted in Figure 5.16. Solution of the governing equation is given for the case of a constant mass velocity to provide simple formulas for the preferred mode and for numerical comparisons with experimental results. Justification for assuming a constant velocity is based on high-speed camera observations of several impacting tubes. It was observed that the mode of buckling was selected early in the deformation process. We also assume that unperturbed motion has the diameter and the wall thickness increasing at a rate proportional to the rate of tube shortening. The buckling is treated as a perturbation of this motion stemming from axisymmetric imperfections of the uniform initial radial displacements and velocities.

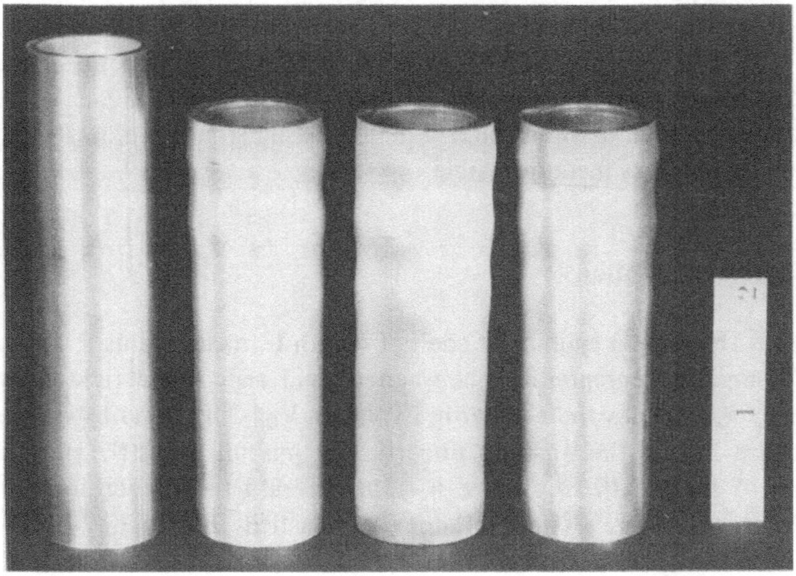

MP-7504-17

FIGURE 5.15 PLASTICALLY BUCKLED TUBES CAUSED BY AXIAL IMPACT (Tubes 13, 14, & 15)

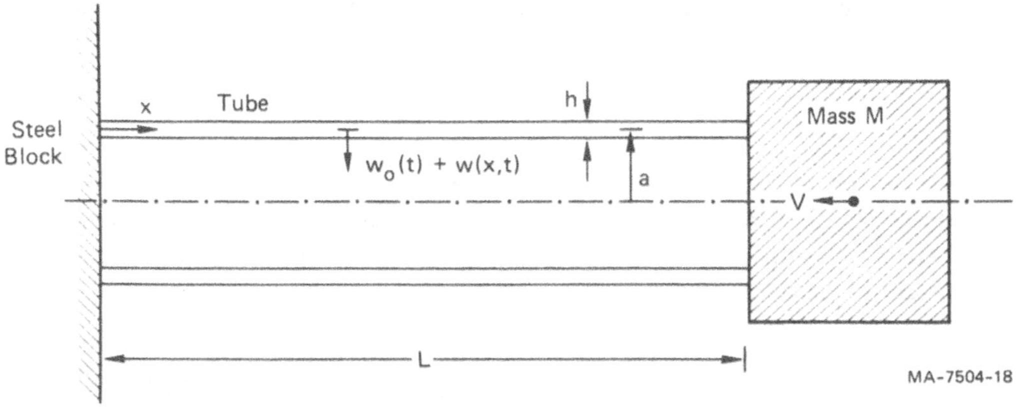


FIGURE 5.16 TUBE COMPRESSED BETWEEN RIGID WALL AND RIGID MASS

It is supposed that the strain rates associated with the unperturbed motion dominate the strain rates introduced by the perturbed motion so that no strain-rate reversal occurs until the buckling is well developed. This supposition ensures that the stress states at every material point remain on the yield surface and therefore have associated strain-rate vectors as outward normals to the yield surface at the stress state. The yielding criterion is that of von Mises and the strain hardening is isotropic. To keep the theory simple, the strain hardening portion of the stress-strain curve from static compression tests is approximated by a straight line. The static data for 6061-T6 and 2024-T3 aluminum alloys, the materials used in the impact experiments, are satisfactory because the stress-strain curves for these materials are insensitive to strain rate. Finally, elastic strains are neglected and the materials are regarded as incompressible.

### 5.2.2 Unperturbed Motion

Figure 5.16 shows a cylindrical shell of length L, mean radius a, and thickness h, about to undergo compression between a rigid slab and a rigid heavy mass, which initially approaches the slab with a velocity $V_0$. The inward displacement of the midsurface during the ensuing unperturbed motion is $w_0(t)$, assumed to be independent of the length coordinate x. The outward radial coordinate measured from the midsurface is z. We again adopt the plasticity Equations (4.1.1) through (4.1.10).

On the midsurface the axial and circumferential strains are

$$\dot{\epsilon}_x = -V/L \qquad \dot{\epsilon}_\theta = -\dot{w}_0/a \tag{5.2.1}$$

where $V = V(t)$ is the current mass velocity. Our assumption that these strain components are proportional to one another may be expressed as

$$\dot{\epsilon}_\theta = -k\dot{\epsilon}_x \tag{5.2.2}$$

where k is the proportionality constant to be determined. If we neglect the effect of strain hardening on the unperturbed motion we may set $\sigma \approx -\sigma_0$ in (4.1.9), the relations giving the stresses in terms of the strain rates. Then, after substituting (5.2.2) in (4.1.9), making use of (4.1.6) for the generalized strain rate, we obtain the constant midsurface stresses

$$\sigma_x = \frac{-(2-k)\sigma_0}{\left[3(1-k+k^2)\right]^{1/2}} \qquad \sigma_\theta = \frac{(2k-1)\sigma_0}{\left[3(1-k+k^2)\right]^{1/2}} \tag{5.2.3}$$

The equations governing the motion of the rigid mass, of magnitude M, and the tube expansion, are

$$M\dot{V} = 2\pi ah\sigma_x \ , \quad \rho a\ddot{w}_0 = \sigma_\theta \tag{5.2.4}$$

in which the stresses are the midsurface values (5.2.3). Integration of Equations

(5.2.4) to obtain velocities, use of the relation $\dot{w}_0/a = -kV/L$ from (5.2.1) and (5.2.2), and substitution of the stress formulas (5.2.3) show that k must satisfy the quadratic equation

$$(2k - 1)/(2 - k)k = ma^2/ML^2 \tag{5.2.5}$$

where $m = 2\pi ahL\rho$ is the mass of the shell. The right-hand side of (5.2.5) is generally small enough to allow the solution to be approximated by

$$k = (1 + ma^2/ML^2)/2 \tag{5.2.6}$$

After we substitute (5.2.6) into the stress formulas (5.2.3), we have

$$\sigma_x = -(1 - 2ma^2/3ML^2)\sigma_0 \qquad \sigma_\theta = (2ma^2/3ML^2)\sigma_0 \tag{5.2.7}$$

We proceed by assuming that the right hand side of (5.2.5) is small enough to let $k \approx 1/2$, which is the case in the experiments. Thus we have the midsurface stress values $\sigma_x \approx -\sigma_0$ and $\sigma_\theta \approx 0$.

With $k = 1/2$ the strain rates are

$$\dot{\epsilon}_x = -V/L \qquad \dot{\epsilon}_\theta = (1 - z/a)(V/2L) \qquad \dot{\epsilon} = V/L \tag{5.2.8}$$

and $\dot{\epsilon}_z$ is determined by the incompressibility condition (4.1.1). In obtaining the generalized strain rate in (5.2.8), we neglected powers of (z/a) higher than one. Substitution of the strain rates (5.2.8) and the approximation $\sigma = \sigma_0$ into the stress formulas (4.1.9) leads to

$$\sigma_x = -(1 + z/3a)\sigma_0 \qquad \sigma_\theta = -(2z/3a)\sigma_0 \tag{5.2.9}$$

A refinement of the approximate generalized stress can be made by letting $\sigma = \bar{\sigma}$, a value averaged over the generalized strain, as was done in Section 4.1

Again with $k = 1/2$, integration of the equation of motion of the mass M, given by (5.2.4) with $\sigma_x = -\sigma_0$, gives

$$V = V_0(1 - t/t_f) \qquad t_f = (M/m)(L\rho V_0/\sigma_0) \tag{5.2.10}$$

where $t_f$ is the duration of motion.

### 5.2.3 Perturbed Motion

Let the axisymmetric perturbation of $w_0(t)$ be $w(x, t)$ so that the total inward displacement is $w_0 + w$. The strain rates become

$$\dot{\epsilon}_x = -V/L + z\dot{w}'' \qquad \dot{\epsilon}_\theta = (1 - z/a)(V/2L - \dot{w}/a) \tag{5.2.11}$$

where primes denote partial differentiation with respect to x. In the results that follow, terms with powers of (z/a) higher than one and terms with products of perturbation quantities are neglected. Thus, by substituting (5.2.11) into (4.1.6) we obtain

$$\dot{\epsilon} = V/L + (2\dot{w}/3a - a\dot{w}'')(z/a) \tag{5.2.12}$$

which can be integrated to give the generalized strain

$$\epsilon = X/L + \left[(2/3a)(w - \tilde{w}) - a(w - \tilde{w})''\right](z/a) \tag{5.2.13}$$

where

$$X = \int_0^t V\,dt \tag{5.2.14}$$

is the mass displacement and $\tilde{w} = w(x, 0)$ are the initial displacement imperfections.

Substitution of the generalized strain (5.2.13) in the hardening law (4.1.10) gives the generalized stress

$$\sigma = \sigma_0 + E_h\left[(2/3a)(w - \tilde{w}) - a(w - \tilde{w})''\right](z/a) \tag{5.2.15}$$

where we have used the approximation $\sigma_0 + E_h X/L \approx \sigma_0$. Substitution of (5.2.15) and the strain rates (5.2.11) and (5.2.12) into the stress formulas (4.1.9) results in

$$\sigma_x = (1 + 2\dot{w}L/3Va)\sigma_0 - \Big[\sigma_0(1 - 4\dot{w}L/Va - \dot{w}''La/V)/3$$

$$+ E_h\Big\{2(w - \tilde{w})/3a - a(w - \tilde{w})''\Big\}\Big](z/a) \tag{5.2.16}$$

$$\sigma_\theta = -(4\dot{w}L/3Va)\sigma_0$$

$$-(2\sigma_0/3)(1 - 2\dot{w}L/Va - \dot{w}''La/V)(z/a) \tag{5.2.17}$$

According to the sign convention of Figure 5.17, stresses (5.2.16) and (5.2.17) have the following resultant membrane forces and moments

$$N_x = -(1 + 2\dot{w}L/3Va)\sigma_0 h \tag{5.2.18}$$

$$N_\theta = -(4\dot{w}L/3Va)\sigma_0 h \tag{5.2.19}$$

$$M_x = \Big[\sigma_0(1 - 4\dot{w}L/Va - \dot{w}''La/V) + E_h\Big\{2(w - \tilde{w})/a - 3a(w - \tilde{w})''\Big\}\Big](h^3/36a) \tag{5.2.20}$$

$$M_\theta = (1 - 2\dot{w}L/Va - \dot{w}''La/V)\sigma_0(h^3/18a) \tag{5.2.21}$$

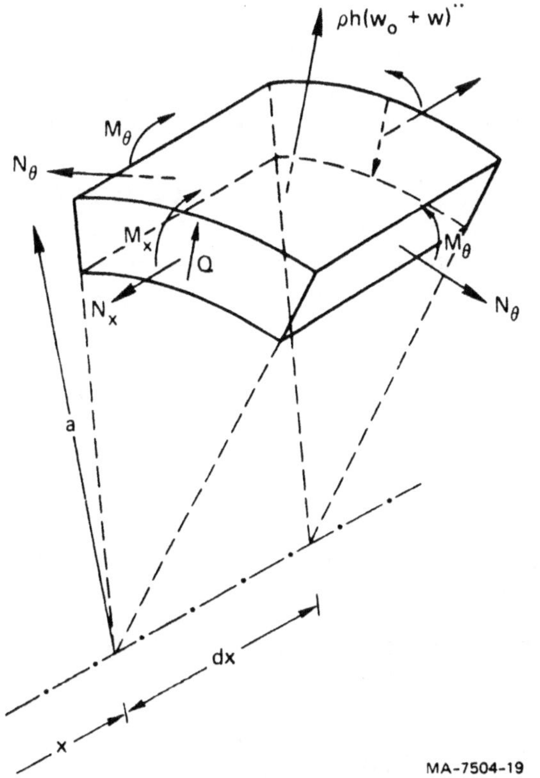


FIGURE 5.17 UNIT SHELL ELEMENT WITH FORCES AND MOMENTS

The restoring moment $M_x$, given by (5.2.20), comprises two groups of terms. The first group is the directional moment that occurs because the stress state on the yield ellipse is linearly dependent on the distance z of the fiber from the midsurface. The second group is the strain-hardening moment. Also resisting the buckling is the circumferential membrane force (5.2.19).

### 5.2.4 Governing Equation

With the aid of Figure 5.17, which shows a unit shell element with its attendant forces and moments, we can derive the following equations of motion or equilibrium:

$$Q - M'_x - N_x w' = 0 \tag{5.2.22}$$

$$Q' + N_\theta / a = \rho h \ddot{w} \tag{5.2.23}$$

The unperturbed displacement $w_0$ does not appear in (5.2.22) and (5.2.23) because $w'_0 = 0$ and the unperturbed equation of radial motion (5.2.4) cancels in (5.2.23). Combining (5.2.22) and (5.2.23) by eliminating the shear force Q and noting that $(N_x w')' \approx N_x w''$ (neglect of product of perturbation terms) gives the equation

$$M''_x + N_x w'' + N_\theta / a = \rho h \ddot{w} \tag{5.2.24}$$

Substitution of expressions (5.2.18), (5.2.19), and (5.2.20) in (5.2.24) gives the governing equation in terms of the perturbed displacements:

$$\left[ (\sigma_0 L / V)(4w + a^2 w'')^{\cdot\,''} + E_h \left\{ 3a^2 (w - \tilde{w})'' - 2(w - \tilde{w}) \right\}'' \right] (h^2 / 36a^2)$$

$$+ \sigma_0 w'' + (4\sigma_0 L / 3Va^2)\dot{w} + \rho \ddot{w} = 0 \tag{5.2.25}$$

The four terms, in order, represent effects of longitudinal bending, axial thrust, hoop force, and inertia.

By letting the radius become infinitely large in (5.2.25), we obtain the corresponding flat plate equation:

$$\left[ (\sigma_0 L / 3V)\dot{w} + E_h (w - \tilde{w}) \right]'''' (h^2 / 12) + \sigma_0 w'' + \rho \ddot{w} = 0 \tag{5.2.26}$$

Equation (5.2.26) governs perturbations that are independent of the lateral y coordinate in a plate with the unperturbed stress condition $\sigma_y = 0$. By omitting the directional longitudinal bending term in (5.2.26), we obtain the rod equation:

$$(E_h h^2/12)(w - \tilde{w})'''' + \sigma_0 w'' + \rho \ddot{w} = 0 \qquad (5.2.27)$$

Equation (5.2.25) governing flow buckling of a tube can be made dimensionless by the quantities

$$u = w/a \qquad \xi = x/L \qquad \tau = V_0 t/2L \qquad \eta = 1 - \tau/\tau_f$$

$$\alpha^2 = h^2/12a^2 \qquad \beta = E_h/\sigma_0 \qquad \gamma = a/L \qquad \mu = M/m \qquad (5.2.28)$$

so that, with the dots and primes now denoting partial differentiation with respect to $\eta$ and $\xi$ we have

$$\ddot{u} - (\mu/3\eta)\left[\alpha^2(4u + \gamma^2 u'')'' + (4/\gamma^2)u\right]^{\cdot}$$

$$+ 2\mu\tau_f\left[\alpha^2\beta\left\{\gamma^2(u - \tilde{u})'' - (2/3)(u - \tilde{u})\right\} + u\right]'' = 0 \qquad (5.2.29)$$

In (5.2.29),

$$\tau_f = V_0 t_f/2L = (\rho V_0^2/2\sigma_0)(M/m) \qquad (5.2.30)$$

which is the energy equation, where $\tau_f$ is the final axial strain.

For the case of sustained axial flow caused by the mass M compressing the tube at a constant velocity $V_0$, let

$$\tau = V_0 t/L \qquad s = \sigma_0/\rho V_0^2 \qquad (5.2.31)$$

Then (5.2.25) can be converted into the dimensionless form

$$\ddot{u} + (s/3)\left[\alpha^2(4u + \gamma^2 u'')'' + (4/\gamma^2)u\right]^{\cdot}$$

$$+ s\left[\alpha^2\beta\left\{\gamma^2(u - \tilde{u})'' - (2/3)(u - \tilde{u})\right\} + u\right]'' = 0 \qquad (5.2.32)$$

where the dots denote differentiation with respect to $\tau$.

### 5.2.5 Modal Solutions

We consider two cases depending on the mass velocity. In the first case, the mass is brought to rest at constant deceleration according to (5.2.10). The equation governing the plastic buckling is (5.2.29). In the second case, the mass velocity is assumed to be constant and the governing equation is (5.2.32).

Let the dimensionless perturbations and initial imperfections be represented by

$$u(\xi,\eta) = \sum_{n=1}^{\infty} u_n(\eta)\sin n\pi\xi \qquad (5.2.33)$$

$$\tilde{u}(\xi) = u(\xi,1) = \sum_{n=1}^{\infty} a_n \sin n\pi\xi \qquad (5.2.34)$$

and substitute into (5.2.29). Each mode amplitude is then found to be governed by the equation

$$\ddot{u}_n - Q_n\dot{u}_n/\eta - R_n^2 u_n = S_n a_n \qquad (5.2.35)$$

where

$$Q_n = (\mu/3)\left[\alpha^2\pi^2 n^2(\gamma^2\pi^2 n^2 - 4) + 4/\gamma^2\right] \qquad (5.2.36)$$

$$R_n^2 = 2\mu\tau_f\pi^2 n^2\left[1 - \alpha^2\beta\,(\gamma^2\pi^2 n^2 + 2/3)\right] \qquad (5.2.37)$$

$$S_n = 2\mu\tau_f\alpha^2\beta\pi^2 n^2(\gamma^2\pi^2 n^2 + 2/3) \qquad (5.2.38)$$

Equation (5.2.35) has the same form as (4.1.59), which describes the plastic buckling of an imploded shell. Amplification occurs only when $R_n^2 > 0$. Formula (5.2.37) therefore gives the maximum value $\bar{n}$ of the mode number for which amplification is possible as

$$\bar{n} \approx 1/\alpha\sqrt{\beta}\,\gamma\pi \qquad (5.2.39)$$

The solution of (5.2.35) is analogous to the solution given by (4.1.65) through

(4.1.68). In particular, the amplification function for initial displacement imperfections is

$$A_n(\zeta) = \eta^{\nu}\Big[I_{\nu}(R_n\eta)K_{\nu-1}(R_n) + K_n(R_n\eta)I_{\nu-1}(R_n)\Big]$$

$$\times (2\mu\tau_f\pi^2n^2/R_n) - S_n/R_n^2 \qquad (5.2.40)$$

which is analogous to (4.1.66); again $\nu = (Q_n + 1)/2$. The terminal value of the amplification factor is

$$A_n(0) \approx \mu\tau_f\pi^2n^2 2^{\nu}\Gamma(\nu)I_{\nu-1}(R_n)/R_n^{\nu-1} \qquad (5.2.41)$$

which is analogous to (4.1.70). The preferred mode number may be taken as the value N that maximizes (5.2.41).

In the second case, where we assume that $V = V_0$ throughout the axial compression, we represent the dimensionless perturbations and initial imperfections by

$$u(\xi,\tau) = \sum_{n=1}^{\infty} u_n(\tau)\sin n\pi\xi \qquad (5.2.42)$$

$$\tilde{u}(\xi) = u(\xi,0) = \sum_{n=1}^{\infty} a_n \sin n\pi\xi \qquad (5.2.43)$$

Substitution of (5.2.42) and (5.2.43) in (5.2.32) shows that each mode amplitude is governed by the equation

$$\ddot{u}_n + Q_n\dot{u}_n - R_n^2 u_n = S_n a_n \qquad (5.2.44)$$

where

$$Q_n = (s/3)\Big[\alpha^2\pi^2n^2(\gamma^2\pi^2n^2 - 4) + 4/\gamma^2\Big] \qquad (5.2.45)$$

$$R_n^2 = s\pi^2n^2\Big[1 - \alpha^2\beta(\gamma^2\pi^2n^2 + 2/3)\Big] \qquad (5.2.46)$$

$$S_n = s\alpha^2\beta\pi^2n^2(\gamma^2\pi^2n^2 + 2/3) \qquad (5.2.47)$$

Growth of initial imperfections occurs when $R_n^2 > 0$, a condition that is satisfied when the mode number is less than $\bar{n}$, again given approximately by (5.2.39).

The solution of (5.2.44) satisfying the initial values of displacement and velocity perturbations

$$u_n(0) = a_n \qquad \dot{u}_n(0) = b_n \tag{5.2.48}$$

is

$$u_n(\tau) = A_n(\tau)a_n + B_n(\tau)b_n \tag{5.2.49}$$

where

$$A_n = (\lambda_n^+ e^{\lambda_n^- \tau} - \lambda_n^- e^{\lambda_n^+ \tau})(1 + S_n/R_n^2)/(\lambda_n^+ - \lambda_n^-) - S_n/R_n^2 \tag{5.2.50}$$

$$B_n = (e^{\lambda_n^+ \tau} - e^{\lambda_n^- \tau})/(\lambda_n^+ - \lambda_n^-) \tag{5.2.51}$$

$$\lambda_n^\pm = \left[\pm(Q_n^2 + 4R_n^2)^{1/2} - Q_n\right]/2 \tag{5.2.52}$$

The functions $A_n$ and $B_n$ given by (5.2.50) and (5.2.51) are the required amplification functions when the velocity of the mass M is taken as the constant $V_0$ in the governing Equation (5.2.32), and hence in the modal Equation (5.2.44). The duration of the sustained axial plastic flow, however, is calculated by assuming that the initial velocity decreases linearly with time according to (5.2.10). Investigations that include this variable velocity in the governing equation are made with formula (5.2.41). The remainder of the analysis in this section treats the velocity as a constant in the governing equation.

### 5.2.6 Amplification Functions

We again illustrate the nature of the amplification functions (5.2.50) and (5.2.51) by an example from a set of experiments in which tubes of 6061-T6 aluminum alloy were subjected to axial impact; the experiments are treated in Sections 5.2.9 and 5.2.10. The example is designated tube 20 and has the following data:

| | | | |
|---|---|---|---|
| midsurface radius | $a$ | = | 0.45 inch (1.145 cm) |
| thickness | $h$ | = | 0.10 inch (0.254 cm) |
| length | $L$ | = | 4.0 inch (10.16 cm) |
| density | $\rho$ | = | 0.097 $lb/in^3$ (2.7 $g/cm^3$) |
| tube mass | $m$ | = | $2\pi ahL\rho$ = 0.112 lb (51 g) |
| compressing mass | $M$ | = | 0.264 lb (120 g) |
| impact velocity | $V_0$ | = | 411 ft/s (125 m/s) |
| yield stress | $\sigma_0$ | = | 44,500 psi (307 MPa) |
| hardening modulus | $E_h$ | = | 105,000 psi (724 MPa) |

These data provide the following values for the dimensionless parameters:

$\alpha^2 = h^2/12a^2 = 0.004115$ $\beta = E_h/\sigma_0 = 2.34$ $\gamma = a/L = 0.113$

$\mu = M/m = 2.353$ $s = \sigma_0/\rho V_0^2 = 7.24$ $\tau_f = \mu/2s = 0.162$

The value of the last parameter $\tau_f$ is a prediction of 16.2% for the final axial strain. These parameters are substituted into formulas (5.2.45), (5.2.46), and (5.2.47), leaving only n as the parameter in $Q_n$, $R_n$, and $S_n$. Formulas (5.2.50), (5.2.51), and (5.2.52) now give the relationship between each amplification function and the mode number n.

These relationships are shown in Figures 5.18 and 5.19. The development

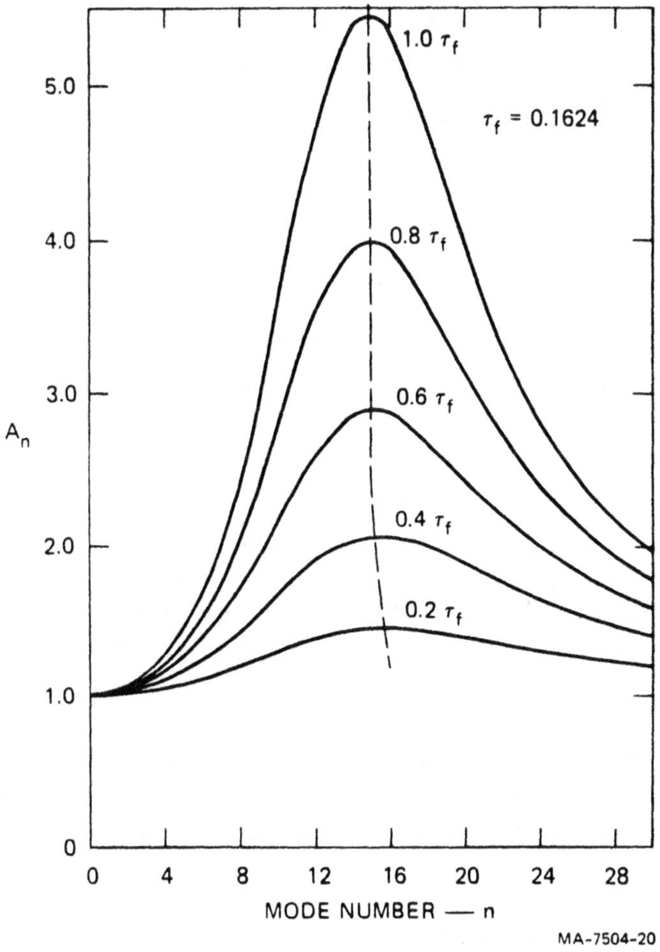


FIGURE 5.18 AMPLIFICATION FUNCTIONS FROM INITIAL DISPLACEMENTS (Tube 20 in Table 5.1)

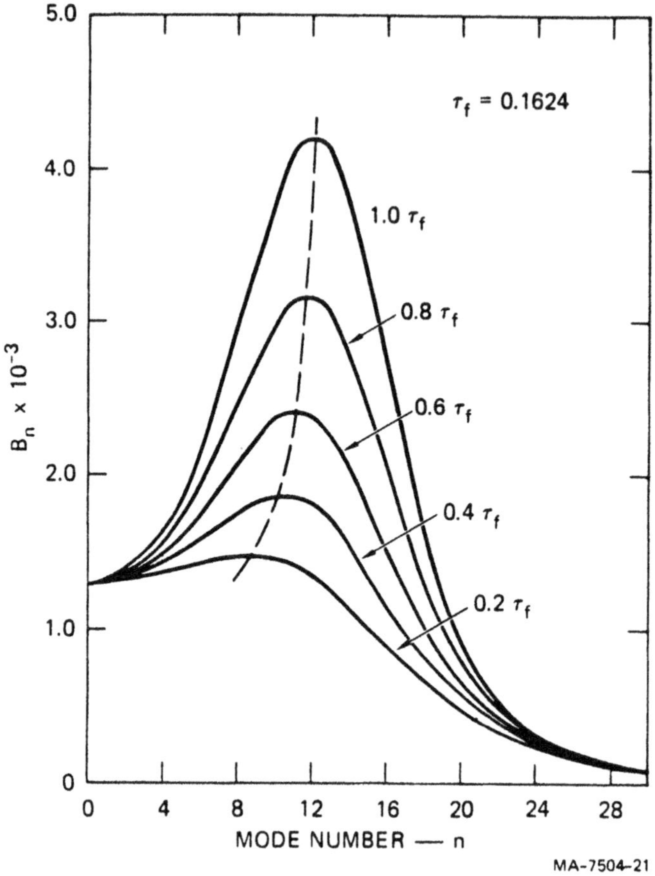


FIGURE 5.19 AMPLIFICATION FUNCTIONS FROM INITIAL VELOCITIES (Tube 20 in Table 5.1)

and shape of the curves are typical of plastic flow buckling, the main feature being the development of a narrow band of harmonics that are eventually amplified much more than is any other harmonic. The dashed lines through the maxima of each constant-time curve give a reasonable prediction of the development of the preferred harmonic.

Figures 5.18 and 5.19 show that the preferred mode is selected during the first half of the motion. This behavior indicates that the assumption of a uniform velocity in the governing equation is probably satisfactory for predicting the preferred mode number.

According to formula (5.2.29), tube 20 has a cutoff mode number of $\bar{n} = 29$. Thus, the preferred mode number N is less than $\bar{n}$. Figures 5.18 and 5.19 show most amplified mode numbers of 15 and 12 stemming from initial displacements and velocities, respectively. The experimental value was 12.

### 5.2.7 Preferred Mode and Critical Velocity Formulas

In many cases of axial impact of tubes, the properties and conditions are such that $Q_n >> 2R_n$ and $R_n^2 >> S_n$ when n is near the preferred mode number N. In these cases we may approximate the final value of the displacement amplification function by

$$A_n(\tau_f) = e^{\lambda_n^+ \tau_f} \tag{5.2.53}$$

Differentiating (5.2.53) with respect to $n^2$ and equating the result to zero then leads to

$$N^2 = 2(1 - 2\alpha\beta) / \pi^2 \alpha \gamma^2 \tag{5.2.54}$$

provided we made the reasonable approximations $2\alpha^2\beta/3$ and $2\alpha\beta\gamma^2 << 1$. In the example of tube 20 in Section 5.2.6, formula (5.2.54) gives N = 13, which compares favorably with the prediction of N = 15 from (5.2.50). Two physical consequences of the assumptions are that the ratio of the hardening modulus to the yield stress should not be large and that the length should be several times the radius. It is interesting to note that for many practical cases formula (5.2.54) is independent of the impact velocity V and does not depend strongly on the hardening parameter $\beta$. This observation is in agreement with the experimental results of Section 5.2.10.

The maximum value of $A_n$ is found by substituting in (5.2.53) the preferred mode number N from (5.2.54), where, according to our approximations, $\lambda_n^+ \approx R_n^2/Q_n$. The impact velocity is introduced through $\tau_f = \mu/2s = \mu\rho V_0^2/2\sigma_0$ (energy equation). This process leads to

$$V_0^2 = \frac{(2\sigma_0/\rho\mu)4\alpha[1 - \alpha(1 - 2\alpha\beta)(2\beta + 1)] \ln A_n}{3[1 - 4\alpha\beta + 8\alpha^2\beta^2(1 - 2\alpha\beta)]} \tag{5.2.55}$$

where, $\alpha, \beta$, and $\mu$ are defined in (5.2.28). Formula (5.2.55) may be regarded as a formula for the critical impact velocity that corresponds to an acceptable amplification factor. For tube 20, if we take $A_n = 10$ the critical velocity is $V_c$ = 535 ft/s (163 m). The experimental impact velocity was 411 ft/s (125 m/s), and it caused moderately large amplitudes (twice the wall thickness).

Figure 5.20 shows the variation of amplification $A_n(\tau_f)$ with the impact velocity $V_0$ for tube 20. One of the two curves shown gives the variation according to formula (5.2.50), with $\tau = \tau_f$, and the other curve gives the variation according to the explicit formula (5.2.55). The comparison provides confidence in the use of (5.2.55) for determining critical velocities.

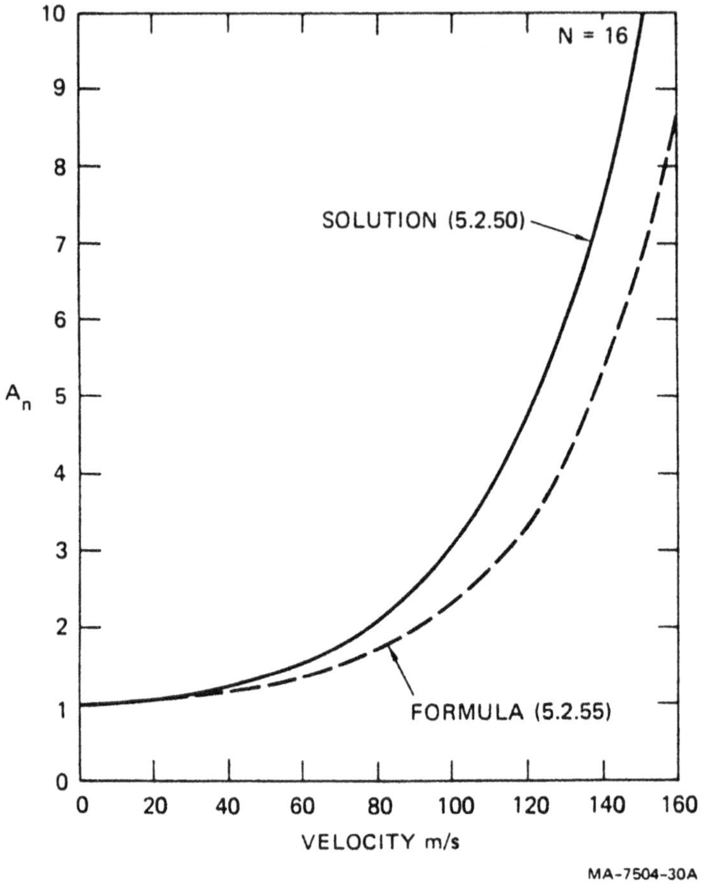


FIGURE 5.20 IMPACT VELOCITY-AMPLIFICATION CURVES FOR TUBE 20

### 5.2.8 Directional and Hardening Moments

From formula (5.2.20), we see that the directional and hardening contributions to the restoring moment $M_x$ are

$$M^{(d)} = -(\sigma_0 L/a)(4w/a + w''a) \cdot (h^3/36a) \tag{5.2.56}$$

$$M^{(h)} = E_h\left[2(w - \tilde{w})/a - 3a(w - \tilde{w})''\right](h^3/36a) \tag{5.2.57}$$

If we convert formulas (5.2.56) and (5.2.57) to our dimensionless quantities (u, $\xi, \tau, \alpha, \beta$, and $\gamma$), select the perferred mode N, and compare the relative magnitudes in a time-integrated manner by means of the ratio

$$\Lambda_a = \int_0^{\tau_f} M_N^{(h)}(\tau)\,d\tau \Big/ \int_0^{\tau_f} M_N^{(d)}(\tau)\,d\tau \tag{5.2.58}$$

we obtain the approximate formula

$$\Lambda_a = \frac{4\alpha\beta}{3} \cdot \frac{3(1-2\alpha\beta)+\alpha}{(1-2\alpha\beta)-2\alpha} \cdot \frac{1-\alpha(2\beta+1)(1-2\alpha\beta)}{1-4\alpha\beta+8\alpha^2\beta^2(1-2\alpha\beta)} \tag{5.2.59}$$

In the derivation of (5.2.59), only growth from initial displacement imperfections is considered and the preferred mode number is assumed to be adequately predicted by formula (5.2.54). We recall that the strain hardening is represented by the parameter $\beta$, so the ratio $\Lambda_a$ tends to zero as $E_h$ tends to zero. For tube 20, the ratio is $\Lambda_a = 1.06$, which implies that for the typical experimental tube the contributions of the directional and hardening moments to the restoring moment are comparable. Comparisons for other tubes can be made by using formula (5.2.59). The ratio of the time-integrated moments depends only on the parameters $\alpha$ and $\beta$, that is, on the radius-to-thickness ratio and the hardening modulus-to-yield stress ratio.

According to (5.2.54), the preferred mode number increases with decreasing hardening modulus. If fact, when $\beta = 0$,

$$N^2 = 2/\pi^2\alpha^2\gamma^2 = 4\sqrt{3}\,L^2/\pi^2 ah \tag{5.2.60}$$

For tube 20, formula (5.2.60) predicts an increase of the mode number from N = 13, from (5.2.54), to N = 16 as the hardening modulus is reduced to zero.

According to (5.2.55), the relationship between the impact velocity and the amplification factor when $\beta = 0$ is

$$V_0^2 = (2\sigma_0/\rho\mu)\left[4\alpha(1-\alpha)/3\right]\ell n\, A_N \tag{5.2.61}$$

For tube 20, formula (5.2.61) predicts a critical impact velocity of $V_0 = 437$ ft/s (133 m/s) based on amplification factor of $A_N = 10$, which is a reduction from $V_0 = 535$ ft/s (163 m/s) predicted by (5.2.55).

### 5.2.9 Description of Experiments

A smooth-bore rifle of 0.46-inch I.D. (1.17 cm) was used to project the tubes against a machined and polished flat face of a heavy steel slab 12 inches long and

12 inches in diameter (30.5 by 30.5 cm). The steel slab and the carriage assembly for the rifle were supported on the same I-beam, and the rifle barrel was aligned optically to be perpendicular to the flat face of the slab. The tubes were attached by adaptor plugs to push rods, as shown in Figure 5.21. Each push rod was a close fit in the barrel and, just before firing, the end of the rod rested against a cartridge. After firing, the tube-plug-rod assemblage traveled along aligned Teflon tracks that ended short of the slab. Installed in the track were accurately spaced electrical contact pins for measuring the tube velocity.

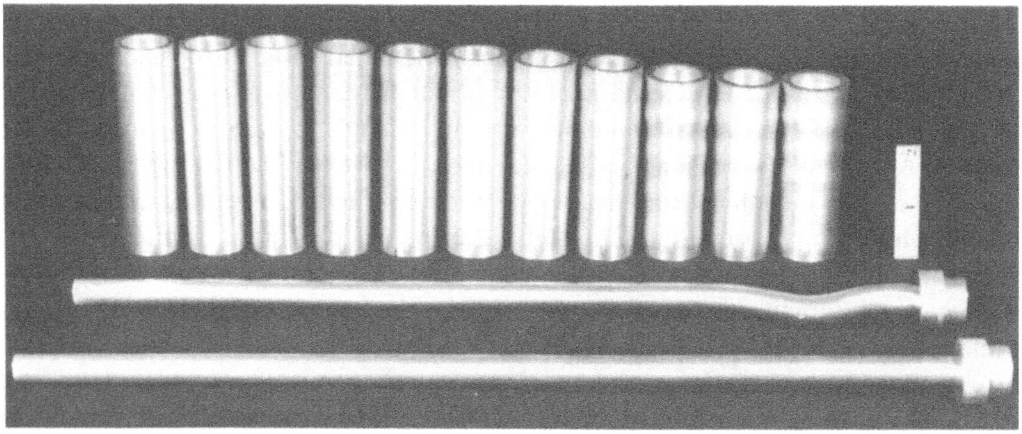

MP-7504-23

FIGURE 5.21 PLASTICALLY BUCKLED TUBES CAUSED BY AXIAL IMPACT AND TWO PUSH ROD-ADAPTER ASSEMBLIES

Along with the rod-plug assemblies, Figure 5.21 shows a series of 6061-T6 aluminum alloy tubes, originally 4 inches (10.2 cm) long (impacted ends shown upwards). From such specimens magnified tube profiles of the type shown in Figure 5.22 were drawn by passing a pointer along a generator, the pointer being an extension of a linear differential transformer, which sent a signal to activate a plotter. This device magnified many inperceptible buckles sufficiently to allow wavelength measurements.

Photographs of six of the tubes were taken during impact using an image converter camera (one photograph per test) to observe the development of plastic buckling. The technique consisted of performing three identical impact tests and taking the photograph at a different time in each test.

Static compression tests were performed on six tubes to provide a comparison of the static and dynamic buckling mode numbers. The ends of the tubes were

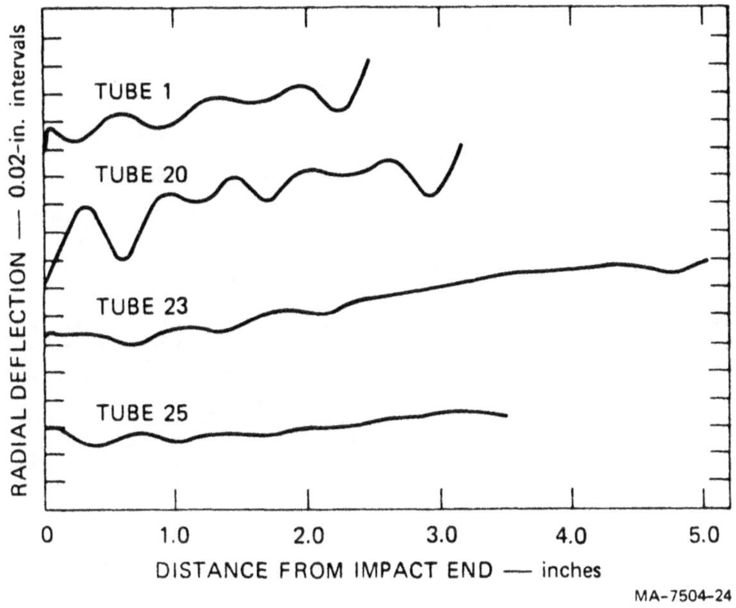


FIGURE 5.22 TUBE PROFILES

lubricated to allow expansion along the flat heavy steel plattens of the compression machine.

Yield stress and strain-hardening values were obtained from bilinear fits of stress-strain curves from compression tests on 3/8-inch-long (9.5 mm) tube specimens. Again, the ends were lubricated, and little barreling was observed to 10% strain.

### 5.2.10 Comparison of Theory and Experiment

Table 5.1 lists the main experimental data. Tubes of 6061-T6 or 2024-T3 aluminum alloy of 1-inch O.D. (2.54 cm), 0.1 or 0.095 inch thick (0.254 or 0.241 cm), and 3, 4, or 6 inches long (7.62, 10.16, or 15.24 cm) were impacted axially at velocities ranging from 130 to 411 ft/s (40 to 125 m/s). The resulting number of half-waves, listed in the final column were obtained from the magnified profiles, examples of which are shown in Figure 5.22.

Table 5.2 lists the preferred mode numbers predicted in three ways, all assuming a constant velocity in the governing equation. The first column of mode numbers headed $A_n(\tau_f)$, was obtained by substituting $\tau_f$ into formula (5.2.50) and numerically finding the value of n that maximized the amplification function. The second column of mode numbers, headed $B_n(\tau_f)$, was obtained similarly using for-

Table 5.1

EXPERIMENTAL DATA FOR TUBES

| Tube Material | Tube Number | Rod-Plug Weight (M) (grams) | Tube Weight (m) (grams) | Length L (cm) | Velocity $V_o$ (m/s) | Shortening X (cm) | Strain X/L (%) | Half-waves N |
|---|---|---|---|---|---|---|---|---|
| | 1,2,3 | 127 | 36 | 7.62 | 101 | 1.283 | 16.8 | 8 |
| | 4 | 127 | 36 | 7.62 | 101 | 1.397 | 18.3 | 8 |
| | 5* | 120 | 51 | 10.16 | 53 | 0.437 | 4.4 | 12 |
| | 6* | 120 | 51 | 10.16 | 58 | 0.516 | 5.2 | 12 |
| | 7* | 120 | 51 | 10.16 | 63 | 0.635 | 6.3 | 12 |
| Aluminum | 8 | 120 | 49 | 10.16 | 78 | 0.833 | 8.3 | 12 |
| | 9* | 120 | 51 | 10.16 | 87 | 0.914 | 9.2 | 12 |
| 6061-T6 | 10* | 120 | 51 | 10.16 | 88 | 0.965 | 9.5 | 12 |
| | 11* | 120 | 51 | 10.16 | 99 | 1.189 | 11.7 | 12 |
| | 12* | 120 | 51 | 10.16 | 100 | 1.194 | 11.9 | 12 |
| | 13,14,15 | 127 | 48 | 10.16 | 104 | 1.397 | 13.8 | 12 |
| | 16* | 120 | 51 | 10.16 | 115 | 1.575 | 15.5 | 11 |
| | 17* | 120 | 51 | 10.16 | 120 | 1.727 | 17.2 | 11 |
| | 18* | 120 | 51 | 10.16 | 121 | 1.740 | 17.4 | 12 |
| | 19* | 120 | 51 | 10.16 | 122 | 1.803 | 17.8 | 11 |
| | 20* | 120 | 51 | 10.16 | 125 | 1.981 | 19.5 | 12 |
| | 21 | 300 | 76 | 15.24 | 40 | 0.518 | 3.4 | † |
| | 22 | 300 | 76 | 15.24 | 75 | 1.549 | 10.2 | 14 |
| | 23,24 | 300 | 76 | 15.24 | 94 | 2.235 | 14.7 | 15 |
| Aluminum | 25 | 127 | 51 | 10.16 | 117 | 1.143 | 11.3 | 11 |
| 2024-T3 | 26 | 127 | 51 | 10.16 | 117 | 1.041 | 10.3 | 11 |
| | 27 | 127 | 51 | 10.16 | 117 | 1.194 | 11.8 | 11 |

* Tubes of thickness 0.254 cm (0.1 in.) (a/h = 4.50); remaining tubes of thickness 0.241 cm (0.095 in.) (a/h = 4.76).

† Insufficient growth of buckling for measurement.

Table 5.1 (concluded)

EXPERIMENTAL DATA FOR TUBES

| Tube Material | Tube Number | Rod-Plug Weight (M) (grams) | Tube Weight (m) (grams) | Length L (inches) | Velocity $V_o$ (ft/s) | Shortening X (inches) | Strain X/L (%) | Half-waves N |
|---|---|---|---|---|---|---|---|---|
| | 1,2,3 | 127 | 36 | 3 | 332 | 0.505 | 16.8 | 8 |
| | 4 | 127 | 36 | 3 | 332 | 0.550 | 18.3 | 8 |
| | 5* | 120 | 51 | 4 | 175 | 0.172 | 4.4 | 12 |
| | 6* | 120 | 51 | 4 | 190 | 0.203 | 5.2 | 12 |
| | 7* | 120 | 51 | 4 | 208 | 0.250 | 6.3 | 12 |
| Aluminum | 8 | 120 | 49 | 4 | 256 | 0.328 | 8.3 | 12 |
| | 9* | 120 | 51 | 4 | 285 | 0.360 | 9.2 | 12 |
| 6061-T6 | 10* | 120 | 51 | 4 | 288 | 0.380 | 9.5 | 12 |
| | 11* | 120 | 51 | 4 | 324 | 0.468 | 11.7 | 12 |
| | 12* | 120 | 51 | 4 | 327 | 0.470 | 11.9 | 12 |
| | 13,14,15 | 127 | 48 | 4 | 342 | 0.550 | 13.8 | 12 |
| | 16* | 120 | 51 | 4 | 378 | 0.620 | 15.5 | 11 |
| | 17* | 120 | 51 | 4 | 395 | 0.680 | 17.2 | 11 |
| | 18* | 120 | 51 | 4 | 397 | 0.685 | 17.4 | 12 |
| | 19* | 120 | 51 | 4 | 400 | 0.710 | 17.8 | 11 |
| | 20* | 120 | 51 | 4 | 411 | 0.780 | 19.5 | 12 |
| | 21 | 300 | 76 | 6 | 130 | 0.204 | 3.4 | † |
| | 22 | 300 | 76 | 6 | 246 | 0.610 | 10.2 | 14 |
| | 23,24 | 300 | 76 | 6 | 310 | 0.880 | 14.7 | 15 |
| Aluminum | 25 | 127 | 51 | 4 | 385 | 0.450 | 11.3 | 11 |
| 2024-T3 | 26 | 127 | 51 | 4 | 385 | 0.410 | 10.3 | 11 |
| | 27 | 127 | 51 | 4 | 385 | 0.470 | 11.8 | 11 |

* Tubes of thickness 0.1 in. (a/h = 4.50); remaining tubes of thickness 0.095 in. (a/h = 4.76).

† Insufficient growth of buckling for measurement.

Table 5.2

THEORETICAL AND EXPERIMENTAL MODE NUMBERS

| Tube Material | Tube Number | Strain $\tau_f$ (%) | $A_n(\tau_f)$ | $B_n(\tau_f)$ | Formula (5.2.54) | Experiment N |
|---|---|---|---|---|---|---|
| | | | Preferred Mode Number N | | | |
| | 1,2,3,4 | 15.89 | 11 | 9 | 10 | 8 |
| | 5 | 2.94 | 15 | 9 | 13 | 12 |
| | 6 | 3.47 | 15 | 9 | 13 | 12 |
| | 7 | 4.16 | 15 | 9 | 13 | 12 |
| | 8 | 6.56 | 16 | 11 | 13 | 12 |
| Aluminum | 9 | 7.18 | 15 | 11 | 13 | 12 |
| 6061-T6 | 10 | 7.98 | 15 | 11 | 13 | 12 |
| | 11 | 10.09 | 15 | 11 | 13 | 12 |
| | 12 | 10.28 | 15 | 11 | 13 | 12 |
| | 13,14,15 | 12.65 | 15 | 12 | 14 | 12 |
| | 16 | 13.74 | 15 | 12 | 13 | 11 |
| | 17 | 15.00 | 15 | 12 | 13 | 11 |
| | 18 | 15.16 | 15 | 12 | 13 | 12 |
| | 19 | 15.38 | 15 | 12 | 13 | 11 |
| | 20 | 16.24 | 15 | 12 | 13 | 12 |
| | 21 | 2.73 | 24 | 13 | 21 | — |
| | 22 | 9.76 | 23 | 17 | 21 | 14 |
| | 23,24 | 15.50 | 23 | 18 | 21 | 15 |
| 2024-T3 | 25,26,27 | 12.67 | 14 | 11 | 10 | 11 |

mula (5.2.51). The third column of mode numbers, headed 'Formula,' was obtained by (5.2.54), which was derived from formula (5.2.50) with $\tau = \tau_f$ under the conditions $Q_n >> 2R_n$ and $R_n^2 >> S_n$ for n near the preferred number N. The experimental values are listed in the final column. Overall, the predictions are satisfactory, especially when they are based on initial velocity imperfections.

In Figure 5.23 there are six photographs of six different tubes taken during impact with an image converter camera. Figures 5.23 (a, b, c) are photographs of the identical tubes 13, 14, and 15 taken 65, 100, and 150 $\mu$s after impact, which occurred at a velocity of 342 ft/s (104 m/s). The longitudinal strains at these times were approximately 5.9%, 7.9%, and 10.5% compared with 5.5%, 7.8%, and 10.3% calculated by assuming a constant force resisting the rod and plug. At 65 $\mu$s, the preferred mode of buckling is under way. At this time the longitudinal strain is 5.9% and since the final strain is $\tau_f = 13.75\%$, the preferred mode is evident at about 0.4 $\tau_f$.

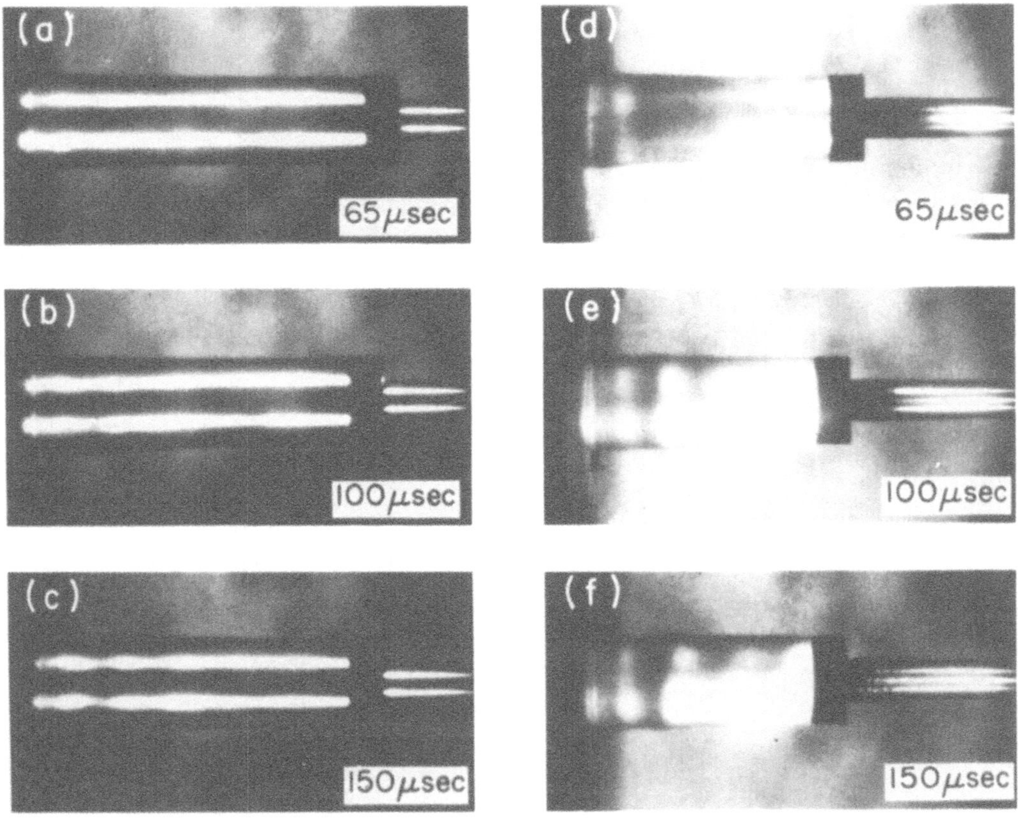


FIGURE 5.23 IMAGE CONVERTER CAMERA PHOTOGRAPHS OF TUBE IMPACT; (a), (b), and (c) are tubes 13, 14, and 15; (d), (e), and (f) are tubes 1, 2, and 3.

A similar behavior is seen in Figures 5.23 (c, d, e) for the identical tubes 1, 2, and 3 subjected to an impact of 332 ft/s (101 m/s). Again, the measured and calculated longitudinal strains agree assuming a constant resistive force. At 65 $\mu$s the preferred buckling is under way. The longitudinal strain is 7.2%, and since the final strain is 16.83%, the preferred mode is evident at about 0.4 $\tau_f$. In both cases the mode numbers observed in Figure 5.23 are the same as those obtained from the final profiles of the type shown in Figure 5.22.

If we examine Figures 5.18 and 5.19 showing the amplification spectra at fixed times or longitudinal strains, we see that the theory predicts a preferred mode number at about 0.4 $\tau_f$, with little change thereafter (dashed lines in the figures), which is in agreement with the photographic observation. This result and the general overall agreement of the experimental and theoretical mode numbers suggest that the assumption of a constant axial velocity is reasonable for predicting mode numbers.

Measurements of radius and thickness increases on several tubes showed that final average hoop and radial strains were within a few percent of each other. This observation reinforces our assumption of Section 5.2.2 that our strain rate proportionality constant may be taken as $k = \frac{1}{2}$.

### 5.2.11 Slow Buckling

When the impact velocity $V_0$ is small, the inertia terms in Equations (5.2.44) governing the amplitudes $u_n$ can be neglected to give

$$Q_n \dot{u}_n - R_n^2 u_n = S_n a_n \tag{5.2.62}$$

where the coefficients are given by (5.2.45), (5.2.46), and (5.2.47). For the case of initial displacement imperfections only, the solution of (5.2.62) is

$$u_n(\tau) = A_n(\tau) a_n = \left[ (1 + S_n/R_n^2) e^{R_n^2 \tau / Q_n} - S_n/R_n^2 \right] a_n \tag{5.2.63}$$

Growth of initial imperfections occurs when $R_n^2 > 0$, a condition that is satisfied when the mode number is less than $\bar{n}$ given approximately by (5.2.39). As in the dynamic case, each tube has a family of spectral curves like those in Figure 5.18. According to (5.2.63), the amplification grows exponentially with time or the imposed axial strain. The exponential term in (5.2.63) depends on the mode number, and the preferred mode number is approximately that which maximizes the exponential. Consequently, we have the same situation as that occurring in axial impact where maximizing (5.2.53) leads to the preferred mode number formula (5.2.54), again with the reasonable provisos that $2\alpha^2\beta/3$ and $2\alpha\beta\gamma^2 << 1$. This result means that the inertial forces do not influence the preferred mode selection. Also, the mode is not strongly dependent on the strain hardening.

Because the theory predicts negligible influence of the inertial forces on the preferred mode selection, static buckling tests were performed on two 3-inch-long (7.62 cm) tubes of 6061-T6 aluminum (Tubes 1-4), two 4-inch-long (10.16 cm) tubes of 6061-T6 aluminum (Tubes 5-20), and two 4-inch-long (10.16 cm) tubes of 2024-T3 aluminum (Tubes 25-27). These tests produced preferred mode numbers of 8, 10, and 10, respectively, as compared with 8, 12, and 11 from the corresponding impact tests. Thus, experimentally, the changes in the number of halfwaves are small.

Figure 5.24 shows the profiles of the two 3-inch-long (7.62 cm) tubes of 6061-T6 aluminum caused by static axial compression to about 0.7% axial strain. Figure 5.25 shows a comparison of one of the profiles from the static test and the profile of Tube 4 caused by axial impact. The preferred mode numbers are identical.

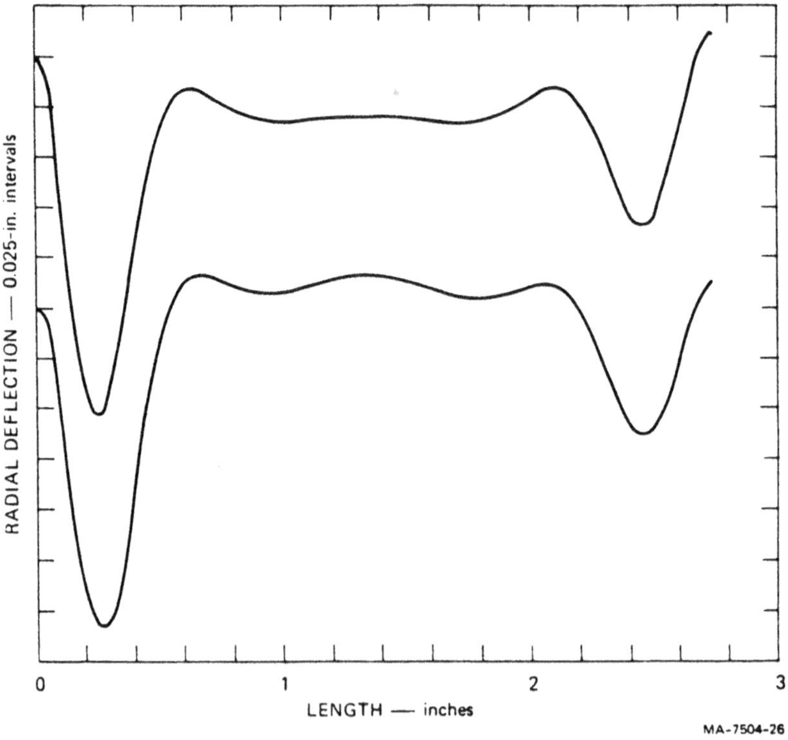


FIGURE 5.24 TUBE PROFILES FROM STATIC COMPRESSION TESTS (Specifications of Tubes 1–4)

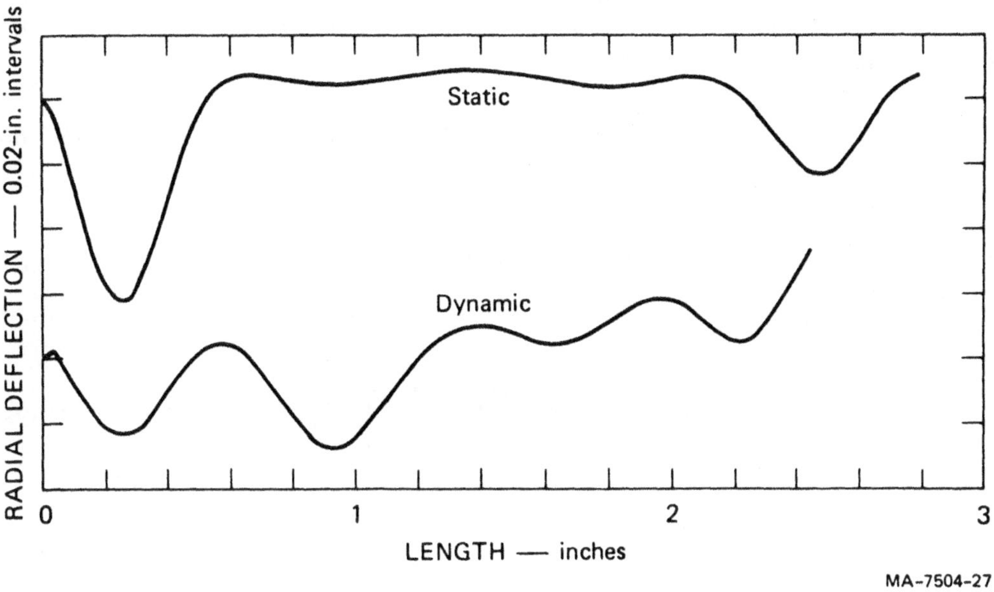


FIGURE 5.25 COMPARISON OF TUBE PROFILES FROM STATIC AND IMPACT TESTS (Tube 4)

### 5.2.12 Axial Impact of Plates

The dynamic buckling of plates is treated here as a special ease of the preceding tube theory. A more general treatment can be found in Section 6.

In Section 5.2.4 we obtained the Equation (5.2.26) which governs the flow buckling of plates compressed uniaxially in a manner similar to the tubes (Figure 5.16). By introducing the dimension quantities

$$u = w/h \qquad \xi = x/L \qquad \tau = V_0 t/2L \qquad \eta = 1 - \tau/\tau_f$$

$$\zeta^2 = h^2/12L^2 \qquad \beta = E_h/\sigma_0 \qquad \mu = M/m \qquad s = \sigma_0/\rho V_0^2 \qquad (5.2.64)$$

in Equation (5.2.26) we obtain

$$\ddot{u} - (\zeta^2\mu/3\eta)\dot{u}'''' + 2\tau_f\mu\left[\zeta^2\beta(u - \tilde{u})'' + u\right]'' = 0 \qquad (5.2.65)$$

In (5.2.64), M is the mass per unit width of plate that is compressing the plate and $m = \rho hL$ is the mass of a unit width of plate. With these modified definitions of the masses M and m, the symbol $\tau_f$ is again given by (5.2.30) and is the final axial strain; thus $\tau_f = \mu/2s$. In (5.2.66), the dots and primes denote partial differentiation with respect to $\eta$ and $\xi$.

The set of equations governing the modal amplitudes is obtained by substituting the Fourier representations (5.2.33) and (5.2.34) into Equation (5.2.65). Thus for each mode we have

$$\ddot{u}_n - Q_n\dot{u}_n/\eta - R_n^2 u_n = S_n \qquad (5.2.66)$$

where

$$Q_n = (\mu\zeta^2/3)(\pi n)^4 \qquad (5.2.67)$$

$$R_n^2 = 2\tau_f\mu(\pi n)^2\left[1 - \zeta^2\beta(\pi n)^2\right] \qquad (5.2.68)$$

$$S_n = 2\tau_f\mu\zeta^2\beta(\pi n)^4 \qquad (5.2.69)$$

Because significant amplification occurs only if $R_n^2 > 0$, only modes with numbers less than $\bar{n}$ are amplified, where

$$\bar{n} = 1/\zeta\sqrt{\beta}\ \pi \qquad (5.2.70)$$

The solution can proceed by the method outlined in Section 4.1.8 for imploded cylindrical shells. However, here we shall retain our simplifying assumption that the velocity of the compressing mass remains equal to the impact velocity $V_0$ for the duration of compression, given by $\tau_f$.

After omitting $\eta$ and replacing $\tau$ by

$$\tau = V_0 t/L \qquad (5.2.71)$$

the dimensionless quantities allow the governing Equation (5.2.26) to be written in the form

$$\ddot{u} + (s\zeta^2/3)\dot{u}'''' + s\left[\zeta^2\beta(u - \tilde{u})'' + u\right]'' = 0 \qquad (5.2.72)$$

where the dots and primes denote partial differentiation with respect to $\tau$ and $\xi$. Substitution in (5.2.72) of the Fourier representations (5.2.33) and (5.2.34) leads to

$$\ddot{u}_n + Q_n\dot{u}_n - R_n^2 u_n = S_n \qquad (5.2.73)$$

where

$$Q_n = (s\zeta^2/3)(\pi n)^4 \tag{5.2.74}$$

$$R_n^2 = s(\pi n)^2\left[1 - \zeta^2\beta(\pi n)^2\right] \tag{5.2.75}$$

$$S_n = s\zeta^2\beta(\pi n)^4 \tag{5.2.76}$$

The solution of (5.2.73) satisfying the initial conditions (5.2.48) is (5.2.49), where $A_n$, $B_n$, and $\gamma_n$ are given by (5.2.50), (5.2.51), and (5.2.52).

Figures 5.26 and 5.27 show the amplification curves for Plate 20, a designa-

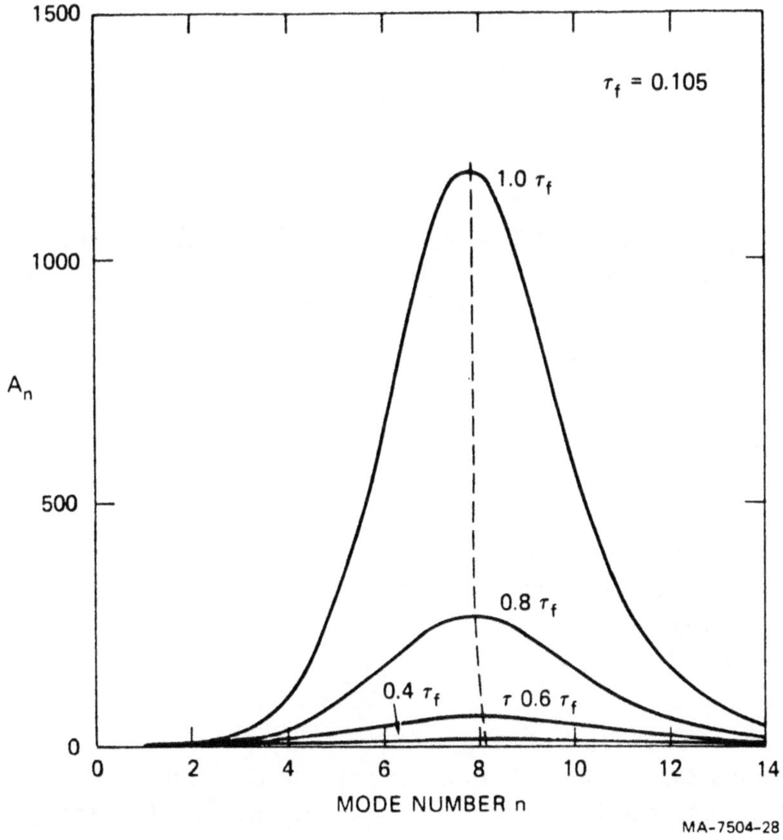


FIGURE 5.26 AMPLIFICATION FUNCTIONS FROM INITIAL DISPLACEMENTS (Plate 20)

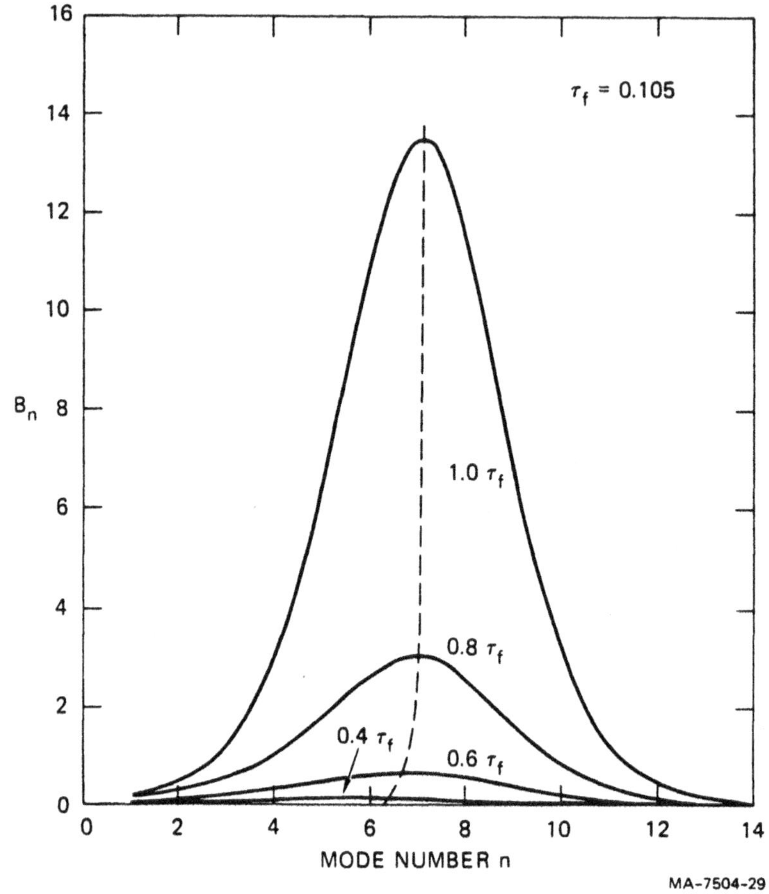


FIGURE 5.27 AMPLIFICATION FUNCTIONS FROM INITIAL VELOCITIES (Plate 20)

tion meaning a plate with the specifications of Tube 20 ($E_h$, $\sigma_0$, $\rho$, h, L, $V_0$, and appropriately modified m and M). The curves are typical of flow buckling. Growth is possible only if $R_n^2 > 0$, which affects only modes with a number less than $\bar{n}$ given by (5.2.75) as

$$\bar{n} = 1/\zeta\sqrt{\beta}\ \pi \tag{5.2.77}$$

A formula can be derived from the amplification function (5.2.50) for $A_n$, with the $Q_n$, $R_n^2$, and $S_n$ of (5.2.74), (5.2.75), and (5.2.76), that provides an approximate prediction for the preferred mode number. If we notice that the predominant quantity in (5.2.50) affecting the variation of $A_n$ with n is the factor $e^{\lambda_n^+ \tau}$, the value of n maximizing $A_n$ is given approximately by $\partial\lambda_n^+/\partial(n^2) = 0$.

With no further approximations this maximization leads to the equation

$$(\pi N)^6 - (18\beta^2/s)(\pi N)^4 + (18\beta/s\zeta^2)(\pi N)^2 - (9/2s\zeta^4) = 0 \qquad (5.2.78)$$

For negligible strain hardening ($\beta \approx 0$), (5.2.78) reduces to

$$N^3 = 3/\pi^3 \sqrt{2s}\ \zeta^2$$

$$= (18\sqrt{2}/\pi^3)(L/h)^2\left[V_0/\sqrt{\sigma_0/\rho}\right] \qquad (5.2.79)$$

For Plate 20, (5.2.79) predicts $N = 7$, which is consistent with the maximum $A_n(\tau_f)$ in Figure 5.26. Predictions for all the plates given by the numerically obtained maximum of $A_n$, by the solution of (5.2.78), and by (5.2.79), are identical. For 6061-T6 aluminum, the 3-, 4-, and 6-inch-long (7.62, 10.16, and 15.24 cm) plates have $N = 6$, 7, and 10 and for the 2024-T3 aluminum plates $N = 7$. Because (5.2.78) and (5.2.79) predict the same values of N, the strain hardening does not influence the selection of the preferred mode number.

The trend from rapid to slow buckling behavior in plates is completely different from that in tubes. Equation (5.2.78) shows that as we consider smaller impact velocities we regard $s \to \infty$, so $N \to 1$, the lowest integer possible in the representation. This decrease in mode number represents a large change, whereas, for tubes the change is small.

A formula giving the amplification resulting from plate impact can be derived if we first made the approximation

$$A_N(\tau_f) \approx \left[\lambda_N^-/(\lambda_N^- - \lambda_N^+)\right]e^{\lambda_N^+\tau_f} \qquad (5.2.80)$$

This approximation requires that $\xi^2\beta(\pi N)^2$ is small enough to make $S_n/R_n^2$ small. If we now employ (5.2.74) and (5.2.75) for $Q_n$ and $R_n$, the result (5.2.78), and $\tau_f = \mu\rho V_0^2/2\sigma_0$, we can convert (5.2.80) into

$$V_0^2 = (4\sigma_0/3\rho\mu)\, f(N)\, \ln(3A_N/4) \qquad (5.2.81)$$

where

$$f(N) = \zeta^2(\pi N)^2 \Big/ \left[1 - 2\beta\zeta^2(\pi N)^2\right] \qquad (5.2.82)$$

As an initial prediction for $V_0$, the preferred mode may be approximated by (5.2.79), that is,

$$\zeta^2(\pi N)^2 = (9\zeta^2/2s)^{1/3} \qquad (5.2.79)$$

Because $s = \sigma_0/\rho V_0^2$ appears in the right-hand side of (5.2.81), iteration is needed to determine $V_0$. However, for many cases we may neglect the term $2\beta\zeta^2(\pi N)^2$ so that (5.2.81) becomes

$$V_0 = 2(2/3)^{1/4}\left[\zeta(\sigma_0/\rho)\right]^{1/2}\left[\mu \ \ell n \ (3A_N/4)\right]^{3/4}$$

$$\approx \left[(\sigma_0/\rho)(h/L)\right]^{1/2}\left[\mu \ \ell n \ (3A_N/4)\right]^{3/4} \qquad (5.2.83)$$

Formula (5.2.83) provides a critical impact velocity for the criterion of a specific value of $A_n$.

## 5.3 FORCES AND ENERGY ABSORPTION DURING AXIAL PLASTIC COLLAPSE OF TUBES

### 5.3.1 Axial Collapse Experiments

Axial collapse experiments were performed with the apparatus shown in Figure 5.28. The shell to be tested is 4 inches (102 mm) in diameter and 6 inches (152 mm) long. A steel aft mass weighing 3.5 pounds (1.59 kg) in the form of a thick, stepped plate is seen at the left end of the shell. The smaller diameter is a slip fit into the shell, and the larger diameter extends outside the shell diameter to provide a flat bearing surface at the step. The front end of the shell is similarly supported by a thin plate that extends up to a guide bushing. The aft mass is connected to the same bushing to form a stiff carriage for accelerating the mass and shell. This is a peg-in-hole connection so that on impact the mass moves forward freely.

The carriage bushing slides on a steel bar to maintain alignment of the cylinder as it moves. The assembly is accelerated by another peg-in-hole connection to a pusher bar, seen at the left of the photograph with its peg not yet inserted into the carriage. The pusher bar is accelerated by a pneumatic piston below the track that supports the pusher carriage and guide bar (not seen in the photograph). Impact is against a 6-inch-thick (152 mm) by 12-inch-square (305 mm) steel target block. The guide bar is secured to the target block by a press fit into a hole in the block accurately machined at a right angle to the impact surface. This arrangement gives impact simultanously around the circumference of the shell to within 5 $\mu$s at impact velocities near 100 ft/s (30 m/s).

A – Energy Absorbing Ring
B – Grooved Rear Support

MP-4750-46

FIGURE 5.28 SHELL, END MASS, AND GUIDE APPARATUS FOR AXIAL COLLAPSE EXPERIMENTS

Figure 5.29 gives post-test photographs of an aluminum 6061-T6 shell with a wall thickness of 0.035 inch (0.89 mm, a/h = 57) tested at an impact velocity of 102 ft/s (31 m/s). It buckled at the ends, where the stress was highest for the longest time. The total crush (change in length) was 1.55 inches (39.4 mm). Four half-buckles (two inward, two outward) collapsed at the front end, and one (outward) collapsed at the aft end. The average half-wavelength, referred to the undeformed state, was 0.308 inch (7.82 mm). The first two half-buckles at the front end were in the symmetric mode and the second two were in the third mode (triangular).

Figure 5.30 gives photographs of an ogival cone of the same wall thickness, length, and aft diameter as in the cylinder, impacted at the same velocity. Its

19

19

MP–4750–35

FIGURE 5.29 PLASTIC COLLAPSE BUCKLING IN 0.035-INCH-THICK (0.89 mm) ALUMINUM CYLINDRICAL SHELL IN 102 ft/sec (31 m/s) IMPACT

(a = 2 inches = 51 mm, L = 6 inches = 152 mm)

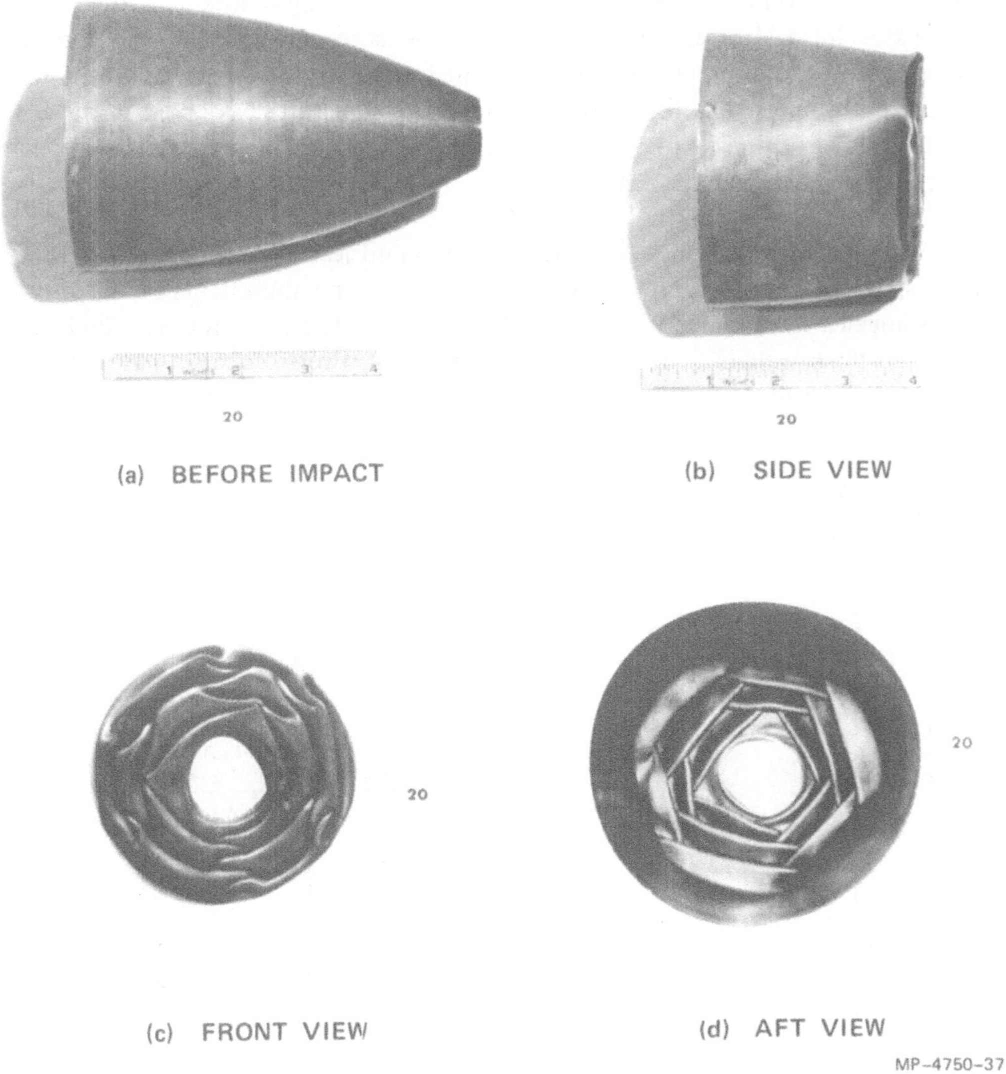


FIGURE 5.30 PLASTIC COLLAPSE BUCKLING IN 0.035-INCH-THICK (0.89 mm) TAPERED ALUMINUM CONE IN 102 ft/sec (31 m/s) IMPACT

(Ogival cone, front radius 0.5 inch (12.7 mm), aft radius 2 inches (51 mm), length 6 inches (152 mm)

front-end diameter is 1 inch (25 mm), so its radius-to-thickness ratio ranged from 14 at the forward end to 57 at the aft end. Total crush was 2.68 inches (68.1 mm), entirely at the front end where axial stress and strain were highest. Seven buckles formed, the first in the symmetric mode, the second in the fourth mode, and the remaining five in the third mode.

Figure 5.31 gives the record from an accelerometer mounted on the aft mass, and a tabular summary of observations made from motion pictures taken at 10,000 frames/s. The overall character of the acceleration is that it gradually increases as the buckles form, because each buckle is at a larger diameter of the cone and requires a larger force to crush. The numbers on the curve correspond to the times at which each buckle was first visible in the motion pictures, as given in the inset. As each buckle begins to form, the acceleration increases, and as each nears complete collapse the acceleration decreases.

A similar observation was made for the cylindrical shell, but the pattern was not as apparent because only a few buckles formed and the effects of the forward and aft buckles were intermingled. A third experiment, on a cylindrical shell with a 0.060-inch-thick wall (1.5 mm, a/h = 33), showed similar collapse but entirely in the symmetric mode.

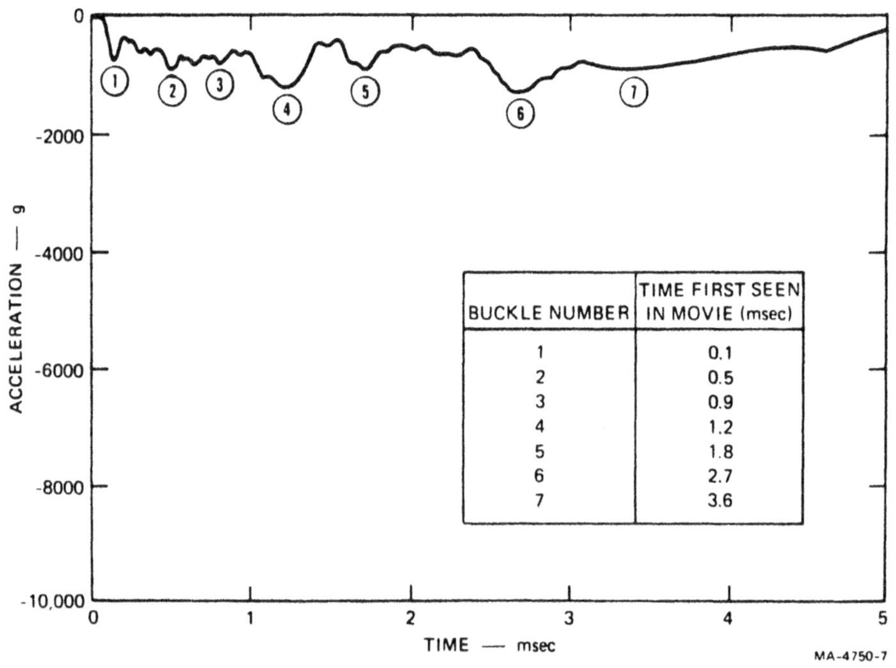


| BUCKLE NUMBER | TIME FIRST SEEN IN MOVIE (msec) |
|---|---|
| 1 | 0.1 |
| 2 | 0.5 |
| 3 | 0.9 |
| 4 | 1.2 |
| 5 | 1.8 |
| 6 | 2.7 |
| 7 | 3.6 |

FIGURE 5.31 ACCELERATION OF AFT MASS IN CONE TEST

### 5.3.2 Theoretical Estimates of Collapse Forces

Shell crushing as in these experiments takes place in two distinct stages, buckling and collapse. In the initial stage, a buckling pattern is established by axial compression, nearly uniform along the shell length, that amplifies some of the small imperfections in the shell. This stage is the subject of Sections 5.1 (thin shells) and 5.2 (thicker shells, as in the tests here) and lasts only a few transit times of an elastic wave along the shell length. During this time, buckles are formed throughout the shell length, sometimes too small to be seen but much larger than the imperfections, so they determine the buckle wavelengths for the second stage. In the second stage, the buckling pattern undergoes the large deformation collapse that was observed in the motion pictures here. This collapse stage can last orders of magnitude longer than the buckling stage and, except for very thick shells, absorbs most of the impact energy.

For shells of intermediate radius-to-thickness ratio, as in Figures 5.29 and 5.30, the buckling modes of the first stage pass reasonably unaltered into the collapse modes of the second stage. For very thin shells, the buckling modes of the first stage are altered by large deformation elastic buckling which, as seen in Figure 5.1, can increase the buckle wavelengths by factors of two to six. This process is extremely complex. The buckle pattern for a collapse analysis of these thin shells is probably best determined by a heuristic argument for patterns that minimize stretching and the length of plastic hinges that form during collapse. These are the Yoshimura[8] diamond patterns.

Shells in which the transfer from buckling to collapse takes place in essentially the same pattern are those in which buckling is initiated by axial plastic flow, as in Section 5.2. Thus, for the shells of Figures 5.29 and 5.30, we use buckle modes determined by the analysis of Section 5.2 as initial conditions for a collapse analysis.

Equations (5.2.51), (5.2.52), and (5.2.54) were used to calculate the theoretical half-wavelengths in Table 5.2. These agree well with the observed half-wavelengths of collapse in the two cylindrical shells. Calculated wavelengths for shells in the range $30 < a/h < 100$ are given to an accuracy of 2% by the formula

$$\lambda_o = 5.6\,h + a/20 \tag{5.3.1}$$

To calculate collapse force, we use the simple symmetric mode shape shown in Figure 5.32 and calculate the energy required to collapse a buckle from zero displacement to the nearly completely folded form in Figure 5.32(b). Approximate extension to polygonal buckles is described later. The simple buckle shape is suggested by the two principal mechanisms of energy absorption, membrane plastic flow and plastic bending. In this idealized buckle shape, membrane stretching and compression take place as hoop strain in the straight sections, and plastic bending takes place as plastic hinge rotation at the apex of each half-buckle.

In this model, the force at the beginning of collapse is infinite, but the energy is finite. A large initial force has a counterpart in the experiments, as seen in Fig-

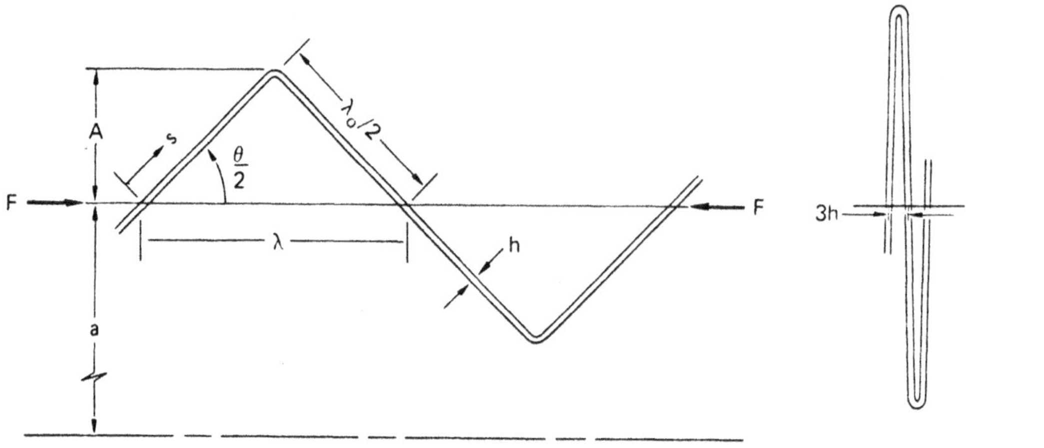


(a) DURING BUCKLING (b) AFTER BUCKLING

FIGURE 5.32 BUCKLING SHAPE USED IN THEORY TO PREDICT PLASTIC COLLAPSE FORCE

ure 5.31, but the model is not expected to reproduce the detailed force history. Instead, we calculate the energy U for complete collapse and divide by the axial displacement to determine an average value for F. Thus, we use

$$F_{ave} = U/(\lambda_o - 3h) \tag{5.3.2}$$

in which 3h is the typical final value $\lambda_f$ of instantaneous half-wavelength $\lambda$ as observed in the experiments. For complete folding to a solid form, $\lambda_f = h$.

During crush, the instantaneous energy absorbed by hoop strain is

$$U_h = 2\pi a h \sigma_o \cdot 2 \int_0^{\lambda_o/2} \epsilon(\lambda, s) \tag{5.3.3}$$

in which $\sigma_o$ is flow stress, and

$$\epsilon(\lambda, s) = 2As/\lambda_o a \tag{5.3.4}$$

is circumferential strain. The instantaneous buckle amplitude, from the geometry in Figure 5.32, is

$$A(\lambda) = (\lambda_o^2 - \lambda^2)^{1/2}/2 \tag{5.3.5}$$

Substitution of Equation (5.3.5) into Equation (5.3.3) and integration as indicated gives for the instantaneous hoop energy

$$U_h = \frac{\pi h \sigma_o \lambda_o}{2} (\lambda_o^2 - \lambda^2)^{1/2}$$

The total hoop strain energy absorbed by each half buckle is found by substitution of the final value $\lambda_f = 3h$, which gives

$$U_h(3h) = \frac{\pi h \sigma_o \lambda_o}{2} (\lambda_o^2 - 9h^2)^{1/2} \qquad (5.3.6)$$

The instantaneous energy absorbed by a plastic hinge is

$$U_b(\lambda) = 2\pi a M_o \theta \qquad (5.3.7)$$

where the plastic moment* per unit of circumference is

$$M_o = \sigma_o h^2/4$$

and, from the geometry in Figure 5.32, the hinge rotation $\theta$ is

$$\theta = 2\cos^{-1}(\lambda/\lambda_o)$$

The total energy absorbed by each plastic hinge is

$$U_b(3h) = \pi a \sigma_o h^2 \cos^{-1}(3h/\lambda_o) \qquad (5.3.8)$$

The strain energies from Equations (5.3.6) and (5.3.8) are added and then equated to the work done by the average axial force as indicated in Equation (5.3.1) to obtain finally

$$F_{ave} = \frac{\pi h \sigma_o}{\lambda_o - 3h} \left[ \frac{\lambda_o}{2} (\lambda_o^2 - 9h^2)^{1/2} + ah \cos^{-1}(3h/\lambda_o) \right] \qquad (5.3.9)$$

With Equations (5.3.1) and (5.3.9), a graph was constructed of the average collapse force versus radius-to-thickness ratio. The result is shown in Figure 5.33. The collapse force is normalized with respect to $F_o = 2\pi a h \sigma_o$, the force that pro-

---

*Bending strain dominates at the hinges, and hoop strain dominates away from the hinges. Therefore, the interaction of hoop and bending deformation is neglected.

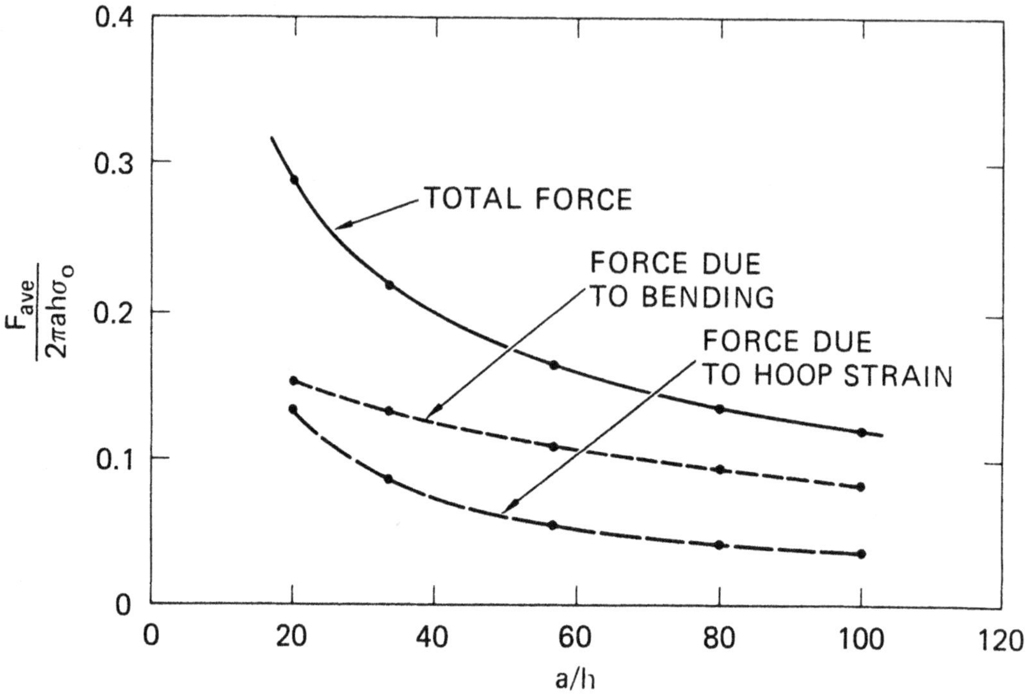


FIGURE 5.33 PREDICTED AVERAGE PLASTIC COLLAPSE FORCE VERSUS RADIUS-TO-THICKNESS RATIO IN 6061-T6 ALUMINUM CYLINDRICAL SHELLS

duces yielding of the shell in axial compression. Also shown are the separate contributions from hoop strain and from bending at the plastic hinges. The contributions from the two mechanisms are comparable.

### 5.3.3 Comparison of Theory and Experiment

Equation (5.3.9) was evaluated for the two cylindrical shells in the experiments. For both cases, $\sigma_o$ = 45,000 psi (310 MPa) and a = 2 inches (51 mm). For the first shell, h = 0.035 inch (0.89 mm), a/h = 57, and $\lambda_o = 5.6h + a/20$ = 0.296 inch (6.83 mm). The calculated average collapse force is $F_{ave}$ = 3250 pounds (1483 kg), and the calculated acceleration of the 3.5-pound (1.60 kg) aft mass is 930 g. This is in excellent agreement with the average acceleration measured during postbuckling collapse, as shown in Figure 5.34(a)* For this case 33% of the energy was absorbed in hoop strain and 67% in bending strain.

---

*The measured acceleration versus time histories have been converted to acceleration versus crush for comparison with the theory. This conversion is particularly useful for the cone discussed later.

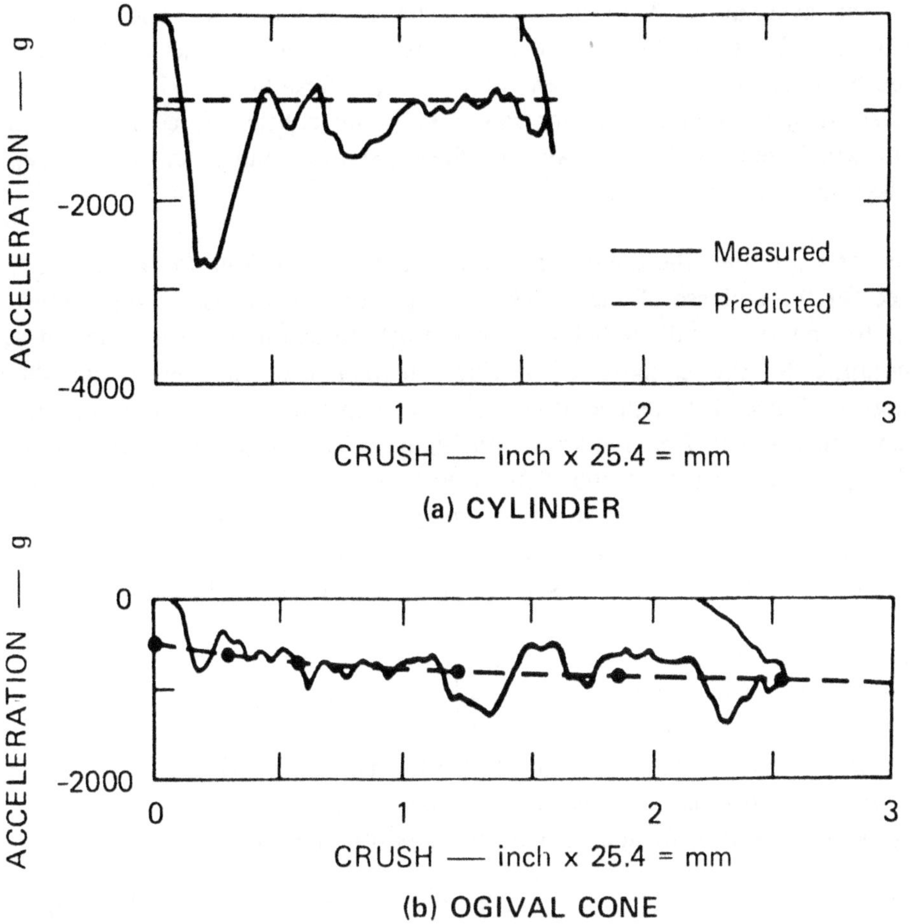


MA-4750-12B

FIGURE 5.34 MEASURED AND PREDICTED AVERAGE ACCELERATIONS FOR TWO SHELLS

For the second shell, h = 0.06 inch (1.52 mm), a/h = 33, and $\lambda_o = 5.6h + a/20 = 0.436$ inch (11.1 mm). The calculated average collapse force is $F_{ave}$ = 7420 pounds (3386 kg), and the predicted acceleration of the 3.5-pound (1.60 kg) aft mass is 2120 g. This is also in good agreement with the average acceleration measured during the postbuckling collapse of a single buckle. For this case, 40% of the energy was absorbed in hoop strain and 60% in bending strain.

Thus, this simple theory can predict the average postbuckling collapse force, even though the assumed axisymmetric buckling modes are not the only ones observed in the experiments. [Recall that, in the plastic collapse test of the 0.035-inch-thick (0.89 mm) shell, the first two buckles formed were axisymmetric, but the other two buckles were in the third (triangular) mode.] This is reasonable because the average hoop strain energy depends primarily on the axial buckle wavelength, and the bending strain energy depends only on the shell circumference, both of which are independent of the circumferential buckling mode. This result also suggests that the theory can predict the collapse force in the tapered cone, in which most of the buckles produced were asymmetric. Therefore, the theory was applied to the cone.

As for a cylinder, the theory applied to a cone can be used to predict the average force for the collapse of each buckle. Thus, for the collapse of more than one buckle, the theory would predict the force-crush function to be a series of steps corresponding to the collapse of buckles at increasing diameters and buckle wavelengths. Since this stepped function is an approximation to the smooth and continuous force-crush function shown in Figure 5.31, no loss in accuracy is introduced by approximating this stepped function by a smooth and continuous function. Specifically, it is assumed that the strain energy density is a continuous function of the coordinate x along the cone axis. Thus, in Equations (5.3.1) and (5.3.9), the wavelength $\lambda_0$, as well as the radius a, are taken as continuous functions of x.

The force-versus-crush relationship for the cone test was then calculated from Equation (5.3.9). For comparison with experimental data, two conversions were made. First, the crush force was converted to acceleration of the aft mass by dividing it by the weight of the 3.5-pound (1.60 kg) mass. Second, the axial coordinate x was converted to crush distance by multiplying it by the factor $(\lambda_0 - 3h)/\lambda_0$, where this factor is also considered a continuous function of x.

The predicted and measured curves of acceleration versus crush distance are shown in Figure 5.34(b). As for the cylinders, good agreement is obtained between the predicted and measured average acceleration versus crush.

## REFERENCES

1. B. Budiansky and J.W. Hutchinson, "Dynamic Buckling of Imperfection-Sensitive Structures," *Proc. of the 11th Intl. Congress of Appl. Mech.*, Springer-Verlag, Berlin, pp. 636-651, 1964.

2. H.E. Lindberg and R.E. Herbert, "Dynamic Buckling of a Thin Cylindrical Shell Under Axial Impact," ASME Trans., Series E, J. Appl. Mech., **33**, pp. 105-112, 1966.

3. Th. von Kármon, L.G. Dunn, and H.S. Tsien, "The Influence of Curvature on the Buckling Characteristic of Structures," J. Aerospace Sci., **7**, pp. 276-289, 1940.

4. B.O. Almroth, A.M.C. Holmes, and D.O. Brush, "An Experimental Study of the Buckling of Cylinders Under Axial Compression," Exptl. Mech., **4**, pp. 263-270, 1964.

5. S.O. Rice, "Mathematical Analysis of Random Noise," in N. Wax, *Selected Papers on Noise and Stochastic Processes,* Dover Pub., Inc., New York, pp. 133-294, 1954.

6. J.W. Hutchinson and B. Budiansky, "Dynamic Buckling Estimates," AIAA J. **4**, 3, pp. 525-530 March 1966.

7. A.L. Florence and J.N. Goodier, "Dynamic Plastic Buckling of Cylindrical Shells in Sustained Axial Compressive Flow," J. Appl. Mech., **35**, pp. 80-86, March 1968.

8. Y. Yoshimura, "On the Mechanism of Buckling of a Circular Cylindrical Shell Under Axial Compression," NACA TM 1390, July 1955.

# 6. PLASTIC FLOW BUCKLING OF RECTANGULAR PLATES

## 6.1 INTRODUCTION

Figure 6.1 shows the plastic buckling that occurs when tubing of square cross section is subjected to axial impact at velocities sufficient to cause moderate plastic strains.[1] Each side of the tube is a laterally supported rectangular plate under uniaxial compression. In the experiments to be discussed later, it was found that the wavelengths of the buckle's like those shown in Figure 6.1 were fairly uniform along the length of each plate and that they were reproducible.

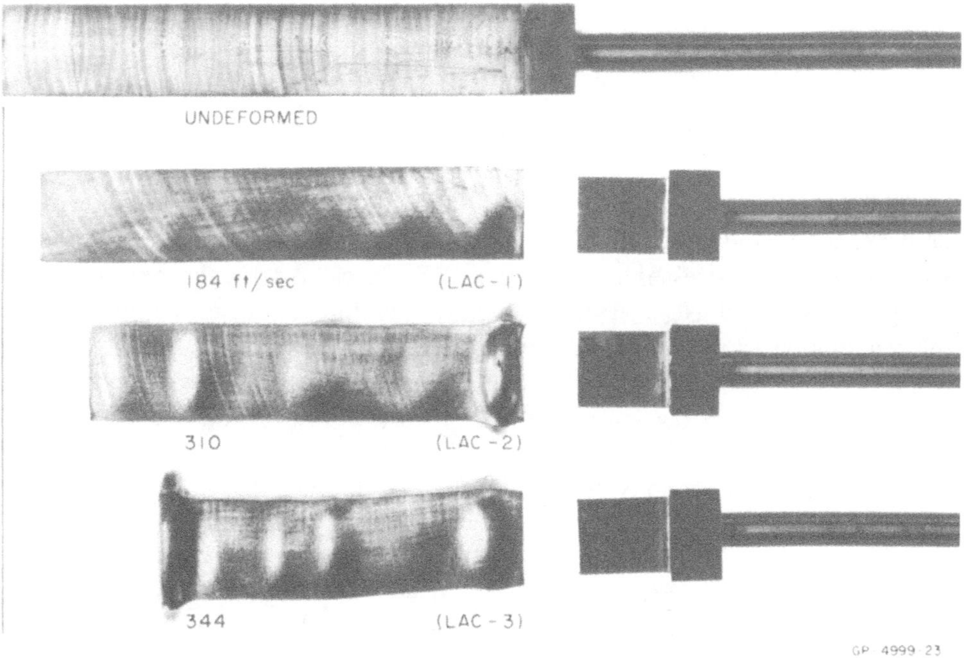


FIGURE 6.1 PLASTIC BUCKLING OF SQUARE TUBES CAUSED BY AXIAL IMPACT

Rather than develop a theory to address the experimental observations directly, we consider first a more general case of imposed in-plane loading. Specifically, we apply as part of the unperturbed state the proportional loading $N_x$, $N_y$, and $N_{xy}$ (see Figure 3.41 for plate nomenclature and coordinates) along the edges of the rectangular plate to create a uniform stress field $\sigma_x^{\circ}$, $\sigma_y^{\circ}$, and $\tau_{xy}^{\circ}$ independent of the depth coordinate z. Buckling is again defined as the growth of plate deflections stemming from initial transverse deflections or velocities. Derivation of the equation governing plate deflections uses rigid-plastic theory with the von Mises yield condition, the associated flow rule, and linear strain hardening. We assume that no strain-rate reversal occurs and that the strain hardening allows us to approximate the unperturbed driving stresses with constant values; that is, the strain hardening should not be excessive.

The governing equation is specialized to treat uniaxial compression and is solved by modal analysis. The solution predicts preferred buckling mode numbers in reasonably good agreement with experimental mode numbers. The solution is also used to derive formulas for preferred mode numbers and the relationship between the plate impact velocity and the amplification of the initial deflections.

## 6.2 PERTURBATIONAL FLEXURE

The nomenclature, sign convention, and coordinate system are shown in Figures 3.39, 3.40, and 3.41. Let $\dot{\epsilon}_x^{\circ}$, $\dot{\epsilon}_y^{\circ}$, and $\dot{\epsilon}_{xy}^{\circ}$ be the midsurface strain rates associated with the stresses $\sigma_x^{\circ}$, $\sigma_y^{\circ}$, and $\tau_{xy}^{\circ}$, caused by the external in-plane loading $N_x$, $N_y$, and $N_{xy}$. Then, as flexure develops because of initial transverse displacements, $w_o(x,y)$ and $\dot{w}_o(x,y)$, the strain rates are

$$\dot{\epsilon}_x = \dot{\epsilon}_x^{\circ} + z\dot{\kappa}_x \qquad \dot{\epsilon}_y = \dot{\epsilon}_y^{\circ} + z\dot{\kappa}_y \qquad \dot{\epsilon}_{xy} = \dot{\epsilon}_{xy}^{\circ} + z\dot{\kappa}_{xy} \tag{6.2.1}$$

assuming that plane sections remain plane. In (6.2.1), the changes of curvature and twist are

$$\kappa_x = -\frac{\partial^2 w}{\partial x^2} \qquad \kappa_y = -\frac{\partial^2 w}{\partial y^2} \qquad \kappa_{xy} = -\frac{\partial^2 w}{\partial x \partial y} \tag{6.2.2}$$

where $w = w(x,y,t)$ is the deflection added to the initial lack of planarity $w_o = w_o(x,y)$.

After eliminating $\dot{\epsilon}_z$ by means of the plastic incompressibility condition

$$\dot{\epsilon}_x + \dot{\epsilon}_y + \dot{\epsilon}_z = 0$$

the equivalent strain rate becomes

$$\dot{\epsilon}^2 = 4(\dot{\epsilon}_x^2 + \dot{\epsilon}_y^2 + \dot{\epsilon}_x\dot{\epsilon}_y + \dot{\epsilon}_{xy}^2)/3 \tag{6.2.3}$$

which, according to (6.2.1), is approximately

$$\dot{\epsilon}^2 = \dot{\epsilon}_o^2 + (8z/3)\left[\dot{\epsilon}_x{}^{\circ}\dot{\kappa}_x + \dot{\epsilon}_y{}^{\circ}\dot{\kappa}_y + (\dot{\epsilon}_x{}^{\circ}\dot{\kappa}_y + \dot{\epsilon}_y{}^{\circ}\dot{\kappa}_x)/2 + \dot{\epsilon}_{xy}{}^{\circ}\dot{\kappa}_{xy}\right] \tag{6.2.4}$$

In (6.2.4), $\dot{\epsilon}_o$ is the equivalent strain rate at the midsurface.

An approximation to the equivalent strain rate is obtained by noting that the perturbation term in (6.2.4) is small relative to the unperturbated term, $\dot{\epsilon}_o^2$. Thus, approximately

$$\dot{\epsilon} = \dot{\epsilon}_o + (4z/3)\left[\dot{\epsilon}_x{}^{\circ}\dot{\kappa}_x + \dot{\epsilon}_y{}^{\circ}\dot{\kappa}_y + (\dot{\epsilon}_x{}^{\circ}\dot{\kappa}_y + \dot{\epsilon}^{\circ}{}_y\dot{\kappa}_x)/2 + \dot{\epsilon}_{xy}{}^{\circ}\dot{\kappa}_{xy}\right]/\dot{\epsilon}_o \tag{6.2.5}$$

We now restrict our problem to cases of proportional loading. In other words, we consider cases in which $\sigma_x{}^{\circ}$, $\sigma_y{}^{\circ}$, and $\tau_{xy}{}^{\circ}$ are maintained in constant ratio throughout the motion. With the usual assumption of plate theory that $\sigma_z = 0$, the equivalent stress is given by

$$\sigma^2 = \sigma_x^2 + \sigma_y^2 - \sigma_x\sigma_y + 3\tau_{xy}^2 \tag{6.2.6}$$

and the proportional loading leads to constant values of the ratios $\sigma_x{}^{\circ}/\sigma^{\circ}$, $\sigma_y{}^{\circ}/\sigma^{\circ}$, and $\tau_{xy}{}^{\circ}/\sigma^{\circ}$, where $\sigma^{\circ}$ is the midsurface value of the equivalent stress. Because of the flow rule

$$\dot{\epsilon}_x = \dot{\lambda}\sigma_x' \qquad \dot{\epsilon}_y = \dot{\lambda}\sigma_y' \qquad \dot{\epsilon}_{xy} = \dot{\lambda}\tau_{xy} \qquad \dot{\epsilon} = 2\dot{\lambda}\sigma/3 \tag{6.2.7}$$

associated with the von Mises yield condition, we have the midsurface strain rate ratios

$$\dot{\epsilon}_x{}^{\circ}/\dot{\epsilon}_o = (2\sigma_x{}^{\circ} - \sigma_y{}^{\circ})/2\sigma^{\circ} \qquad \dot{\epsilon}_y{}^{\circ}/\dot{\epsilon}_o = (2\sigma_y{}^{\circ} - \sigma_x{}^{\circ})/2\sigma^{\circ}$$

$$\dot{\epsilon}_{xy}{}^{\circ}/\dot{\epsilon}_o = 3\tau_{xy}/2\sigma^{\circ} \tag{6.2.8}$$

and proportional loading keeps these ratios constant. In (6.2.7)

$$\sigma_x' = (2\sigma_x - \sigma_y)/3 \quad \text{and} \quad \sigma_y' = (2\sigma_y - \sigma_x)/3 \tag{6.2.9}$$

are the deviatoric stresses when $\sigma_z = 0$. We can now integrate (6.2.5) to give the equivalent strain

$$\epsilon = \epsilon_o + (4z/3)\left[\dot{\epsilon}_x{}^{\circ}\kappa_x + \dot{\epsilon}_y{}^{\circ}\kappa_y + (\dot{\epsilon}_x{}^{\circ}\kappa_y + \dot{\epsilon}_y{}^{\circ}\kappa_x)/2 + \dot{\epsilon}_{xy}{}^{\circ}\kappa_{xy}\right]/\dot{\epsilon}_o \tag{6.2.10}$$

The flow rule in the form (6.2.8) allows us to express (6.2.10) as

$$\epsilon = \epsilon_o + z(\sigma_x^{\circ}\kappa_x + \sigma_y^{\circ}\kappa_y + 2\tau_{xy}^{\circ}\kappa_{xy})/\sigma^{\circ} \tag{6.2.11}$$

which we can introduce in the linear strain-hardening law

$$\sigma = \sigma_o + E_h\epsilon \tag{6.2.12}$$

to give

$$\sigma = \sigma^{\circ} + z(\sigma_x^{\circ}\kappa_x + \sigma_y^{\circ}\kappa_y + 2\tau_{xy}^{\circ}\kappa_{xy})E_h/\sigma^{\circ} \tag{6.2.13}$$

From the flow rule (6.2.7) and the deviatoric stress (6.2.9), we have

$$\sigma_x = (2\dot{\epsilon}_x + \dot{\epsilon}_y)(2\sigma/3\dot{\epsilon}) \qquad \sigma_y = (2\dot{\epsilon}_y + \dot{\epsilon}_x)(2\sigma/3\dot{\epsilon})$$

$$\tau_{xy} = \dot{\epsilon}_{xy}(2\sigma/3\dot{\epsilon}) \tag{6.2.14}$$

If we substitute in (6.2.14) the strain rates (6.2.1) and (6.2.5) and the stress (6.2.13), we obtain the stress formulas

$$\sigma_x = \sigma_x^{\circ} + z\Big[\{(2 - 3\alpha^2/2)\dot{\kappa}_x + (1 - 3\alpha\beta/2)\dot{\kappa}_y - 3\alpha\gamma\dot{\kappa}_{xy}\}/\dot{\lambda}_o$$
$$+ E_h\,\alpha(\alpha\kappa_x + \beta\kappa_y + 2\gamma\kappa_{xy})\Big] \tag{6.2.15}$$

$$\sigma_y = \sigma_y^{\circ} + z\Big[\{(1 - 3\alpha\beta/2)\dot{\kappa}_x + (2 - 3\beta^2/2)\dot{\kappa}_y - 3\beta\gamma\dot{\kappa}_{xy}\}/\dot{\lambda}_o$$
$$+ E_h\,\beta(\alpha\kappa_x + \beta\kappa_y + 2\gamma\kappa_{xy})\Big] \tag{6.2.16}$$

$$\tau_{xy} = \tau_{xy}^{\circ} + z\Big[\{-(3\alpha\gamma/2)\dot{\kappa}_x - (3\beta\gamma/2)\dot{\kappa}_y + (1 - 3\gamma^2)\dot{\kappa}_{xy}\}/\dot{\lambda}_o$$
$$+ E_h\,\gamma(\alpha\kappa + \beta\kappa_y + 2\gamma\kappa_{xy})\Big] \tag{6.2.17}$$

where

$$\alpha = \sigma_x^{\circ}/\sigma^{\circ} \qquad \beta = \sigma_y^{\circ}/\sigma^{\circ} \qquad \gamma = \tau_{xy}^{\circ}/\sigma^{\circ} \tag{6.2.18}$$

and

$$\dot{\lambda}_o = 3\dot{\epsilon}_o/2\sigma^{\circ} \tag{6.2.19}$$

From the yield surface relation (6.2.6) the constants are related by

$$\alpha^2 + \beta^2 - \alpha\beta + 3\gamma^2 = 1 \tag{6.2.20}$$

which means that only two of the three possible loading forces $N_x$, $N_y$, and $N_{xy}$ are independent when plastic straining is occurring in all three components. The loading is specified by two of the constants $\alpha$, $\beta$, and $\gamma$ of (6.2.18) and the flow rule (6.2.8) requires the corresponding strain rate ratios to be

$$\dot{\epsilon}_x^{\circ}/\dot{\epsilon}_o = (2\alpha - \beta)/2 \qquad \dot{\epsilon}_y^{\circ}/\dot{\epsilon}_o = (2\beta - \alpha)/2$$

$$\dot{\epsilon}_{xy}^{\circ}/\dot{\epsilon}_o = 3\gamma/2 \tag{6.2.21}$$

We still have to specify the loading rate through the midsurface equivalent strain rate, $\dot{\epsilon}_o$, before the loading specification is complete. Another way of looking at loading rate specification is to specify one of the strain rate components while keeping their ratios constant, in accordance with (6.2.21). This procedure determines the equivalent strain rate, $\dot{\epsilon}_o$. Time integration determines $\epsilon_o$ and hence, by (6.2.12), the equivalent stress $\sigma^{\circ}$. Thus the $\dot{\lambda}_o$ of (6.2.19) is determined.

The forces and moments in a plate of thickness, h, resulting from the stress distribution (6.2.15), (6.2.16), and (6.2.17) are

$$N_x = \sigma_x^{\circ} h \qquad N_y = \sigma_y^{\circ} h \qquad N_{xy} = \tau_{xy}^{\circ} h \tag{6.2.22}$$

$$M_x = \left[\{(2 - 3\alpha^2/2)\dot{\kappa}_x + (1 - 3\alpha\beta/2)\dot{\kappa}_y - 3\alpha\gamma\dot{\kappa}_{xy}\}/\dot{\lambda}_o + E_h\,\alpha(\alpha\kappa_x + \beta\kappa_y + 2\gamma\kappa_{xy})\right] h^3/12 \tag{6.2.23}$$

$$M_y = \left[\{(1 - 3\alpha\beta/2)\dot{\kappa}_x + (2 - 3\beta^2/2)\dot{\kappa}_y - 3\beta\gamma\dot{\kappa}_{xy}\}/\dot{\lambda}_o + E_h\,\alpha(\alpha\kappa_x + \beta\kappa_y + 2\gamma\kappa_{xy})\right] h^3/12 \tag{6.2.24}$$

$$M_{xy} = \left[\{-(3\alpha\gamma/2)\dot{\kappa}_x - (3\beta\gamma/2)\dot{\kappa}_y + (1 - 3\gamma^2)\dot{\kappa}_{xy}\}/\dot{\lambda}_o + E_h\,\gamma(\alpha\kappa_x + \beta\kappa_y + 2\gamma\kappa_{xy})\right] h^3/12 \tag{6.2.25}$$

Once again we see that each moment has two components, one directional and the other hardening.

## 6.3 GOVERNING EQUATION

### 6.3.1 General Loading

The equation of motion of a plate was derived in Section 3.5. Allowing for the deflection being the sum of the initial and buckling deflections, we have

$$\frac{\partial^2 M_x}{\partial x^2} + 2\frac{\partial^2 M_{xy}}{\partial x \partial y} + \frac{\partial^2 M_y}{\partial y^2}$$

$$+ \left(N_x \frac{\partial^2}{\partial x^2} + 2N_{xy}\frac{\partial^2}{\partial x \partial y} + N_y \frac{\partial^2}{\partial y^2}\right)(w + w_o) = \rho h \frac{\partial^2 w}{\partial t^2} \qquad (6.3.1)$$

After substituting expressions (6.2.22) through (6.2.25) for the forces and moments, with (6.2.2) giving the curvature and twist changes, (6.3.1) becomes

$$\left[(4 - 3\alpha^2)\frac{\partial^4 \dot{w}}{\partial x^4} - 12\alpha\gamma \frac{\partial^4 \dot{w}}{\partial x^3 \partial y} + 2(4 - 3\alpha\beta - 6\gamma^2)\frac{\partial^4 \dot{w}}{\partial x^2 \partial y^2}\right.$$

$$\left. - 12\beta\gamma \frac{\partial^4 \dot{w}}{\partial x \partial y^3} + (4 - 3\beta^2)\frac{\partial^4 \dot{w}}{\partial y^4}\right]\frac{h^2}{24\dot{\lambda}_o}$$

$$+\left[\alpha^2\frac{\partial^4 w}{\partial x^4} + 4\alpha\gamma\frac{\partial^4 w}{\partial x^3 \partial y} + 2(\alpha\beta + 2\gamma^2)\frac{\partial^4 w}{\partial x^2 \partial y^2} + 4\beta\gamma\frac{\partial^4 w}{\partial x \partial y^3} + \beta^2\frac{\partial^4 w}{\partial y^4}\right]\frac{h^2 E_h}{12}$$

$$-\left(\sigma_x^\circ \frac{\partial^2}{\partial x^2} + 2\tau_{xy}^\circ \frac{\partial^2}{\partial x \partial y} + \sigma_y^\circ \frac{\partial^2}{\partial y^2}\right)(w + w_o) + \rho \ddot{w} = 0 \qquad (6.3.2)$$

The constants $\alpha$, $\beta$, and $\gamma$ are given by the midsurface stress ratios (6.2.18).

### 6.3.2 Uniaxial Compression

To represent uniaxial compression, we take

$$\sigma_x^\circ = -\sigma^\circ \ (\sigma^\circ > 0), \qquad \sigma_y^\circ = 0, \qquad \tau_{xy}^\circ = 0$$

so that from (6.2.18)

$$\alpha = -1 \qquad \beta = 0 \qquad \gamma = 0$$

Consequently, the flow rule formulas (6.2.21) give the strain rates

$$\dot{\epsilon}_x^{\circ} = -\dot{\epsilon}_o \; (\dot{\epsilon}_o > 0) \; , \quad \dot{\epsilon}_y^{\circ} = \dot{\epsilon}_o/2 \; , \quad \dot{\epsilon}_{xy} = 0 \; .$$

and (6.2.19) gives the plastic flow parameter

$$\dot{\lambda}_o = 3\dot{\epsilon}_x^{\circ}/2\sigma^{\circ}{}_x$$

We specify $\dot{\epsilon}_x^{\circ}$, and hence $\dot{\epsilon}_o$, from which we obtain $\epsilon_o$. The linear strain-hardening law (6.2.12) then gives $\sigma^{\circ}$, and hence $\sigma_x^{\circ}$. Thus $\dot{\lambda}_o$ is determined.

The governing equation (6.3.2) now reduces to

$$\frac{\sigma_x^{\circ} h^2}{24\dot{\epsilon}_x}\left(\frac{\partial^4 \dot{w}}{\partial x^2} + 8\frac{\partial^4 \dot{w}}{\partial x^2 \partial y^2} + 4\frac{\partial^4 \dot{w}}{\partial y^4}\right)$$

$$+ \frac{h^2 E_h}{12}\frac{\partial^4 w}{\partial x^4} - \sigma_x^{\circ}\frac{\partial^2 w}{\partial x^2} + \rho\ddot{w} = \sigma_x^{\circ}\frac{\partial^2 w_o}{\partial x^2} \tag{6.3.3}$$

As an approximation we let the uniaxial stress $\sigma_x^{\circ}$ be a constant equal to the average value from the stress-strain curve up to the strain at which the deflections are to be evaluated. Also, we shall restrict our solutions to cases of constant strain rate. Thus, we let

$$\sigma_x^{\circ} = -\bar{\sigma} \; (\bar{\sigma} > 0) \qquad \dot{\epsilon}_x = -V/a \tag{6.3.4}$$

where V is rate of decrease of a plate of length a.

By introducing in Equation (6.3.3) the dimensionless quantities

$$u = w/h \qquad u_o = w_o/h \qquad \xi = x/a \qquad \eta = y/b$$

$$\tau = Vt/a \qquad \alpha^2 = h^2/12a^2 \qquad \beta = E_h/\bar{\sigma} \qquad \gamma = a/b$$

$$s = \bar{\sigma}\, 2\rho V^2 \tag{6.3.5}$$

in which b is the width of the plate, we obtain the equation governing uniaxial compression in the form

$$\ddot{u} + s\alpha^2 \left( \frac{\partial^4 \dot{u}}{\partial \xi^4} + 8\gamma^2 \frac{\partial^4 \dot{u}}{\partial \xi^2 \partial \eta^2} + 4\gamma^4 \frac{\partial^4 \dot{u}}{\partial \eta^4} \right) + 2s\alpha^2 \beta \frac{\partial^4 u}{\partial \xi^4} + 2s \frac{\partial^2 u}{\partial \xi^2} = -2s \frac{\partial^2 u_o}{\partial \xi^2}$$

$$(\ )^{\cdot} \equiv \partial/\partial\tau \qquad (6.3.6)$$

The first term represents inertial resistance to buckling, the second group involving $\dot{u}$ represents the directional moment resistance, the third term represents the strain-hardening moment resistance, the fourth term represents the axial thrust causing buckling, and the final term represents the influence of the initial deflection imperfections.

## 6.4 UNIAXIAL COMPRESSION OF SIMPLY SUPPORTED PLATES

### 6.4.1 Modal Solution

In terms of the dimensionless space variables $\xi$, $\eta$, u (replacing x, y, w) the boundary conditions for a simply supported rectangular plate are

$$u = 0 \qquad \frac{\partial^2 u}{\partial \xi^2} = 0 \qquad \xi = 0,1$$

$$u = 0 \qquad \frac{\partial^2 u}{\partial \eta^2} = 0 \qquad \eta = 0,1 \qquad (6.4.1)$$

The initial plate deflection and transverse velocity may be represented by

$$u_o(\xi, \eta) = \sum_m \sum_n a_{mn} \sin m\pi\xi \, . \sin n\pi\eta \qquad (6.4.2)$$

$$\dot{u}(\xi, \eta, 0) = \sum_m \sum_n b_{mn} \sin m\pi\xi \, . \sin n\pi\eta \qquad (6.4.3)$$

Physically, according to the dimensionless groups (6.3.5), the amplitudes $a_{mn}$ and $b_{mn}$ are fractions of the thickness h and the quantity hV/a, respectively.

For the deflection representation, we let

$$u(\xi, \eta, \tau) = \sum_m \sum_n u_{mn}(\tau) \, . \sin m\pi\xi \ \sin n\pi\eta \qquad (6.4.4)$$

in which the coefficients satisfy the initial conditions

$$u_{mn}(0) = 0 \qquad \dot{u}_{mn}(0) = b_{mn} \tag{6.4.5}$$

Substitution of (6.4.2) and (6.4.4) in the governing equation (6.3.6) leads to the requirement that

$$\ddot{u}_{mn} + Q_{mn}\,\dot{u}_{mn} - R_m^2\,u_{mn} = S_m\,a_{mn} \tag{6.4.6}$$

where the coefficients are

$$Q_{mn} = s\alpha^2\left[(m\pi)^4 + 8\gamma^2(m\pi)^2(n\pi)^2 + 4\gamma^4(n\pi)^4\right] \tag{6.4.7}$$

$$R_m^2 = 2s\,(m\pi)^2\left[1 - \alpha^2\beta\,(m\pi)^2\right] \tag{6.4.8}$$

$$S_m = 2s\,(m\pi)^2 \tag{6.4.9}$$

The solution of (6.4.6) satisfying initial conditions (6.4.5) is

$$u_{mn}(\tau) = A_{mn}(\tau)\,a_{mn} + B_{mn}(\tau)\,b_{mn} \tag{6.4.10}$$

where

$$A_{mn} = \left[(\lambda_{mn}^{+}e^{\lambda_{mn}^{-}\tau} - \lambda_{mn}^{-}e^{\lambda_{mn}^{+}\tau})/(\lambda_{mn}^{+} - \lambda_{mn}^{-}) - 1\right](1 + S_m/R_m^2) \tag{6.4.11}$$

$$B_{mn} = (e^{\lambda_{mn}^{+}\tau} - e^{\lambda_{mn}^{-}\tau})/(\lambda_{mn}^{+} - \lambda_{mn}^{-}) \tag{6.4.12}$$

$$\lambda_{mn}^{\pm} = \left[\pm\,(Q_{mn}^2 + 4R_m^2)^{1/2} - Q_{mn}\right]/2 \tag{6.4.13}$$

The solution (6.4.10) shows that each coefficient in the modal representation (6.4.4) is obtained by an amplification of the corresponding initial displacement and velocity coefficients. The magnification factors or amplification functions $A_{mn}$ and $B_{mn}$, given by (6.4.11) and (6.4.12), contain exponential terms that suggest growth with time. From (6.4.13), only $\lambda_{mn}^{+}$ can be positive to provide exponential amplification and $\lambda_{mn}^{+}$ is positive if $R_m^2 > 0$, that is, if $m < 1/\pi\alpha\sqrt{\beta}$. Thus $\overline{m} = 1/\pi\alpha\sqrt{\beta}$ may be considered as a cutoff mode number in that only modes with $m < \overline{m}$ are amplified.

The partial derivative of $\lambda_{mn}^{+}$ with respect to $(n\pi)^2$ is negative for all positive values of n. Hence the value of n that maximizes $\lambda_{mn}^{+}$ is n = 1, and we may rewrite the solution in the form

$$u(\xi,\eta,\tau) = \sin\pi\eta \sum_{m} u_m(\tau) \sin m\pi\xi \tag{6.4.14}$$

where the coefficients satisfy the equation

$$\ddot{u}_m + Q_m \dot{u}_m - R_m^2 u_m = S_m \tag{6.4.15}$$

with

$$Q_m = s\alpha^2\left[(m\pi)^4 + 8(\gamma\pi)^2(m\pi)^2 + 4(\gamma\pi)^4\right] \tag{6.4.16}$$

$$R_m^2 = 2s(m\pi)^2\left[1 - \alpha^2\beta(m\pi)^2\right] \tag{6.4.17}$$

$$S_m = 2s(m\pi)^2 \tag{6.4.18}$$

The solution of (6.4.15) satisfying the initial conditions

$$u_m(0) = 0 \qquad \dot{u}_m(0) = b_m \tag{6.4.19}$$

is

$$u_m(\tau) = A_m(\tau)\, a_m + B_m(\tau)\, b_m \tag{6.4.20}$$

where

$$A_m = \left[(\lambda_m^+ e^{\lambda_m^- \tau} - \lambda_m^- e^{\lambda_m^+ \tau})/ (\lambda_m^+ - \lambda_m^-) - 1\right](1 + S_m/R_m^2) \tag{6.4.21}$$

$$B_m = (e^{\lambda_m^+ \tau} - e^{\lambda_m^- \tau})/ (\lambda_m^+ - \lambda_m^-) \tag{6.4.22}$$

$$\lambda_m^{\pm} = \left[\pm (Q_m^2 + 4R_m^2)^{1/2} - Q_m\right]/2 \tag{6.4.23}$$

### 6.4.2 Amplification Functions

The dependence on the mode number m at a given time of the amplification functions (6.4.21) and (6.4.22) is illustrated by plate LAC-2 of the experiments described in Section 6.6. The data for LAC-2 are as follows:

| | | | |
|---|---|---|---|
| length | $a$ | = | 5 inches (12.70 cm) |
| width | $b$ | = | ¾ inch (1.91 cm) |
| thickness | $h$ | = | 1/16 inch (0.16 cm) |
| density | $\rho$ | = | 0.097 lb/in.$^3$ (2.7 g/cm$^3$) |
| impact velocity | $V$ | = | 310 ft/s (94.45 m/s) |
| yield stress | $\sigma$ | = | 30,000 psi (207 MPa) |
| hardening modulus | $E_h$ | = | 48,300 psi (333 MPa) |

The corresponding values of the dimensionless parameters are

$$\alpha^2 = h^2/12a^2 = 1/76800 \quad \beta = E_h/\bar{\sigma} = 1.61 \qquad \gamma = a/b = 6.67$$

$$s = \bar{\sigma}/2\rho V^2 = 4.29 \qquad \tau_f = 0.16$$

in which $\tau_f$ is the fractional shortening of the plate caused by uniaxial compression. The parameter values are substituted in (6.4.16) through (6.4.18), leaving $Q_m$, $R_m$, and $S_m$, and hence $\lambda_m^+$ and $\lambda_m^-$, as functions only of the mode number m. The resulting amplification functions are shown in Figures 6.2 and 6.3. The development and shape of the curves are typical of plastic flow buckling showing the band of harmonics with the largest amplification. The value of the preferred mode number, corresponding to the maximum amplification, remains the same during the second half of the motion.

### 6.4.3 Preferred Mode and Critical Velocity Formulas

In many cases, including the experiments discussed in Section 6.6, the properties and conditions are such that $Q_m >> 2R_m$ when m is near the preferred mode number. In these cases we may approximate the final value of the amplification function by

$$A_m(\tau_f) = e^{R_m^2 \tau_f / Q_m} \tag{6.4.24}$$

Differentiation of (6.4.24) with respect to $m^2$ and equating the result to zero leads to the equation

$$\left[1 + 8\alpha^2\beta(\gamma\pi)^2\right](m\pi)^4 + 8\alpha^2\beta(\gamma\pi)^2(m\pi)^2 - 4(\gamma\pi)^4 = 0 \tag{6.4.25}$$

for the preferred mode. This result may be simplified even further for cases in which $\alpha^2\beta(\gamma\pi)^2 < 2$ or, with physical quantities, $E_h/\bar{\sigma} < 2(b/h)^2$. In these cases,

$$m^2 = 2\gamma^2\left[\{1 + 8\alpha^2\beta(\gamma\pi)^2\}^{1/2} - 2\alpha^2\beta(\gamma\pi)^2\right] \qquad (6.4.26)$$

Finally, if $8\alpha^2\beta(\gamma\pi)^2$ is small enough to allow binomial approximations, we have

$$m^2 \approx 2\gamma^2\left[1 + 2\alpha^2\beta(\gamma\pi)^2\right] \approx 2\gamma^2 \qquad (6.4.27)$$

Let us examine briefly the conditions we have introduced. The condition $Q_m^2 >> 4R_m^2$, assuming that the preferred mode is comparable to the value of $\gamma =$

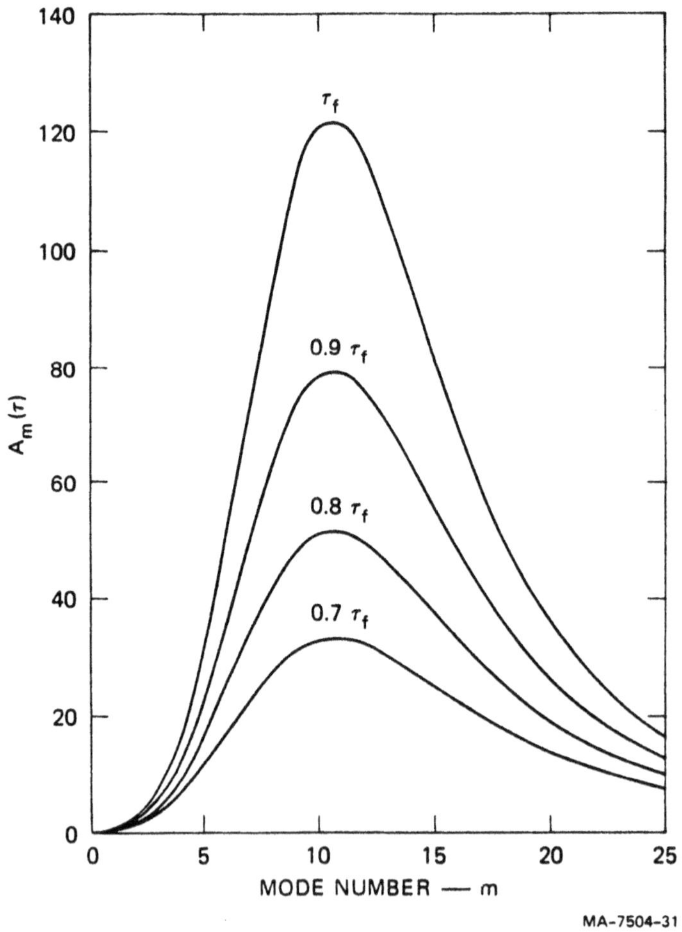


FIGURE 6.2 AMPLIFICATION FUNCTIONS FROM INITIAL DEFLECTIONS

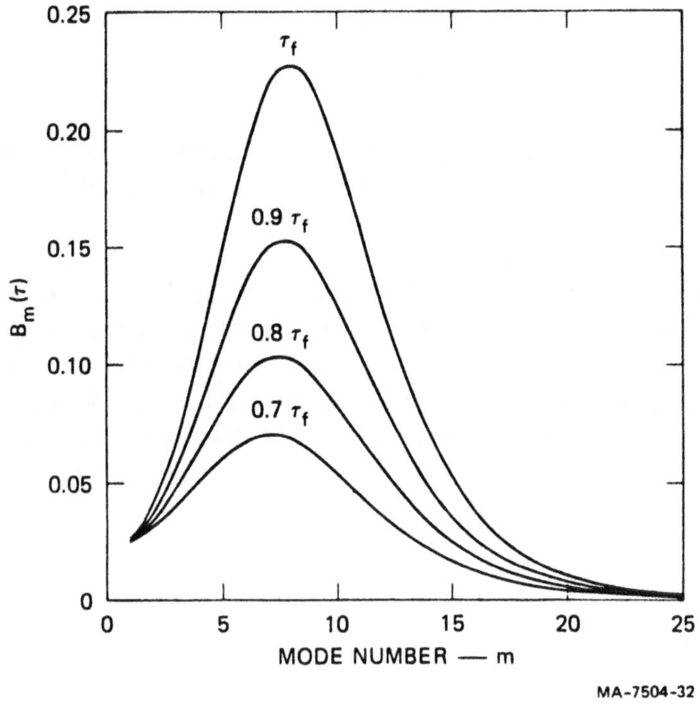


FIGURE 6.3 AMPLIFICATION FUNCTIONS FROM INITIAL VELOCITIES

a/b, is satisfied if $\pi^4(\bar{\sigma}/\rho V^2)(h/b)^4(a/b)^2 >> 1$. If we consider a plate between two rigid walls that approach each other at velocity V satisfying $V/c << \pi^2(h/b)^2(a/b)/\beta^{1/2}$, where $c = (E_h/\rho)^{1/2}$, the preferred mode by (6.4.25), is independent of V. For plate SAC-4 of the experiments, we require $V << 5600$ ft/s (1707 m/s). In the experiment, V = 578 ft/s (176 m/s) produced a shortening of 29%, so we predict that the preferred mode is independent of impact velocity for this plate geometry and material.

The next condition, $E_h/\bar{\sigma} < 2(b/h)^2$, also applies to the experimental plates. For SAC-4, we have b/h = 4 and $\bar{\sigma}$ = 30,000 psi (207 MPa), so if the average hardening modulus is such that $E_h <$ 960,000 psi (6614 MPa), we can employ (6.4.26) to predict the preferred mode. The final condition in physical terms is that $(2\pi^2/3)(h/b)^2\beta$ should be small enough compared with unity to allow binomial approximations. For plate SAC-4, which has the highest value of the thickness-to-width ratio of all the plates tested, we obtain 0.66. Formula (6.4.26) gives $m = 1.06\sqrt{2}\gamma = 15$, whereas (6.4.27) gives $m = \sqrt{2}\gamma = 14$.

If we approximate the preferred mode number by $m = \sqrt{2}\gamma$ and substitute this value in (6.4.16) and (6.4.17) for $Q_m$ and $R_m^2$, we obtain from (6.4.24) the corresponding amplification function

$$A_m(\tau_f) = e^{\tau_f/6\alpha^2(\gamma\pi)^2} \qquad (6.4.28)$$

where $\tau_f = Vt_f/a$. We are now in a position to obtain an approximate relationship between V, the constant rate of plate shortening, and the amplification function for the preferred mode, at time $t_f$.

Let the physical event correspond to the experiments of Section 6.6, as depicted in Figure 6.4. The plate with a mass M at the rear impacts a rigid surface at normal incidence with a velocity V. Because for simplicity the theory assumes a constant velocity throughout the shortening $\Delta a$, we take the duration of shortening to be $t_f = \Delta a/V$; therefore, $\tau_f = \Delta a/a$ is the final value of the strain $\epsilon_x$. We now equate the plastic work to the initial kinetic energy of the mass to obtain

$$abh\bar{\sigma}\tau_f = MV^2/2 \qquad (6.4.29)$$

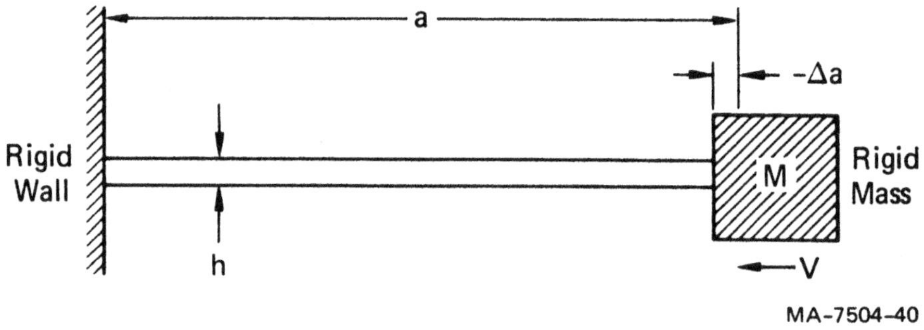


FIGURE 6.4 PLATE UNDERGOING COMPRESSION

Introducing in (6.4.29) the plate mass and the mass ratio

$$m = \rho abh \qquad \mu = M/m \qquad (6.4.30)$$

gives

$$\tau_f = \rho\mu V^2/2\bar{\sigma} \qquad (6.4.31)$$

Substitution of $\tau_f$ from (6.4.31) into (6.4.28) then leads to the required relationship between impact velocity and amplification function

$$V^2 = (2\bar{\sigma}/\rho\mu)\, 6\alpha^2(\gamma\pi)^2 \ln A_m$$

$$= \pi^2(\bar{\sigma}/\rho\mu)\,(h/b)^2 \ln A_m \qquad (6.4.32)$$

Formula (6.4.32) may be regarded as a formula for the critical impact velocity that produces the maximum acceptable amplification factor. For plate SAC-4, if we choose $A_m = 10$ the critical velocity is $V_c = 670$ ft/s (204 m/s). In the experiment, an impact velocity of 578 ft/s (176 m/s) caused moderately large buckling amplitudes.

### 6.4.4 Directional and Hardening Moments

When formula (6.2.23) for $M_x$ is specialized for application of uniaxial compression of simply supported plates, it becomes

$$M_x = \left[(\dot{\kappa}_x/2 + \dot{\kappa}_y)/\dot{\lambda}_o + E_h \kappa_x\right] h^3/12 \qquad (6.4.33)$$

where $\dot{\lambda}_o = 3V/2a\overline{\sigma}$, and the curvature changes are the derivatives (6.2.2). In terms of the dimensionless variables (6.3.5) the moment expression (6.4.33) is

$$M = -\left(\frac{\partial^2 \dot{u}}{\partial \xi^2} + 2\gamma^2 \frac{\partial^2 \dot{u}}{\partial \eta^2} + 3\beta \frac{\partial^2 u}{\partial \xi^2}\right) \alpha^2/3 \qquad (6.4.34)$$

where $M = M_x/\overline{\sigma}h^2$.

In Section 6.4.1 we showed that the preferred lateral mode number is $n = 1$, so we represent u by (6.4.14) and M by

$$M = \sin \pi \eta \sum_m M_m \sin m\pi\xi \qquad (6.4.35)$$

Thus, Equation (6.4.34) gives

$$M_m = \left[\{(m\pi)^2 + 2(\gamma\pi)^2\}\, \dot{u}_m + 3\beta (m\pi)^2 u_m\right] \alpha^2/3 \qquad (6.4.36)$$

The amplification function is $A_m$ given by (6.4.28) when we approximate the preferred mode number by $m = \sqrt{2}\gamma$. For displacement imperfections alone, $u_m = A_m a_m$ so (6.4.36) becomes

$$M_m = \left[2/3\alpha^2 + 6\beta\,(\gamma\pi)^2\right] \alpha^2 A_m a_m/3 \qquad (6.4.37)$$

giving as the ratio of the directional and hardening moments

$$\Lambda = M_m^{(d)}/M_m^{(h)} = 1/9\alpha^2\beta\,(\gamma\pi)^2 = (4/3\pi^2)\,(b/h)^2\overline{\sigma}/E_h \qquad (6.4.38)$$

Among the simply supported plates of the experiments, SAC-4 gives the lowest value of the ratio $\Lambda$ (1.34), which indicates that for this plate the directional moment is more important than the hardening moment. For SAC-1, 2, and 3, which have half the thickness of SAC-4, we obtain $\Lambda = 5.36$, so in these plates the directional moment dominates.

## 6.5 UNIAXIAL COMPRESSION OF UNSUPPORTED PLATES

In the CSC series of experiments, the plates were not supported along the edges. It was observed that the plate buckling deformations were independent of the transverse coordinate. Consequently, we eliminate the y coordinate from our theory to predict the preferred buckling mode and to obtain a critical velocity formula.

### 6.5.1 Governing Equation, Modal Solution, and Amplification Functions

If the dimensionless deflection u is taken to be a function of dimensionless length and time, $\xi$ and $\tau$, the governing equation (6.3.6) becomes

$$\ddot{u} + s\alpha^2\dot{u}'''' + 2s\alpha^2\beta u'''' + 2su'' = -2su_o'' \qquad (6.5.1)$$

$$(\;) \equiv \partial/\partial\tau \qquad (\;)' \equiv \partial/\partial\xi$$

where the dimensionless quantities are those of (6.3.5). The boundary conditions are

$$u = 0 \quad \text{and} \quad u'' = 0 \quad \text{at} \quad \xi = 0,\ 1 \qquad (6.5.2)$$

The initial plate deflection and transverse velocity may be represented by

$$u_o(\xi) = \sum_m a_m \sin m\tau\xi \qquad (6.5.3)$$

$$\dot{u}(\xi, 0) = \sum_m b_m \sin m\tau\xi \qquad (6.5.4)$$

The buckling deflection may be represented by

$$u(\xi,\tau) = \sum_{m} u_m(\tau) \sin m\pi\xi \tag{6.5.5}$$

in which the coefficients satisfy the initial conditions

$$u_m(0) = 0 \qquad \dot{u}_m(0) = b_m \tag{6.5.6}$$

Substitution of (6.5.3) and (6.5.5) in the governing equation (6.5.1) leads to the requirement that

$$\ddot{u}_m + Q_m\dot{u}_m - R_m^2 u_m = S_m a_m \tag{6.5.7}$$

where the coefficients are

$$Q_m = s\alpha^2(m\pi)^4 \tag{6.5.8}$$

$$R_m^2 = 2s(m\pi)^2[1 - \alpha^2\beta(m\pi)^2] \tag{6.5.9}$$

$$S_m = 2s(m\pi)^2 \tag{6.5.10}$$

The solution of (6.5.7) satisfying the initial conditions (6.5.6) is

$$u_m(\tau) = A_m(\tau)\, a_m + B_m(\tau)\, b_m \tag{6.5.11}$$

where the amplification functions are given by

$$A_m = \left[(\lambda_m^+ e^{\lambda_m^- \tau} - \lambda_m^- e^{\lambda_m^+ \tau}) / (\lambda_m^+ - \lambda_m^-) - 1\right](1 + S_m/R_m^2) \tag{6.5.12}$$

$$B_m = (e^{\lambda_m^+ \tau} - e^{\lambda_m^- \tau}) \;/\; (\lambda_m^+ - \lambda_m^-) \tag{6.5.13}$$

$$\lambda_m^{\pm} = \left[\pm (Q_m^2 + 4R_m^2)^{1/2} - Q_m\right]/2 \tag{6.5.14}$$

### 6.5.2 Preferred Mode and Critical Velocity Formulas

In formulas (6.5.12) and (6.5.13) for the amplification functions, the dominant factor is the exponential term with the positive power $\lambda_m^+\tau$. We can therefore base a prediction for the preferred mode on the value of m that maximizes $\lambda_m^+$. From (6.5.14), (6.5.8), and (6.5.9), $\lambda_m^+$ is maximized by the value of m satisfying

$$s\alpha^4(m\pi)^6 - \left[1 - 2\alpha^2\beta(m\pi)^2\right]^2 = 0 \qquad (6.5.15)$$

In many cases, as in the CSC series of experiments, we may neglect $2\alpha^2\beta(m\pi)^2$ in (6.5.15) to obtain

$$m = 1/\pi\,(s\alpha^4)^{1/6} \qquad (6.5.16)$$

for the preferred mode number.

If (6.5.16) is substituted in formula (6.5.12) for the amplification function $A_m$ we obtain

$$A_m(\tau_f) \approx (4/3)\,e^{(s/\alpha^2)^{1/3}\tau_f} \qquad (6.5.17)$$

for its value at the final compressive strain $\tau_f$.

We again consider the plate being compressed between a rigid wall and a mass M that approaches the wall at a constant velocity V, as shown in Figure 6.4. Substitution of the relationship (6.4.31) for the final strain in (6.5.17) leads to the relationship between the impact velocity and the amplification function

$$V^2 = (4\bar{\sigma}/\rho\mu^{3/2})\,\alpha\,[\ln(4A_m/3)]^{3/2} \qquad (6.5.18)$$

Formula (6.5.18) may be regarded as a formula for the critical impact velocity that produces the maximum acceptable amplification factor. For plates 4CSC-1, 2, if we choose $A_m = 10$ the critical velocity is $V_c = 70$ ft/s (21 m/s). In the experiments, the results of Figure 6.5 show that distinct amplitudes occurred at $V = 59$ ft/s (18 m/s) and that excessively large amplitudes occured at $V = 100$ ft/s (30 m/s), indicating that the critical velocity formula provides reasonable predictions.

### 6.5.3 Directional and Hardening Moments

If we follow the procedure of Section 6.4.4 for comparing the magnitudes of

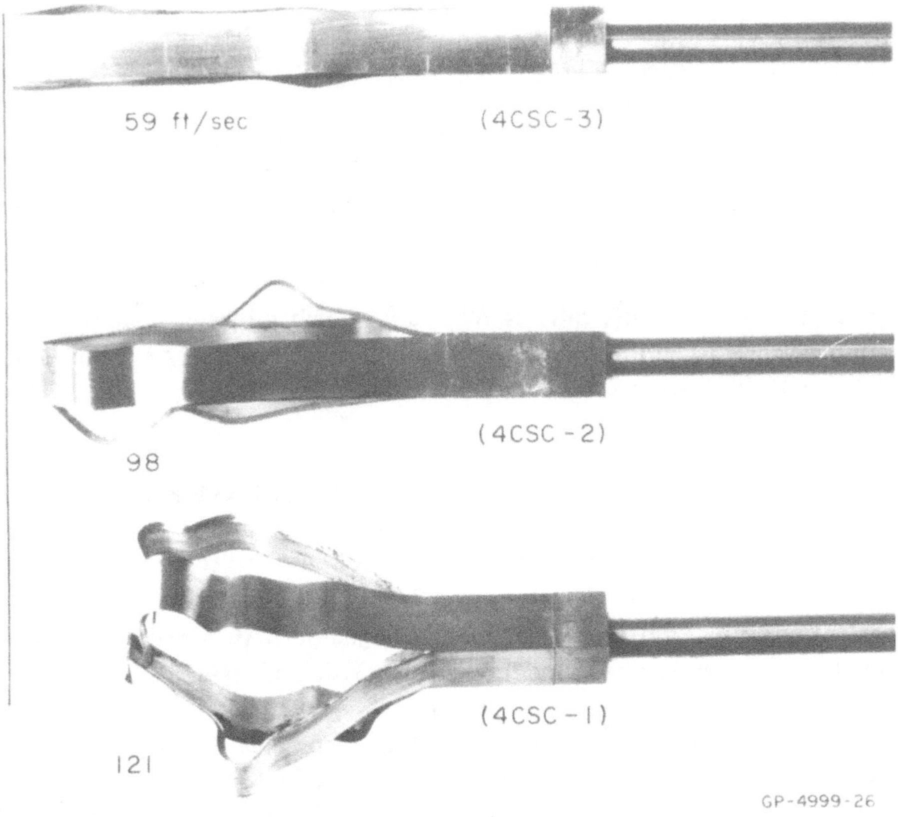


FIGURE 6.5 PLASTIC BUCKLING OF RECTANGULAR PLATES CAUSED BY INPLANE IMPACT

the directional and hardening moments, we find that the dimensionless moment is

$$M = -(\dot{u} + 3\beta u)''\alpha^2/3 \tag{6.5.19}$$

which is (6.4.34) when u is independent of $\eta$. When M is represented by

$$M = \sum_m M_m \sin m\pi\xi \tag{6.5.20}$$

and u is represented by (6.5.5) we have

$$M_m = (\dot{u}_m + 3\beta u_m)(m\pi)^2\alpha^2/3 \tag{6.5.21}$$

If we let m be the preferred mode number given approximately by (6.5.16) and confine our investigation to the amplification function $A_m$ given approximately by (6.5.17), we obtain from (6.5.21)

$$M_m = \left[(s/\alpha^2)^{1/3} + 3\beta\right] u_m (m\pi)^2 \alpha^2/3 \qquad (6.5.22)$$

Hence (6.5.22) gives the ratio of the directional and hardening moments as

$$\Lambda = M_m^{(d)}/M_m^{(h)} = (s/\alpha^2)^{1/3}/3\beta \qquad (6.5.23)$$

For plate 4CSC-3 this ratio is $\Lambda = 5.9$, which indicates that the directional moment is more important than the hardening moment. The values of $\Lambda$ for 4CSC-2 and 4CSC-1 are 9.8 and 8.9.

## 6.6 COMPARISON OF THEORETICAL AND EXPERIMENTAL RESULTS

Square tubes of aluminum alloy 6063-T5 were projected against a massive steel target slab as described in Section 5.2.9 for circular tubes. Figure 6.1 shows an undeformed tube and three specimens designated LAC-1, -2, and -3 that were subjected to axial impact at velocities of 184, 310, and 344 ft/s (56, 94, and 105 m/s). Buckling is evident in LAC-2 and LAC-3. Each side of the tube is regarded as a rectangular plate. In the theory, the edges are regarded as simply supported because rotation occurred there. The transverse mode is a single halfwave ($n = 1$), and several waves appeared along the length. The stress-strain relationship was obtained from standard tensile tests, the results from which are shown in Figure 6.6

Table 6.1 lists the data for 12 plate experiments. The last three specimens listed (4CSC-3, -2, -1) were from tubes of square cross section but with cuts along the edges to provide free supports. Figure 6.5 shows the buckled state of these plates caused by impact.

Table 6.2 compares the predicted and experimentally obtained preferred mode of buckling. The numbers refer to the number of halfwaves along the length of the plate. In the column under $A_m(\tau_f)$, the numbers were obtained from the maxima of formula (6.4.21) when $\tau = \tau_f$. These are the mode numbers based on the growth of the initial deflection imperfections. Similarly, the numbers in the column headed $B_m(\tau_f)$ were obtained from the maxima of formula (6.4.22) when $\tau = \tau_f$. These predictions are based on buckling arising from initial velocity distribution. The column headed formulas gives the preferred mode numbers predicted by (6.4.27) for simply supported plates or by (6.5.16) for plates with free edges, both for initial deflection imperfections. The final column gives the numbers obtained in the experiments.

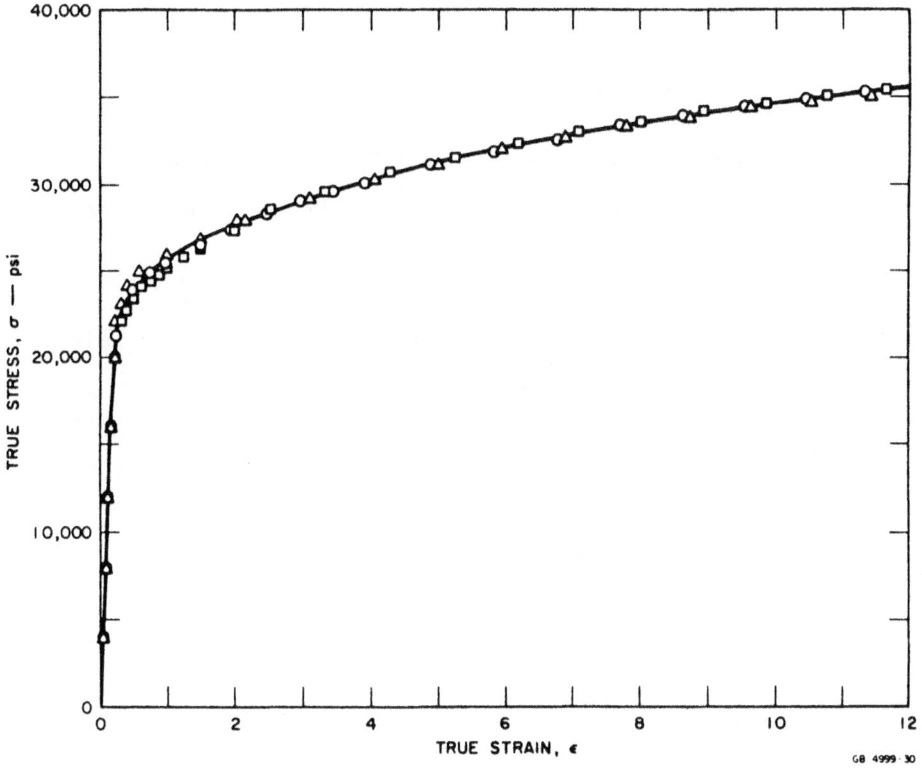


FIGURE 6.6    STRESS-STRAIN CURVE FOR 6063-T5 ALUMINUM ALLOY

A comparison of the predicted and experimental mode numbers in Table 6.2 indicates that the theory employing the growth of initial deflection imperfections leads to satisfactory predictions. For the simply supported plates, the predictions based on the influence of the initial transverse velocity distribution are low, especially for the plates in which the buckling is not well developed. In these cases the $B_m$ vs m curves are still fairly flat over a wide range of m; therefore, selecting the maximum to give the preferred mode is not necessarily realistic. The predictions improve for plates with well developed buckling (for example, see the trend for SAC-3, -2, and -1). For the free supported plates (4CSC-3, -2, -1), the predictions based on the influence of the initial transverse velocity distribution are satisfactory.

## 6.7 SLOW BUCKLING

When the impact velocity is small the inertia terms in Equation (6.3.6) can be neglected; the parameter s, which contains the velocity, then cancels throughout. Equation (6.4.15) for the modal amplitudes becomes

$$Q_m \dot{u}_m - R_m^2 u_m = S_m a_m$$

Table 6.1

EXPERIMENTAL DATA FOR RECTANGULAR PLATES

| Specimen* | Width b (cm) | Thickness h (cm) | Yield $\bar{\sigma}$ (MPa)§ | Modulus† $E_h$ (MPa) | Velocity V (m/s) | Shortening $\Delta a/a$ (%) | Mode‡ m |
|---|---|---|---|---|---|---|---|
| SAC-3 | 1.27 | 0.159 | 207 | 333 | 61 | 10 | 19 |
| SAC-2 | 1.27 | 0.159 | 207 | 333 | 91 | 23 | 19 |
| SAC-1 | 1.27 | 0.159 | 207 | 333 | 122 | 36 | 16 |
| LAC-1 | 1.91 | 0.159 | 198 | 404 | 56 | 7 | 13 |
| LAC-2 | 1.91 | 0.159 | 207 | 333 | 94 | 16 | 12 |
| LAC-3 | 1.91 | 0.159 | 207 | 333 | 105 | 30 | 14 |
| LAC-4 | 1.91 | 0.318 | 207 | 333 | 115 | 15 | 14 |
| LAC-5 | 1.91 | 0.318 | 207 | 333 | 140 | 20 | 10 |
| SAC-4 | 1.27 | 0.318 | 207 | 333 | 176 | 29 | 13 |
| 4CSC-3 | 1.91 | 0.159 | 164 | 1792 | 18 | 1 | 8 |
| 4CSC-2 | 1.91 | 0.159 | 177 | 839 | 30 | 3 | 10 |
| 4CSC-1 | 1.91 | 0.159 | 177 | 839 | 35 | 3 | 10 |

* All plates of length a = 12.7 cm (5 inches).

† Value at $\epsilon$ = 10% when $\epsilon$ > 10%.

‡ Number of halfwaves.

§ 0.1 MPa = 14.5 psi = 1 bar.

Table 6.1 (concluded)

EXPERIMENTAL DATA FOR RECTANGULAR PLATES

| Specimen* | Width b (inch) | Thickness h (inch) | Yield $\bar{\sigma}$ (psi) | Modulus† $E_h$ (psi) | Velocity V (ft/s) | Shortening $\Delta a/a$ (%) | Mode‡ m |
|---|---|---|---|---|---|---|---|
| SAC-3 | 1/2 | 1/16 | 30,000 | 48,300 | 199 | 10 | 19 |
| SAC-2 | 1/2 | 1/16 | 30,000 | 48,300 | 300 | 23 | 19 |
| SAC-1 | 1/2 | 1/16 | 30,000 | 48,300 | 400 | 36 | 16 |
| LAC-1 | 3/4 | 1/16 | 28,700 | 58,600 | 184 | 7 | 13 |
| LAC-2 | 3/4 | 1/16 | 30,000 | 48,300 | 310 | 16 | 12 |
| LAC-3 | 3/4 | 1/16 | 30,000 | 48,300 | 344 | 30 | 14 |
| LAC-4 | 3/4 | 1/8 | 30,000 | 48,300 | 377 | 15 | 14 |
| LAC-5 | 3/4 | 1/8 | 30,000 | 48,300 | 460 | 20 | 10 |
| SAC-4 | 1/2 | 1/8 | 30,000 | 48,300 | 578 | 29 | 13 |
| 4CSC-3 | 3/4 | 1/16 | 23,800 | 260,000 | 59 | 1 | 8 |
| 4CSC-2 | 3/4 | 1/16 | 25,700 | 121,700 | 100 | 3 | 10 |
| 4CSC-1 | 3/4 | 1/16 | 25,700 | 121,700 | 115 | 3 | 10 |

* All plates of length a = 5 inches.

† Value at $\epsilon$ = 10% when $\epsilon$ > 10%.

‡ Number of halfwaves.

Table 6.2

THEORETICAL AND EXPERIMENTAL MODES

| Specimen | $A_m(\tau_f)$ | $B_m(\tau_f)$ | Formulas | Experiment |
|---|---|---|---|---|
| | Preferred Mode Numbers | | | |
| SAC-3 | 14 | 6 | 14 | 19 |
| SAC-2 | 14 | 9 | 14 | 19 |
| SAC-1 | 14 | 11 | 14 | 16 |
| LAC-1 | 10 | 5 | 9 | 13 |
| LAC-2 | 11 | 8 | 9 | 12 |
| LAC-3 | 11 | 9 | 9 | 14 |
| LAC-4 | 9 | 3 | 9 | 14 |
| LAC-5 | 9 | 4 | 9 | 10 |
| SAC-4 | 12 | 4 | 14 | 13 |
| 4CSC-3 | 7 | 5 | 6 | 8 |
| 4CSC-2 | 8 | 6 | 7 | 10 |
| 4CSC-1 | 8 | 7 | 8 | 10 |

where

$$Q_m = \alpha^2\left[(m\pi)^4 + 8(\gamma\pi)^2(m\pi)^2 + 4(\gamma\pi)^4\right]$$

$$R_m^2 = 2(m\pi)^2\left[1 - \alpha^2\beta(m\pi)^2\right]$$

$$S_m = 2\,(m\pi)^2$$

which determines the amplification factor of the initial deflection imperfections

$$A_m(\tau) = (e^{R_m^2\tau/Q_m} - 1)\,(S_m/R_m^2)$$

If we choose the preferred mode number as the value of m that maximizes the ratio $R_m^2/Q_m$, we obtain formulas (6.4.25), (6.4.26), and (6.4.27) according to relative magnitudes of properties occurring in the formulas. Thus we obtain the same buckling modes as those for a large class of dynamic cases, that is, for cases where $Q_m >> 2R_m$, the $Q_m$ and $R_m^2$ being (6.4.16) and (6.4.17).

## REFERENCE

1. J.N. Goodier, "Dynamic Buckling of Rectangular Plates and Sustained Plastic Compressive Flow," in *Engineering Plasticity,* Cambridge University Press, 1968, Proceedings of an International Conference of Plasticity held in Cambridge, England, March 1968.

# BIBLIOGRAPHY


1. A. Marston, "Correspondence on the Theory of the Ideal Column," ASCE Trans., **39**, pp. 108-120, 1898.

2. C. Jensen, "Quebec Bridge Disaster," Engineering, **85**, London, pp. 433-434, 1908.

3. W.E. Lilly, "The Strength of Columns," ASCE Trans., **76**, pp. 258-274, 1913.

4. R.V. Southwell, "On the Analysis of Experimental Observations and Problems of Elastic Stability," Proc. Royal Soc., London, Series A, **135**, pp. 601-616, 1932.

5. H. Grauers, "Buckling of Straight Rods by Impact," Arkiv för Matematik, Astronomi och Fysic (Stockholm), **22**, 26, pp. 1-19, 1932.

6. J. Taub and C. Koning *Impact Buckling of Thin Bars in the Elastic Range Hinged at Both Ends,* NACA TM 748, 1934.

7. J. Taub, *Impact Buckling of Thin Bars in the Elastic Range for Any End Condition,* NACA TM 749, 1934.

8. Th. von Kármon, L.G. Dunn, and H.S. Tsien, "The Influence of Curvature on the Buckling Characteristic of Structures," J. Aerospace Sci., **7**, pp. 276-289, 1940.

9. N.W. McLachlan, *Theory and Application of Mathieu Functions,* University Press, Oxford, 1941.

10. R.V. Churchill, *Fourier Series and Boundary Value Problems,* McGraw-Hill Book Co., New York, pp. 85-90, 1941.

11. M.J. Manjoine, "Influence of Rate of Strain and Temperature on Yield Stresses in Mild Steel," ASME Trans., **66**, Series E, J. Appl. Mech., **11**, pp. 211-218, 1944.

12. J.H. Meier, "On the Dynamics of Elastic Buckling," J. Aero. Sci., **12**, pp. 433-440, 1945.

13. T. Davidson and J.H. Meier, "Impact on Prismatical Bars," Proc. Soc. Exptl. Stress Anal., **4**, 1, pp. 88-11, 1946.

14. F.R. Shanley, "Inelastic Column Theory," J. Aero. Sci., **14**, 5, pp. 261-268, May 1947.

15. S.B. Batdorf, *A Simplified Method of Elastic Stability Analysis for Thin Cylindrical Shells,* NACA Rep. No. 874, U.S. Govt. Printing Off., Washington, DC, 1947.

16. M.A. Lavrentiev and A.V. Ishlinski, "Dynamic Forms of Loss of Stability of Elastic Systems," Doklady Akademiya Nauk, USSR, **64**, pp. 779-782, 1949.

17. J.J. Stoker, *Nonlinear Vibration,* Interscience Pub., Inc., New York, 1950.

18. A. Nadai, *Theory of Flow and Fracture of Solids,* Vol. 1, 2nd ed., McGraw-Hill Book Co., Inc., New York, p. 313, 1950.

19. N.J. Hoff, "The Dynamics of the Buckling of Elastic Columns," ASME Trans., **73**, Series E, J. Appl. Mech., **18**, 1, pp. 68-74, March 1951.

20. G. Gerard and H. Becker, "Column Behavior Under Conditions of Impact," J. Aero. Sci., **19**, pp. 58-65, 1952.

21. J.F. Davidson, "Buckling of Struts Under Dynamic Loading," J. Mech. Phys. Solids, **2**, pp. 54-56, 1953.

22. S.O. Rice, "Mathematical Analysis of Random Noise," in N. Wax, *Selected Papers on Noise and Stochastic Processes,* Dover Publications, New York, 1954.

23. Y. Yoshimura, "On the Mechanism of Buckling of a Circular Cylindrical Shell Under Axial Compression," NACA TM 1390, July 1955.

24. V.V. Bolotin, *The Dynamic Stability of Elastic Systems,* Holden-Day, Inc., San Francisco, 1964, translated from Russian edition, 1956.

25. A. Petre, "Buckling of Bars Subject to Axial Shock," Studii si Cercetari de Mecanica Applicata (Roumania), **7**, 1, pp. 173-178, January 1956.

26. A.F. Schmitt, "A Method of Stepwise Integration in Problems of Impact Buckling," ASME Trans., **78**, Series E, J. Appl. Mech., **23**, 2, pp. 291-294, June 1956.

27. E. Procopovici, "Transverse Deformation of an Elastic Bar Subjected to an Axial Impulsive Force," Studii si Ceretari de Mecanica Applicata, **8**, 3, pp. 839-845, 1957.

28. E. Sevin, "On the Elastic Bending of Columns Due to Dynamic Axial Forces

Including Effects of Axial Inertia," ASME Trans., **82**, Series E, J. Appl. Mech., **27**, pp. 125-131, March 1960.

29. S.P. Timoshenko and J.M. Gere, *Theory of Elastic Stability,* McGraw-Hill Book Co., New York, 1961.

30. R.G. Payton, "Dynamic Membrane Stress in a Circular Elastic Shell," ASME Trans., **83**, Series E, J. Appl. Mech., **28**, 3, pp. 417-420, September 1961.

31. G.W. Housner and W.K. Tso, "Dynamic Behavior of Supercritically Loaded Struts," ASCE Proc., J. Engr. Mech. Div., pp. 41-65, 1962.

32. G.R. Abrahamson and J.N. Goodier, "Dynamic Plastic Flow Buckling of a Cylindrical Shell from Uniform Radial Impulse," Proc. Fourth U.S. Nat. Congress of Appl. Mech., Berkeley, California, pp. 939-950, June 1962.

33. J.N. Goodier and I.K. McIvor, *Dynamic Stability and Nonlinear Oscillations of Cylindrical Shells (Plane Strain) Subjected to Impulsive Pressure,* Stanford University, Div. of Eng. Mech., Tech. Rep. No. 132, Stanford, California 94305, June 1962.

34. G.R. Abrahamson and A.L. Florence, *Investigation of Response of Simplified ICBM-Type Structures to Impulsive Loading,* SRI Project 3570 Report to AFSWC, Kirtland AFB, Albuquerque, New Mexico, Vol. II, Contract No. AF 29(601)-4329, August 1962.

35. S.R. Bodner and P.S. Symonds, "Experimental and Theoretical Investigations of the Plastic Deformation of Cantilever Beams Subjected to Impulse Loading," ASME Trans., **84**, Series E, J. Appl. Mech., **29**, pp. 719-728, December 1962.

36. H. Kolsky, *Stress Waves in Solids,* Dover Publications, New York, pp. 41-47, 1963.

37. D. Slepian, "Contributions to Signal and Noise Theory," IEEE Trans. on Information Theory **IT-9**, pp. 229-233, 1963.

38. P. Perzyna, "The Constitutive Equations for Rate Sensitive Plastic Materials," Quarterly of Applied Mechanics, **20**, pp. 321-332, January 1963.

39. B. Budiansky and J.W. Hutchinson, "Dynamic Buckling of Imperfection-Sensitive Structures," *Proc. of the 11th Intl. Congress of Appl. Mech.,* Springer-Verlag, Berlin, pp. 636-651, 1964.

40. B.O. Almroth, A.M.C. Holmes, and D.O. Brush, "An Experimental Study of the Buckling of Cylinders Under Axial Compression," Exptl. Mech., **4**, pp. 263-270, 1964.

41. H.E. Lindberg, "Buckling of a Very Thin Cylindrical Shell Due to an Impulsive Pressure," ASME Trans., **86**, Series E, J. Appl. Mech., **31**, 2, pp. 267-272, June 1964.

42. J.N. Goodier and I.K. McIvor, "The Elastic Cylindrical Shell Under Uniform Radial Impuse," ASME Trans., **86**, Series E, J. Appl. Mech., **31**, 2, pp. 259-266, June 1964.

43. R.S. Roth and J.M. Klosner, "Nonlinear Response of Cylindrical Shells With Imperfections Subjected to Dynamic Axial Loads," AIAA J., **2**, 10, pp. 1788-1794, October 1964.

44. A.P. Coppa and W.A. Nash, *Dynamic Buckling of Shell Structures Subject to Longitudinal Impact,* Report to Air Force Flight Dyn. Lab., Dayton, Ohio, prepared by General Electric Co., Philadelphia, Pennsylvania 19101, FDL-TDR-64-65, December 1964.

45. H.E. Lindberg, "Impact Buckling of a Thin Bar," ASME Trans., **87**, Series E, J. Appl. Mech., **32**, 2, pp. 315-322, June 1965.

46. H.E. Lindberg, D.L. Anderson, R.D. Firth, and L.V. Parker, *Response of Reentry-Vehicle-Type Shells to Blast Loads,* Lockheed Missiles and Space Co. Rep. No. LMSC-B130200, Vol. IV-C, prepared by SRI International, Menlo Park, California 94025, September 1965.

47. H.E. Lindberg, "Dynamic Plastic Buckling of a Thin Cylindrical Shell Containing an Elastic Core," ASME Trans., **87**, Series E, J. Appl. Mech., **32**, 4, pp. 803-812, December 1965.

48. B.M. Malyshev, "Stability of Columns Under Impact Compression," Mekhanika Tverdogo, Tela, **1**, 4, pp. 137-142, 1966.

49. J.W. Hutchinson and B. Budiansky, "Dynamic Buckling Estimates," AIAA J., **4**, 3, pp. 525-530, March 1966.

50. H.E. Lindberg and R.E. Herbert, "Dynamic Buckling of a Thin Cylindrical Shell Under Axial Impact," ASME Trans., **88**, Series E, J. Appl. Mech., **33**, 1, pp. 105-112, March 1966.

51. G.R. Abrahamson and J.N. Goodier, "Dynamic Flexural Buckling of Rods

Within an Axial Plastic Compression Wave," ASME Trans., \fB88\fR, Series E, J. Appl. Mech., \fB33,\fB 2, pp. 241-247, June 1966.

52. M. Abramovitz and L.A. Stegun, *Handbook of Mathematical Functions*, Applied Mathematics Series 55, National Bureau of Standards, August 1966.

53. G.R. Abrahamson, A.L. Florence, and H.E. Lindberg, *Radiation Damage Study (RADS), Vol. XIII--Dynamic Response of Beams, Plates, and Shells to Pulse Loads,* SRI Project 5733 Report to Avco Missile Systems Division, Wilmington, Massachusetts, AVMSD-0339-66-RR, Vol. XIII, September 1966.

54. J.N. Goodier, "Dynamic Plastic Buckling," In *Dynamic Stability of Structures,* G. Herrmann, Ed., Proc. Intl. Conf. at Northwestern University, October 1965, Pergamon Press, New York, pp. 189-211, 1967.

55. *Dynamic Stability of Structures*, G. Herrmann, Ed., Proceedings of International Conference, Northwestern University, Evanston, Illinois, October 1965, Pergamon Press, New York, 1967.

56. A. Florence and H. Vaughan, "Dynamic Plastic Flow Buckling of Short Cylindrical Shells Due to Impulsive Loading," Int. J. Solids Structures, **4**, pp. 741-756, 1968.

57. A.L. Florence, "Buckling of Visco-Plastic Cylindrical Shells Due to Impulsive Loading," AIAA J., **6**, pp. 532-537, 1968.

58. V.M. Kornev, "Modes of Stability Loss in an Elastic Rod Under Impact," Zhurnal Prikladnoi Mekhaniki i Tecknicheskoi, **9**, 3, pp. 63-68, 1968.

59. J.N. Goodier, "Dynamic Buckling of Rectangular Plates in Sustained Plastic Compressive Flow," in *Engineering Plasticity*, Cambridge University Press, 1968, Proceedings of an International Conference of Plasticity held in Cambridge, England, March 1968.

60. A.L. Florence and J.N. Goodier, "Dynamic Plastic Buckling of Cylindrical Shells in Sustained Axial Compressive Flow," ASME Trans., **90**, Series E, J. Appl. Mech., **35**, 1, pp. 80-86, March 1968.

61. A.L. Florence, "Buckling of Viscoplastic Cylindrical Shells Due to Impulsive Loading," AIAA J., **6**, 3, pp. 532-537, March 1968.

62. D.L. Anderson and H.E. Lindberg, "Dynamic Pulse Buckling of Cylindrical Shells Under Transient Radial Pressures," AIAA J., **6**, pp. 589-598, April 1968.

63. G.R. Abrahamson, A.L. Florence, J.N. Goodier, H.E. Lindberg, and H. Vaughan, *Response of Simplified ICBM-Type Structures to Impulsive Loading,* SRI Project 6057 Report to Air Force Weapons Laboratory, Albuquerque, New Mexico, AFWL-TR-67-70, Vol. II, August 1968.

64. H. Vaughan and H.E. Lindberg, "Dynamic Plastic Buckling of Sandwich Shells," ASME Trans., **90**, Series E, J. Appl. Mech., **35**, 3, pp. 539-546, September 1968.

65. H. Vaughan and H.E. Lindberg, "Impulse Buckling of an Elastic-Plastic Cylinder Containing an Elastic Core," ASME Trans., **90**, Series E, J. Appl. Mech., **35**, 4, pp. 827-829, December 1968.

66. H.E. Lindberg and G.E. Sliter, *Response of Reentry-Vehicle-Type Shells to Transient Surface Pressures,* SRI Project 6397 Report to Air Force Weapons Laboratory, Albuquerque, New Mexico, AFWL-TR-68-56, June 1969.

67. A.L. Florence, "Dynamic Buckling of Viscoplastic Cylindrical Shells," 4th Battelle Colloqium, *Inelastic Behavior of Solids,* McGraw-Hill Book Co., Inc., New York, pp. 471-499, 1970.

68. H. Vaughan and A.L. Florence, "Plastic Flow Buckling of Cylindrical Shells Due to Impulsive Loading," ASME Trans., **92**, Series E, J. Appl. Mech., **37**, 1, pp. 171-179, March 1970.

69. H.E. Lindberg, *Dynamic Pulse Buckling of Cylindrical Shells,* SRI International, Poulter Lab. Tech. Rep. No. 001-70, Menlo Park, California 94025, April 1970.

70. H. Schwieger and C. Spuida, *Investigation of the Dynamic Stability of Axially Impacted Cylindrical Shells,* (in German), Deutsche Luft- und Raumfahrt, Rep. DLR FB 70-19, Mülheim (Ruhr), West Germany, June 1970.

71. G.R. Abrahamson and H.E. Lindberg, "Peak Load-Impulse Characterization of Critical Pulse Loads in Structural Dynamics," in *Dynamic Response of Structures,* G. Herrmann and N. Perrone, Eds., Pergamon Press, Inc., New York, pp. 31-52, 1972.

72. V.M. Kornev and V.N. Solodovnikov, "Axially Symmetrical Instability Modes in a Cylindrical Shell Under Impact," Zhurnal Prikladnoi Mekhaniki i Tekhnicheskoi Fiziki, pp. 95-100, March-April 1972.

73. V.M. Kornev, "Development of Dynamic Forms of Stability Loss of Elastic Systems Under Intensive Loading Over a Finite Time Interval," Zhurnal Prikladnoi Mekhaniki i Tekhnicheskoi Fiziki, pp. 122-128, July-August 1972.

74. D.L. Wessenberg, "*Dynamic Buckling of Thin Aluminum, Circular Cylindrical Shells Subjected to a Uniform Impulse Loading,*" Sandia Lab. Rep. SLL-73-0045, Albuquerque, New Mexico, August 1973.

75. N. Jones and C.S. Ahn, "Dynamic Elastic and Plastic Buckling of Complete Spherical Shells," Int. J. Solids Structures, **10**, pp. 1357-1374, 1974.

76. H.E. Lindberg, "Stress Amplification in a Ring Caused by Dynamic Instability," ASME Trans., **96**, Series E, J. Appl. Mech., **41**, 2, pp. 392-400, June 1974.

77. G.R. Abrahamson, "Critical Velocity for Collapse of a Shell of Circular Cross Section Without Buckling," ASME Trans., **96**, Series E, J. Appl. Mech., **41**, 2, pp. 407-411, June 1974.

78. N. Jones and C.S. Ahn, "Dynamic Buckling of Complete Rigid-Plastic Spherical Shells," ASME Trans., **96**, J. Appl. Mech., **41**, 3, pp. 609-614, September 1974.

79. N. Jones and D.M. Okawa, *Dynamic Plastic Buckling of Rings and Cylindrical Shells,* Massachusetts Institute of Technology, Dept. Ocean Eng. Rep. No. 74-18, Cambridge, Massachusetts 02139, October 1974.

80. H.E. Lindberg and T.C. Kennedy, "Dynamic Plastic Pulse Buckling Beyond Strain-Rate Reversal," ASME Trans., **97**, Series E, J. Appl. Mech., **42**, 2, pp. 411-416, June 1975.

81. V.M. Kornev and A.V. Markin, "Natural Motion Density of Elastic Shells Under Intensive Dynamic Loading," Zhurnal Prikladnoi Mekhaniki i Tekhnicheskoi Fiziki, pp. 173-178, September-October 1975.

82. A.L. Florence and G.R. Abrahamson, *A Theory for Critical Loads to Damage a Cylindrical Shell by a Large Underwater Explosion,* Defense Nuclear Agency Rep. No. DNA 4047F, prepared by SRI International, Menlo Park, California, Project PYU-1714, June 1976.

83. A.L. Florence and G.R. Abrahamson, "Critical Velocity for Collapse of Viscoplastic Cylindrical Shells Without Buckling," ASME Trans., **99**, Series E, J. Appl. Mech., **44**, 1, pp. 89-94, March 1977.

84. G. Maymon and A. Libai, "Dynamics and Failure of Cylindrical Shells Subjected to Axial Impact," AIAA J., **15**, 11, pp. 1624-1630, November 1977.

85. I. Elishakoff, "Axial Impact Buckling of a Column With Random Imperfec-

tions," ASME Trans., **100**, Series E, J. Appl. Mech., **45**, 2, pp. 361-365, June 1978.

86. I. Elishakoff, "Impact Buckling of Thin Bar via Monte Carlo Method," ASME Trans., **100**, Series E, J. Appl. Mech., **45**, 3, pp. 586-590, September 1978.

87. D.G. Zincik and R.C. Tennyson, "Stability of Circular Cylindrical Shells Under Transient Axial Impulsive Loading," AIAA J., **18**, 6, pp. 691-699, June 1980.

# INDEX

GPSR Compliance

*The European Union's (EU) General Product Safety Regulation (GPSR) is a set of rules that requires consumer products to be safe and our obligations to ensure this.*

*If you have any concerns about our products, you can contact us on ProductSafety@springernature.com*

In case Publisher is established outside the EU, the EU authorized representative is:

Springer Nature Customer Service Center GmbH
Europaplatz 3
69115 Heidelberg, Germany

**Batch number: 10809439**

Printed by Printforce, the Netherlands